Cereal Crops

Cereal Crops: Genetic Resources and Breeding Techniques provides the reader practical tools for understanding relationships and challenges of successful farming; improvements to genetic modifications; and environmentally sound methods of production of bulk and quality cereals including wheat, maize, rice, barley, and millets. It explores the trait mapping, cropping systems, genome engineering, and identification of specific germplasms needed for the more effective development of biotic and abiotic stress resistant cereals within the framework of ensuring future food supplies around the world.

Features:

- Focuses on cropping systems, genetics and genome engineering for higher crop production at a global level.
- Features information on specific prebiotic formulas to ward off adverse effects of antibiotics.
- Covers mechanistic as well as practical approaches for enhancing crop production in a sustainable way.
- Includes further in-depth analysis of various topics following each chapter.

This is a vital resource for researchers, crop biologists, and students working with crop production and climate changes that have a significant impact on crop production, spanning basic to advanced level discussions of plant breeding, molecular genetics, and agronomy. Covering mechanistic and practical approaches for enhancing crop production in a sustainable way, this text is beneficial to intensive farmers and stakeholders in the field of crop production.

Cereal Crops

Genetic Resources and Breeding Techniques

Edited by
Tariq Shah, Lixiao Nie,
Marcelo Carvalho Minhoto Texeira Filho, and
Rabia Amir

CRC Press
Taylor & Francis Group
Boca Raton London New York

CRC Press is an imprint of the
Taylor & Francis Group, an **informa** business

First edition published 2023
by CRC Press
6000 Broken Sound Parkway NW, Suite 300, Boca Raton, FL 33487-2742

and by CRC Press
4 Park Square, Milton Park, Abingdon, Oxon, OX14 4RN

CRC Press is an imprint of Taylor & Francis Group, LLC

ISBN: 9781032164496 (hbk)
ISBN: 9781032168968 (pbk)
ISBN: 9781003250845 (ebk)

DOI: 10.1201/9781003250845

Typeset in Times
by Deanta Global Publishing Services, Chennai, India

Writing a book is harder than I thought and more rewarding than I could have ever imagined. None of this would have been possible without my ALLAH (Subhanahu wa Taqadus). His substantial blessing and applause flourished my thoughts and boomed my ambition to have the valued fruit of my modest efforts in the form of this piece of literature from the flourishing spring of growing knowledge. He stood by me during every struggle and all my successes. I'm eternally grateful to my father, who took in an extra mouth to feed when he didn't have to. He taught me discipline, tough love, manners, respect, and so much more that has helped me succeed in life. I truly have no idea where I'd be if he hadn't given me a roof over my head or became the father figure whom I desperately needed at that age. To my family, Wahid Shah, Muzammil Shah, and Niaz Ali Shah: for always being the people I could turn to during those dark and desperate years. They sustained me in ways that I never knew that I needed. Finally, to all those who have been a part of my getting there.

Tariq Shah

Contents

Editors

Tariq Shah is a research associate at the University of Agriculture, Peshawar, Pakistan. He completed his master's degree at the Department of Agronomy, University of Agriculture and also visited China for research purposes (8 months). His research work involves plant stress physiology, plant–microbe interaction coping with weather adversity, and adaptation to climatic variability. He has published 16 articles in peer-reviewed journals and 14 chapters. He is currently editing a book titled *Brassica Species Production under Climatic Extremes* (Springer, Beijing). He is a reviewer for ten peer-reviewed international journals and was a recipient of the Elsevier Peer Review Award 2020. He has also organized two national and three international conferences in Pakistan. He has ongoing collaborations with different countries including China, Taiwan, Germany, and the USA in plant stress physiology. He is also the recipient of multiple awards from the Higher Education Commission.

Lixiao Nie is a professor and PhD supervisor in the Crop Cultivation and Farming System, College of Tropical Crops, Hainan University, Haikou, China. He received his PhD in crop cultivation and farming systems from Huazhong Agricultural University, China, in 2008. In 2005–2007, he conducted his PhD thesis research at the Philippines-based International Rice Research Institute (IRRI) under the supervision of Drs. Shaobing Peng and Bas Bouman subsequently joining Huazhong Agricultural University as a crop physiologist from 2008 to 2019. In 2019, he moved to Hainan University. His research mainly focuses on crop simplified and green planting and cultivation. He teaches courses related to plant physiology and crop cultivation and has published more than 90 papers/chapters in international journals and books. His H-index and RG score are 28 and 35.83, respectively. He serves as an editorial board member for international journals, *Field Crops Research* (2018–present) and *Scientific Reports* (2015–present), and deputy editor-in-chief for a Chinese national journal (2020–present). In recent years, he presided over four National Natural Science Foundation of China (NSFC) projects and two national key research and development projects.

Marcelo Carvalho Minhoto Teixeira Filho has a master's degree (2008) and a PhD (2011) in agronomy from the Faculty of Engineering of Ilha Solteira (FEIS) of São Paulo State University (UNESP). He completed a short-term internship (2017) as visiting professor at Plant Nutrition Group, ETH Zürich, Switzerland. Since 2013, he has been an assistant professor in the Department of Plant Protection, Rural Engineering and Soils (DEFERS) at UNESP where he teaches the discipline of plant nutrition. Since July 2018, he has served as editor-in-chief of the agronomic crop journal *Revista Cultura Agronômica*. He is an associate member of the Brazilian Association of Scientific Editors (ABEC) and the current president of the Permanent Research Commission (CPP) of FEIS-UNESP. His research interests include plant nutrition, fertilization, soil fertility, improved efficiency fertilizers, plant growth-promoting bacteria associated with fertilization reduction, *Azospirillum brasilense*, wheat, corn, soybean, sugarcane, and eucalyptus.

Rabia Amir earned her PhD from the University of Edinburg, UK, and master's degree from Arid Agriculture University Rawalpindi, Pakistan. She received a gold medal during her master's and bachelor's degrees. Her research focuses on abiotic stress signaling in plants, mainly working on flavonoid synthesis pathways in peanut plants under diverse environmental conditions. She has published many research articles in reputed journals including *Nature Biotechnology*, *Frontiers in Plant Science*, and *Plant Genetics Resources*. She has completed five projects concerning plant stress physiology. She is currently assistant professor at the National University of Sciences and Technology, Pakistan and is in charge of the plant cell signaling group. She has published more than ten chapters with reputed publishers. She has received multiple awards from the Higher Education Commission of Pakistan.

Contributors

Faiza Abbas
Department of Plant Biotechnology
Atta-ur-Rahman School of Applied
  Biosciences
National University of Sciences and
  Technology
Islamabad, Pakistan

Waleed Fouad Abobata
Horticulture Research Institute
Agriculture Research Center
Giza, Egypt

Arzoo Ahad
Department of Plant Biotechnology
Atta-ur-Rahman School of Applied
  Biosciences
National University of Sciences and
  Technology
Islamabad, Pakistan

Murtaz Aziz Ahmad
Department of Plant Biotechnology
Atta-ur-Rahman School of Applied
  Biosciences
National University of Sciences and
  Technology
Islamabad, Pakistan

Namrah Ahmad
Department of Plant Biotechnology
Atta-ur-Rahman School of Applied
  Biosciences
National University of Sciences and
  Technology
Islamabad, Pakistan

Ahmad Ali
Department of Plant Biotechnology
Atta-ur-Rahman School of Applied
  Biosciences
National University of Sciences and
  Technology
Islamabad, Pakistan

Rabia Amir
Department of Plant Biotechnology
Atta-ur-Rahman School of Applied
  Biosciences
National University of Sciences and
  Technology
Islamabad, Pakistan

Habib-ur-Rehman Athar
Institute of Pure and Applied Biology
Bahauddin Zakariya University
Multan, Pakistan

Masood Iqbal Awan
Department of Agronomy
University of Agriculture
Faisalabad Pakistan

Murat Aycan
Laboratory of Biochemistry
Faculty of Agriculture
Niigata University
Niigata, Japan

Saba Azeem
Department of Plant Biotechnology
Atta-ur-Rahman School of Applied
  Biosciences
National University of Sciences and
  Technology
Islamabad, Pakistan

Rimsha Azhar
Department of Plant Biotechnology
Atta-ur-Rahman School of Applied
    Biosciences
National University of Sciences and
    Technology
Islamabad, Pakistan

Clara R. Azzam
Cell Research Department
Field Crop Research Institute
Agriculture Research Center
Giza, Egypt

Sadia Banaras
Department of Biotechnology
Biological Sciences
Quaid-i-Azam University
Islamabad, Pakistan

Marouane Baslam
Laboratory of Biochemistry
Faculty of Agriculture
Niigata University
Niigata, Japan

Sherien Bukhat
Institute of Molecular Biology and
    Biotechnology
Bahauddin Zakariya University
Multan, Pakistan

Justin Chipomho
Faculty of Agricultural Sciences and
    Technology
Marondera University of Agricultural
    Sciences and Technology
Marondera, Zimbabwe

Zhicong Dai
Institute of Environment and Ecology
School of The Environment and Safety
    Engineering
Jiangsu University
Zhenjiang, China

and

Jiangsu Collaborative Innovation Center
    of Technology and Material of Water
    Treatment
Suzhou University of Science and
    Technology
Suzhou, China

Daolin Du
Institute of Environment and Ecology
School of the Environment and Safety
    Engineering
Jiangsu University
Zhenjiang, China

Muhammad Ansar Farooq
Institute of Environmental Sciences and
    Engineering
School of Civil and Environmental
    Engineering
National University of Sciences and
    Technology
Islamabad, Pakistan

Nosheen Fatima
Department of Plant Biotechnology
Atta-ur-Rahman School of Applied
    Biosciences
National University of Sciences and
    Technology
Islamabad, Pakistan

Abdul Ghaffar
Department of Agronomy
Muhammad Nawaz Sharif University of
    Agriculture
Multan, Pakistan

Maria Gillani
Department of Plant Biotechnology
Atta-ur-Rahman School of Applied
    Biosciences
National University of Sciences and
    Technology
Islamabad, Pakistan

Alvina Gul
Department of Plant Biotechnology
Atta-ur-Rahman School of Applied
  Biosciences
National University of Sciences and
  Technology
Islamabad, Pakistan

Ghulam Haider
Department of Plant Biotechnology
Atta-ur-Rahman School of Applied
  Biosciences
National University of Sciences and
  Technology
Islamabad, Pakistan

Qurat Ul Ain Ali Hira
Department of Plant Biotechnology
Atta-ur-Rahman School of Applied
  Biosciences
National University of Sciences and
  Technology
Islamabad, Pakistan

Sajid Hussain
State Key Laboratory of Rice Biology
China National Rice Research Institute
Hangzhou, China

and

Soil and Water Testing Laboratory
Multan, Pakistan

Sami Ullah Jan
Department of Bioinformatics &
  Biosciences
Capital University of Science and
  Technology
Islamabad, Pakistan

S. Jeyaraj
Department of Botany
University of Kerala
Thiruvananthapuram, India

Riya Johnson
Department of Botany
University of Calicut
Malappuram, India

Kiran Khurshid
Department of Plant Biotechnology
Atta-ur-Rahman School of Applied
  Biosciences
National University of Sciences and
  Technology
Islamabad, Pakistan

G. S. Anil Kumar
Department of Botany
Govt. Degree College Chaubattakhal
Pauri Garhwal, India

Midhat Mahboob
Department of Plant Biotechnology
Atta-ur-Rahman School of Applied
  Biosciences
National University of Sciences and
  Technology
Islamabad, Pakistan

Saadatullah Malghani
College of Biology and Environment
Nanjing Forestry University
Nanjing, China

Noor-ul-Ain Malik
Department of Plant Biotechnology
Atta-ur-Rahman School of Applied
  Biosciences
National University of Sciences and
  Technology
Islamabad, Pakistan

Hamid Manzoor
Institute of Molecular Biology and
  Biotechnology
Bahauddin Zakariya University
Multan, Pakistan

Nyamande Mapope
Faculty of Agricultural Sciences and
   Technology
Marondera University of Agricultural
   Sciences and Technology
Marondera, Zimbabwe

Shomaila Mehmood
Anhui Key Laboratory of Modern
   Biomanufacturing
School of Life Sciences
Anhui University
Hefei, China

Toshiaki Mitsui
Laboratory of Biochemistry
Faculty of Agriculture
Niigata University
Niigata, Japan

Faiza Munir
Department of Plant Biotechnology
Atta-ur-Rahman School of Applied
   Biosciences
National University of Sciences and
   Technology
Islamabad, Pakistan

Nida Mushtaq
Department of Plant Biotechnology
Atta-ur-Rahman School of Applied
   Biosciences
National University of Sciences and
   Technology
Islamabad, Pakistan

Aneela Mustafa
Department of Plant Biotechnology
Atta-ur-Rahman School of Applied
   Biosciences
National University of Sciences and
   Technology
Islamabad, Pakistan

Salman Nawaz
Department of Plant Biotechnology
Atta-ur-Rahman School of Applied
   Biosciences
National University of Sciences and
   Technology
Islamabad, Pakistan

Misbah Naz
Institute of Environment and Ecology
School of the Environment and Safety
   Engineering
Jiangsu University
Zhenjiang, China

and

State Key Laboratory of Crop Genetics
   and Germplasm Enhancement
Nanjing Agricultural University
Nanjing, China

Aswathy Raj K.P.
Department of Botany
University of Calicut
Malappuram, India

Cosmas Parwada
Faculty of Agricultural Sciences
Department of Agricultural
   Management
Zimbabwe Open University
Hwange, Zimbabwe

Jos T. Puthur
Department of Botany
University of Calicut
Malappuram, India

Khola Rafique
Pest Warning and Quality Control of
   Pesticides
Agriculture Department
Punjab, Pakistan

P.E. Rajasekharan
Indian Institute of Horticultural
    Research
Bangalore, India

Sumaira Rasul
Institute of Molecular Biology and
    Biotechnology
Bahauddin Zakariya University
Multan, Pakistan

Fozia Saeed
Institute of Molecular Biology and
    Biotechnology
Bahauddin Zakariya University
Multan, Pakistan

Nair G. Sarath
Department of Botany
University of Calicut
Malappuram, India

A.M. Shackira
Department of Botany
Sir Syed College
Kerala, India

Sahar A. Sherif
Crop Intensification Research
    Department
Field Crop Research Institute
Agriculture Research Center
Giza, Egypt

Muhammad Tariq
Department of Pharmacology
Lahore Pharmacy College
Lahore, Pakistan

Mathew Veena
Department of Botany
University of Calicut
Malappuram, India

Mustafa Yildiz
Department of Field Crops
Faculty of Agriculture
Ankara University
Ankara, Turkey

Sameera Zafar
Department of Plant Biotechnology
Atta-ur-Rahman School of Applied
    Biosciences
National University of Sciences and
    Technology
Islamabad, Pakistan

Sania Zaib
Department of Biochemistry
Biological Sciences
Quaid-i-Azam University
Islamabad, Pakistan

Akmal Zubair
Department of Biochemistry
Quaid-i-Azam University
Islamabad, Pakistan

1 Farming Systems Improvements in Different Regions

Cosmas Parwada, Justin Chipomho, and Nyamande Mapope

CONTENTS

1.1 INTRODUCTION

Smallholder farmers in Africa are characterized by poor crop productivity and perennial food insecurity, especially in the semi-arid tropics where the majority of smallholder farmers live (Anderson and D'Souza, 2014). The farmers are challenged by poor rainfall, especially in the larger parts of Southern Africa, pests and diseases, and poor soil fertility, resulting from inherently infertile soils that are worsened by inappropriate soil management practices, e.g., nutrient mining (Parwada and Chinyama, 2021). In particular, soils in the semi-arid regions of Zimbabwe are characteristically infertile with low potential to sustain agricultural production under continuous cultivation (Nyathi et al., 2003). The soils are extremely deficient in nitrogen (N), phosphorus (P), and sulfur, and soil fertility on smallholder farms in Zimbabwe continues to decline (Nyathi et al., 2003; Parwada et al., 2022). Therefore, the maintenance of soil fertility is key to sustaining the productivity of smallholder agriculture in sub-Saharan Africa (Nadwa et al., 2011).

Soil infertility is the major limiting factor in crop production among smallholder farmers in Zimbabwe (Nyathi et al., 2003). Smallholder farmers are experiencing perennial food shortages for complex reasons that are exacerbated by the current economic crisis (Parwada et al., 2022). The soils are predominantly sandy and have limited ability to store organic matter and nutrients, thus soil fertility declines rapidly under crop production (Anderson and D'Souza, 2014). Efforts to curb declining soil fertility in the semi-arid regions of Zimbabwe are hampered by a number of

DOI: 10.1201/9781003250845-1

challenges, e.g., increased pressure on land use and poverty, among other factors that are common to much of Southern Africa.

Additionally, concerning specific soil fertility management problems, the semi-arid areas (natural farming regions IV and V) of Zimbabwe also face natural disasters such as perennial droughts and frequent crop pests outbreaks. The regions receive low (<450 mm year^{-1}), erratic, and unreliable rainfall, and are usually affected by mid-season droughts which can cause low crop yields, making farmers vulnerable to food insecurity (Nyathi et al., 2003). Regrettably, at least 70% of communal farmers in Zimbabwe are located in these dry and agriculturally marginalized areas, suggesting that many are food insecure.

The unpredictably low rainfalls in the natural regions IV and V result in frequent failure of rain-fed crop production. Smallholder farmers usually grow cereal crops such as maize (*Zea mays* L.) in areas with more reliable rainfall, and sorghum (*Sorghum bicolor* L. Moench) and pearl millet (*Pennisetum glaucum* L. R. Br.) in very dry areas (regions IV and V). Pearl millet and sorghum are strategic crops for food security in the dry regions because they are more tolerant to drought than maize. Maize production is commonly failing in the natural regions IV and V because of poor rainfall (Parwada and Chinyama, 2021). Crops such as beans, cowpeas, and groundnuts are marginalized in commercial terms; they are considered crops for women and little attention is given to their production. There is also a wide imbalance in production area allocations between cereals and marginalized legumes, resulting in no systematic crop rotations among farmers. Marginalized legumes are grown in small areas and receive less than 5% of soil fertility inputs (Mapfumo and Giller, 2001; Nyathi et al., 2003). The communal areas are characterized by rampant land degradation, and farmers' practices of monoculture cropping of cereals continue to deplete nutrients and have contributed to the continued decline in cereal yields (Parwada et al., 2020). Fortunately, the majority (>60%) of communal farmers in Zimbabwe are women who have adopted crop diversification on their farms through intercropping cereals with the marginalized legumes (Nyathi et al., 2003).

Several methods for soil fertility management such as the use of synthetic fertilizers and animal manure, and crop rotation are available to farmers; nevertheless, use of synthetic fertilizers is not compatible with resource-constrained communal farmers (Anderson and D'Souza, 2014). Synthetic fertilizers are out of reach to the majority of these farmers, hence they usually grow crops with little or no fertilizer application causing rapid soil degradation. In this regard, targeting nutrients within crop rotations and intercrops can enhance crop productivity and allow farmers to use the limited inputs of N and P fertilizers and manure that are available (Gitari et al., 2018). Apart from their direct provision of food and cash, grain legumes play an important role in soil fertility management (Addo-Quaye et al., 2011). It is agreed that growing legumes leads to an improvement in soil fertility; however, Nyathi et al. (2003) observed that the yield of cereal crops is better when they are grown in rotation with legumes than in continuous monocultures. This suggests that it is important to have methods of increasing the contribution of legume–cereals rotations to achieve food security in semi-arid environments. There is also a need to understand the magnitude of the yield benefits associated with the inclusion of legumes, which can be achieved by quantifying the N fixed under certain prevailing environmental conditions, for example.

There is increasing awareness of the importance of food legumes in improving the soil they grow in and in mitigating greenhouse gases (GHG) (Parwada et al., 2020). Nevertheless, food legumes remain poor cousins to major cereal crops such as maize, sorghum, and millet due to the ever-increasing global demand for cereals from burgeoning human populations (Anderson and D'Souza, 2014). Additionally, priorities for the cultivation of and research in food legumes remain secondary to those for cereals in most cropping systems. This chapter examines the role of marginalized small grain legumes in cereal–legume cropping systems on cereal productivity in the semi-arid regions of Zimbabwe.

1.2 CROPPING SYSTEMS IN SEMI-ARID AREAS OF ZIMBABWE

The main methods of growing crops are mixed cropping, intercropping, and rarely sole cropping in the semi-arid regions of Zimbabwe (Nyathi et al., 2003). Notably, legumes are grown in very small areas within the system. Minor crops include cowpea (*Vigna unguiculata* L. Walp), Bambara groundnut (*V. subterranea* L. Verdc), sunflower (*Helianthus annuus* L.), and cotton (*Gossypium hirsutum* L.) (Parwada and Chinyama, 2021). Other minor crops include watermelons (*Citrullus lanatus* Thunb.) and pumpkins (*Cucurbita maxima* L.) that are also planted as intercrops. Farmers are not concerned with the yield of the small grain legume in these intercrops, thus legume yields are very low. Estimated crop yields in one of Zimbabwe's dry regions were 0.40 t ha^{-1} (cowpea), 0.5 t ha^{-1} (pearl millet), 0.70 t ha^{-1} (sorghum), and 0.80 t ha^{-1} (maize) in a normal rainy season (Rohrbach, 2001). The national average yields in smallholder farming areas are 0.30 t ha^{-1} for both cowpea and groundnut and 0.6 t ha^{-1} for cereals (Nyathi et al., 2003). In this chapter, we address the role of minor food legumes such as chickpea (*Cicer arietinum* L.), cowpea (*V. unguiculata* L.), mung bean (*V. radiata* L. Wilczek), and pigeon pea (*Cajanus cajan* L.) on the yield of cereals. These crops are grown as secondary crops in many farming systems, mostly in subsistence agriculture and less so in commercial production systems in Zimbabwe.

1.3 EFFECTS OF LEGUMES ON CEREAL PRODUCTIVITY

The inclusion of legumes in crop rotations can reduce the use of synthetic fertilizers and energy in arable systems, consequently lowering greenhouse gases emissions without lowering cereal yield. The N fertilizer savings observed in legume–cereal rotations range around 277 kg ha^{-1} of CO_2 per year (1 kg N = 3.15 kg CO_2; Kandhro et al., 2007). In recent years, several studies have focused on the role of legumes in the reduction of GHG emissions (Blanchart et al., 2006; Addo-Quaye et al., 2011). Nadwa et al. (2011) demonstrated that legume crops emit around five to seven times less GHG per unit area compared with cereal crops. Zougmore et al. (2000) showed that peas emitted 69 kg N_2O ha^{-1}, far less than winter wheat (368 kg N_2O ha^{-1}) and rape (534 kg N_2O ha^{-1}). In general, N_2O losses from soils covered with legumes are certainly lower than those from both N_2O fertilized grasslands and non-legume crops, as also indicated by Addo-Quaye et al. (2011).

Smallholder farmers rely on inorganic fertilizers which most of them are unable to purchase. Additionally, these types of fertilizers have negative effects on the soil and the environment (Parwada and Chinyama, 2021). The unaffordability of these fertilizers in the context of the smallholder communal farmer results in no or under application of fertilizers causing reduced grain yield. However, including legume crops in the cropping systems can mitigate the soil fertility challenge. The benefits of grain legumes can be divided into a nitrogen effect component and a break crop effect component. The nitrogen effect component is a result of the N provision from biological nitrogen fixation (BNF) (Gitari et al., 2018), which is highest in situations of low N fertilization to subsequent crop cycles (Blanchart et al., 2006). The break crop effect includes non-legume-specific benefits, such as improvements in soil organic matter and structure, phosphorus mobilization, soil water retention and availability, and reduced pressure from diseases and weeds (Anderson and D'Souza, 2014; Parwada et al., 2020).

1.4   EFFECTS OF LEGUMES ON SOIL PROPERTIES

Soil cultivation may cause significant soil organic carbon (SOC) losses through the decomposition of humus (Parwada et al., 2020). The change in land use by shifting from pasture to cropping systems results in the loss of soil C stocks of between 25% and 43% (Nadwa et al., 2011). Legume-based systems improve several aspects of soil fertility, such as SOC and humus content, and N and P availability (Vesterager et al., 2008). Grain legumes can increase SOC by supplying biomass, organic C, and N (Gitari et al., 2018), as well as releasing hydrogen gas as a by-product of biological nitrogen fixation, which promotes the development of bacterial legume nodules in the rhizosphere (Addo-Quaye et al., 2011). Legumes also significantly affect soil N availability when they are grown as winter crops in rice–bean and rice–vetch combinations; the rice residue N content is enhanced 9.7%–20.5%, with values ranging from 1.87 to 1.93 g N kg^{-1} soil (Gitari et al., 2019). These quantities of fixed N can suffice the N demands of most cereals grown in Zimbabwe.

In a study by Parwada and Chinyama (2021), cowpea–cereals relay intercrops modified the chemical properties of sandy soils. The nitrogen, phosphorus, potassium, and carbon content increased under cowpea–sorghum and cowpea–millet intercrops compared with sole cropping of cereals (Table 1.1).

It was observed that growing grain legumes adds more soil organic carbon to sandy soils compared with soils with oats (7.21 g kg^{-1} dry matter, on average) (Gitari et al., 2019). Specifically, the cultivation of pea exerted the most positive action on the organic carbon content (7.58 g kg^{-1}, after harvest, on average), whereas the narrow-leaved lupin had the least effect (7.23 g kg^{-1}, on average) (Nadwa et al., 2011). It is estimated that about 88% of the legume species examined to date can form nitrogen-fixing nodules with rhizobia, which is responsible for up to 80% of the biological nitrogen fixation that takes place in agricultural settings. Thus, legumes may play a key role in sustainable agriculture.

TABLE 1.1

Average Chemical Properties of the Soil Recorded after Three Years under Different Legume–Cereal Intercrop

Parameter	Sorghum	Cowpea-sorghum	Cowpea-millet	Millet sole crop
pH (H$_2$O)	4.5 ± 1.2	6.96 ± 0.3	6.98 ± 0.3	4.4 ± 1.2
EC (dSm^{-1})	4.1 ± 0.03	9.12 ± 0.1	8.99 ± 0.1	4.2 ± 0.03
CEC (cmol$_{(+)}$ kg^{-1})	7.0 ± 0.5	317.2 ± 0.8	319.2 ± 0.8	7.3 ± 0.5
Total C (%)	0.5 ± 0.04	4.5 ± 0.4	3.5 ± 0.4	0.6 ± 0.04
Total N (%)	0.6 ± 0.03	5.16 ± 0.2	4.96 ± 0.2	0.6 ± 0.03
C:N ratio	0.2 ± 0.01	6.8 ± 0.7	6.4 ± 0.7	0.3 ± 0.01
Olsen extractable P (mg kg^{-1})	45.0 ± 7.3	420.4 ± 17.8	320.4 ± 17.8	42.0 ± 7.3
Extractable NH$_4$ (mg kg^{-1})	65.4 ± 0.8	286.3 ± 2.8	270.3 ± 2.8	68.4 ± 0.8
K (mg kg^{-1})	5.4 ± 0.6	8.2 ± 0.5	8.4 ± 0.5	4.8 ± 0.6
Ca (cmol$_{(+)}$ kg^{-1})	0.2 ± 0.05	22.1 ± 2.5	21.1 ± 2.5	0.2 ± 0.05
Mg (cmol$_{(+)}$ kg^{-1})	11.7 ± 1.9	19.8 ± 2.1	18.8 ± 2.1	12.5 ± 1.9
Na (cmol$_{(+)}$ kg^{-1})	0.25 ± 0.03	3.6 ± 0.7	4.2 ± 0.7	0.22 ± 0.03
Cu (cmol$_{(+)}$ kg^{-1})	79.1 ± 36.1	107.2 ± 38.6	105.8 ± 38.6	77.5 ± 36.1
Zn (cmol$_{(+)}$ kg^{-1})	63.2 ± 6.9	312.8 ± 0.6	300.7 ± 0.6	61.8 ± 6.9

EC, electrical conductivity; CEC, cation exchange capacity.
Data are means ± standard error of the means for three replicates.

Legume-based pastures can rehabilitate degraded land by enhancing soil aggregation and promoting the activity of different soil microbes that modify the soil structure (Parwada and Van Tol, 2018). Characteristically, some legume crops are broad-leaved which makes them useful in preventing soil erosion and runoff during a storm (Vesterager et al., 2008). This indicates that legumes can increase water infiltration, thereby increasing the available moisture especially in semi-arid regions. In addition, some legumes, such as alfalfa, are deep-rooted and are therefore able to absorb nutrients and water that are not available to other crop plants (Zhang et al., 2011). This allows the efficient utilization of resources in all soil zones if such crops are included in various intercropping systems.

Soils in semi-arid regions of Zimbabwe are deficient in phosphorus. Symbiotic nitrogen fixation and ammonium assimilation induce soil acidification that can benefit alkaline soils by solubilizing phosphorus from rock phosphates (Nyathi et al., 2003). Other legumes such as lupin can secrete organic anions into the soil, thereby inducing phosphorus solubilization in phosphorus-deficient soils (Zougmore et al., 2000). When cowpea–maize were intercropped, Vesterager et al. (2008) found an increase in P availability at the rhizosphere level associated with significant acidification (−0.73 U) than in sole cropping. Less-labile organic P pools (i.e., NaOH-extractable P pools and acid-extractable P pools) significantly accumulated more in the rhizosphere of legumes in a legume–cereal intercrop than in the cereal monocropping system (Tittonell et al., 2005). This enhanced the P uptake by the plants, thereby increasing the growth and yield performance under intercropping.

There is agreement that cereals produce higher protein grains and higher yields when grown after or in conjunction with legumes (Gitari et al., 2019; Parwada and Chinyama, 2021). In that regard, legumes are frequently rotated with non-legume crops such as cereals. The intercropping of legumes and cereals annually may be beneficial to the environment because it permits more efficient exploitation of the available nutrients, leading to better use of nitrogen in the agro-system and reducing post-harvest nitrogen availability and nitrate leaching (Tajudeen, 2010). The increased nitrogen from cowpea–sorghum and cowpea–millet relay intercrops was observed to enhance the grain yields of the cereals (Parwada and Chinyama, 2021). The grain yields of sorghum and millet were increased by at least 45% under intercropping compared to sole cropping (Figure 1.1).

Several researchers have investigated the yield benefits of legumes for subsequent cereal crops (Gitari et al., 2018, 2019). Courty et al. (2015) reported higher yields of wheat after legumes (field peas, lupins, faba beans, chickpeas, and lentils) than yields of wheat after wheat. In comparison, the wheat–wheat yield was 4.0 t ha^{-1} while the mean grain yield for legume–wheat was 5.2 t ha^{-1}, representing a greater than 30% increase on average. In Botswana, many food legumes are adapted to drought-prone, low-nutrient environments and are used in rotation or as intercrops with cereals. Malawian smallholder farmers have adopted legumes, mainly edible legume intercrops such as pigeon pea and groundnut, to improve both human nutrition and soil fertility (Parwada and Chinyama, 2021).

In temperate environments, the cereal yield is on average 17% and 21% higher in grain–legume-based systems than wheat mono-cropping, under standard and moderate fertilization levels, respectively (Nyathi et al., 2003). However, it should be noted that the yield advantage to subsequent cereal crops provided by legumes also depends on the species and the amounts of fixed N (Gitari et al., 2018). Differences in BNF patterns are also found between the same species. Courty et al. (2015) compared 25 groundnut varieties for plant BNF in three different agro-ecologies in South Africa, highlighting an N-fixed range of between 76 and 188 kg ha^{-1}, depending also on the soil and environmental conditions and on N uptake. Other factors influencing

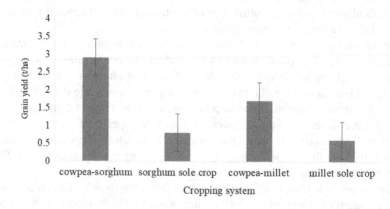

FIGURE 1.1 Average grain yield (t ha^{-1}) of sorghum and millet recorded after three years of cowpea intercropping (adapted from Parwada and Chinyama, 2021).

BNF include the salinity and sodicity (alkalinity) of soils, as observed in chickpea, common bean, and faba bean (Nadwa et al., 2011).

Grain legumes as break crops can also contribute to weed control (Blanchart et al., 2006) by contrasting their specialization and helping stabilize the crop–weed community composition (Gitari et al., 2019). However, concerns remain on the introduction of grain legumes into cropping sequences. Cropping systems that include legumes in rotations should be supported by best crop management practices (e.g., N fertilization rates and timing, soil management, weeding, and irrigation), which often do not match standard techniques normally applied by farmers (Parwada and Chinyama, 2021). For example, some possible risks in terms of nitrate leaching associated with grain legume cultivation can be counteracted by including cover crops in the system (Kandhro et al., 2007). Additional reasons may explain why grain legumes are not very common in high-input cropping systems. These include their low and unstable yields, inadequate policy support, and lack of proper quantification of the long-term benefits of legumes within cropping systems (Gitari et al., 2018). However, other efforts could be addressed, for example, to better sustain livelihoods and increase the economic return to farmers.

1.5 INTERCROPPING OF LEGUMES AND CEREALS

Intercropping is a widely used cropping system in low-input and low-yield farming systems in Zimbabwe (Parwada and Chinyama, 2021). Intercropping is the growing (or mixing) of more than one crop on one piece of land in the same season (Zougmore et al., 2000). Intercropping increases the return on investment from one crop in case the other fails due to poor climatic conditions. It is of immense benefits to farmers who do not have large pieces of land, as it assists the farmers in growing more than one crop type on one piece of land (Tajudeen, 2010). In comparison to other cropping systems such as monoculture cropping, intercropping is less labor and capital intensive because it maximizes returns per unit area and labor unit.

Intercropping is beneficial in increasing cereal yields with reduced inputs, pollution mitigation, and more stable aggregate food yields per unit area. Nevertheless, many constraints make intercropping uncommon in modern agriculture because there should be a single and standardized product that is suitable for mechanization (Courty et al., 2015). It is therefore necessary to optimize intercropping systems to enhance resource-use efficiency and crop yield simultaneously (Egbe, 2010), while also promoting multiple ecosystem services. The direct mutual benefits of cereal–legumes intercropping include below-ground processes in which cereals benefit from legumes-fixed N and increased Fe and Zn bio-availability from the companion legumes (Addo-Quaye et al., 2011). However, the role of marginalized legumes in cereal yield is also ignored such that their inclusion in the intercropping systems is not specifically based on their role in improving soil fertility among communal farmers. Crop physiology, agronomy, and ecology can simultaneously contribute to the improvement of intercropping systems, enhancing crop productivity and resource-use efficiency, thereby making intercropping a viable approach to sustainable intensification, particularly in regions with impoverished soils (Blanchart et al., 2006). However, more research is required to realize these benefits.

1.6 CONCLUSION

Legume crops can directly increase available macro plant nutrients, e.g., nitrogen through BNF and P availability at the rhizosphere level associated with significant acidification. Legumes can recover unavailable forms of soil phosphorus which could be major assets in future cropping systems in the semi-arid tropics. This has directly improved the cereal yield in dry areas. Additionally, growing broad-leaved legumes can also indirectly increase the productivity of cereal crops in dry areas by modifying the soil micro-environments around the crops, e.g., minimizing evapotranspiration and conserving moisture. However, in order to enhance the roles and importance of grain legumes in the context of sustainable agriculture, there is a need for appropriately designed and managed cropping systems such as intercropping. Consequently, legume crops with the ability to accumulate phosphorus from forms that are normally unavailable need to be further studied, because phosphorus is an expensive and limited resource in several cropping systems in the tropics. The inclusion of legume crops in various cropping systems enhances both the quantity and the quality of the cereal grain, which is advantageous to most rural poor in marginalized areas of Zimbabwe. There is a need to quantify the yield benefits accrued from various legume–cereal intercrops or rotations under different soil, climatic, and plant characteristics and management conditions to find a suitable approach to achieve the best improvements.

REFERENCES

Addo-Quaye AA, Darkwa AA, Ocloo GK. 2011. Yield and productivity of component crops in a maize-soybean intercropping system as affected by time of planting and spatial arrangement. *J Agric Biol Sci.* 9: 50–57.

Anderson JA, D'Souza S. 2014. From adoption claims to understanding farmers and contexts: A literature review of conservation agriculture (CA) adoption among smallholder farmers in Southern Africa. *Agric Ecosyst Environ.* 187: 116–132.

Blanchart E, Villenave C, Viallatoux A, Barthès B, Girardin C, Azontonde A, Fellera C. 2006. Long-term effect of a legume cover crop (*Mucuna pruriens* var. utilis) on the communities of soil macrofauna and nematofauna, under maize cultivation, in southern Benin. *Eur J Soil Biol.* 42: S136–S144.

Courty PE, Smith P, Koegel S, Redecker D, Wipf D. 2015. Inorganic nitrogen uptake and transport in beneficial plant root-microbe interactions. *Crit Rev Plant Sci.* 34(1–3): 4–16.

Egbe OM. 2010. Effects of plant density of intercropped soybean with tall sorghum on competitive ability of soybean and economic yield at Otobi, Benue State, Nigeria. *J Cer Oilseeds.* 1(1): 1–10.

Gitari HI, Gachene CKK, Karanja NN, Kamau S, Sharma K, Schulte-Geldermann E. 2018. Optimizing yield and economic returns of rain-fed potato (*Solanum tuberosum* L.) through water conservation under potato-legume intercropping systems. *Agric Water Manage.* 208: 59–66.

Gitari HI, Nyawade SO, Gachene CKK, Karanja NN, Kamau S, Sharma K, Schulte-Geldermann E. 2019. Increasing potato equivalent yield increases returns to investment under potato-legume intercropping system. *Open Agric.* 4: 623–629.

Kandhro MN, Tunio SD, Memon HR, Ansari MA. 2007. Growth and yield of sunflower under influence of mungbean intercropping. *Pak J Agric Res.* 23: 9–13.

Mapfumo P, Giller KE. 2001. *Soil Fertility Management Strategies and Practices by Smallholder Farmers in Semi-Arid Areas of Zimbabwe*. International Crops Research Institute for the Semi-Arid Tropics (ICRISAT) with permission from the Food and Agricultural Organization (FAO). Bulawayo, Zimbabwe and Rome, Italy, 60p.

Nadwa SM, Bationo A, Obanyi SN, Rao IM, Sanginga N, Vanlauwe B. 2011. Inter and intra-specific variation of legumes and mechanisms to access and adapt to less available soil phosphorus and rock phosphate. In: A Bationo et al. (eds.), *Fighting Poverty in Sub-Saharan Africa: The Multiple Roles of Legumes in Integrated Soil Fertility Management*. Springer Science+Business Media B.V, Dordrecht, pp. 47–83. https://doi.org/10.1007/978-94-007-1536-3.

Nyathi P, Kimani SK, Jama B, Mapfumo P, Murwira HK, Okalebo JR, Bationo A. 2003. Soil fertility management in semi-arid areas of East and Southern Africa. In: Gichuru MP, Bationo A, Bekunda MA, Goma HC, Mafongoya PL, Mugendi DN, Murwira HM, Nandwa SM, Nyathi P, Swift MJ (eds.), *Soil Fertility Management in Africa: A Regional Perspective*. Academy Science Publishers, Nairobi, Kenya, pp. 219–252.

Parwada C, Chinyama AT. 2021. Land equivalent ratio (LER) of cowpea-sorghum relay intercrops as affected by different cattle manure application rates under smallholder farming systems. *Front Sustain Food Syst*. https://doi.org/10.3389/fsufs.2021.778144.

Parwada C, Chipomho J, Mandumbu R. 2022. In-field soil conservation practices and crop productivity in marginalized farming areas of Zimbabwe. In: Mupambwa HA, Nciizah DA, Nyambo P, Muchara B, Ndakalimwe NG (eds.), *Food Security for African Smallholder Farmers*. Springer; 1st ed. 2022 edition, pp. 75–87.

Parwada C, Van Tol J. 2018. Effects of litter quality on macroaggregates reformation and soil stability in different soil horizons. *Environ, Dev Sustain*. https://doi.org/10.1007/s10668-018-0089-z.

Parwada C, Van Tol J, Tibugari H, Mandumbu R. 2020. Characterisation of soil physical properties and resistance to erosion in different areas of soil associations. *Afr Crop Sci J*. 28: 93–109.

Rohrbach DD. 2001. Zimbabwe baseline: Crop management options and investment priorities in Tsholotsho. In: Twomlow SJ, Ncube B (eds.), *Improving Soil Management Options for Women Farmers in Malawi and Zimbabwe: Proceedings of a Collaborators Workshop on DFID-supported project Will Women Farmers Invest in Improving their Soil Fertility? Participatory Experimentation in a Risky Environment*, 13–15 September 2000, ICRISAT Bulawayo, Zimbabwe. International Crops Research Institute for the Semi-Arid Tropics, Bulawayo, Zimbabwe, pp. 57–64.

Tajudeen OO. 2010. Evaluation of sorghum-cowpea intercrop productivity in savanna agro-ecology using competition indices. *J Agric Sci*. 2(3): 229–234.

Tittonell P, Vanlauwe B, Leffelaar PA, Rowe EC, Giller KE. 2005. Exploring diversity in soil fertility management of smallholder farms in western Kenya - I. Heterogeneity at region and farm scale. *Agric Ecosyst Environ*. 110: 149–165.

Vesterager JM, Nielsen NE, Hogh-Jensen H. 2008. Effect of cropping history and phosphorous source on yield and nitrogen fixation in sole and intercropped cowpea-maize systems. *Nutr Cycl Agroecosyst*. 80: 61–73.

Zhang G, Yang Z, Dong S. 2011. Interspecific competitiveness affects the total biomass yield in an alfalfa and corn intercropping system. *Field Crop Res*. 124: 66–73.

Zougmore R, Kambou FN, Ouattara K, Guillobez S. 2000. Sorghum-cowpea intercropping: An effective technique against runoff and soil erosion in the Sahel (Saria, Burkina Faso). *Arid Land Res Manage*. 14: 329–334.

2 Stepwise Intensification Option for Enhancing Cereal-Based Cropping Systems

Waleed Fouad Abobata, Clara R. Azzam, and Sahar A. Sherif

CONTENTS

2.1 INTRODUCTION

Agriculture is characterized by the methods used in managing water resources, soil, and energy to provide the food and clothing needs of humans. Throughout history, agriculture has been the foundation of economic, social, political, and cultural development worldwide. One of the main requirements of each dynamic activity is planning within the general objectives of the activity. In order to develop and deal with crises, the agricultural sector, as one of the most important economic activities in many communities, requires coordinated planning. Sustainable agriculture is a

DOI: 10.1201/9781003250845-2

type of agriculture that is more resource efficient, helps people, and is environmentally friendly. To put it another way, sustainable agriculture must be environmentally sound, economically viable, and socially acceptable. The goals of sustainable agriculture are strongly linked with its definitions. The goals of a successful sustainable agricultural program are to

- provide food security along with increased quality and quantity, while considering the needs of future generations;
- conserve water, soil, and natural resources; conserve energy resources inside and outside the farm;
- maintain and improve farmers' profitability; maintain the vitality of rural communities; to conserve biodiversity (Eskandari, 2011).
- Although high yield per unit area has been able to meet the dietary needs of expanding populations in some locations using conventional farming and monocropping systems, these operations require direct and indirect access to expensive energy derived from fossil fuels. In terms of ecology and the environment, monocropping has caused a series of problems. Due to excessive use of resources such as water, soil, forests, pastures, and natural resources, humans not only put them at risk of extinction, but also cause pollution because of industrial activities, chemical fertilizers, and pesticides, which threaten the earth (Reganold, 1992). If farming activities are conducted based on ecological principles, in addition to preventing the destruction of natural ecosystems, the result is a stable condition (Dariush et al., 2006). Additionally, agricultural systems must provide for the needs of people today and those of future generations; therefore, it is essential to achieve sustainable agriculture. One of the key strategies in sustainable agriculture is to restore diversity to agricultural ecosystems and their effective management. Intercropping is a method of increasing diversity in an agricultural ecosystem.
- Intercropping is an example of a sustainable agricultural system by following objectives such as ecological balance and greater utilization of resources, thereby increasing productivity. Intercropping (growing two or more crops simultaneously in the same field) is a way of ameliorating the productivity of land and other inputs (Andrews & Kassam, 1976). Small farmers use intercropping to diversify their products and boost the consistency of their annual output by making better use of land and other resources (Faris et al., 1983). According to Enyi (1973), small farmers across the world are severely hampered by low productivity and inadequate land resources. A preliminary study has revealed that intercropping may be a viable technique for enhancing output. Intercropping is widespread with its highest use in tropical Africa. It is practiced in African counties that have different cultural, economic, colonial, and political backgrounds and experiences. In addition, technological developments and the availability of resources constitute a basis for diversifying the farming systems and the number of crops grown. Sometimes, double and triple cropping and alternate strip cropping systems are practiced in market gardens with certain

vegetable crops. Among the acclaimed crops grown in the tropical zone of Africa is cotton as a cash crop, and rice, maize, sorghum, and millet. Legumes, such as groundnut and cowpea, and cassava are often found in mixed cropping systems.

- This chapter reviews the progress in implementing different types of intercropping systems in Egypt.

2.2 TYPES OF INTERCROPPING SYSTEMS

Growing two or more crops at the same time in the same field is usually described as an "intercropping system", which can be subdivided into four different subsystems (Kugbe et al., 2018):

1. *Mixed intercropping*, in which several mixed crops are grown and randomly distributed in the same space; these crops may be planted and harvested at different times according to their specific characteristics (Kugbe et al., 2018).
2. *Row intercropping*, in which two or more crops are grown simultaneously in a regular arrangement with a well-defined planting pattern, usually consisting of one or more rows of a short duration crop in parallel rows between rows of a long duration crop (Kugbe et al., 2018).
3. *Relay intercropping* is when one or more crops are planted within an existing crop so that the first crop's final stage coincides with the initial development of the subsequent crop (Kugbe et al., 2018).
4. *Strip cropping* is a method of growing two or more crops in the same field, but in distinct and alternating strips that are wide enough to allow autonomous cultivation yet narrow enough to obtain good yields. The results from intercropping studies indicate that crop diversity may improve the quality of an ecosystem. Greater species richness may be associated with nutrient cycling characteristics that can often regulate soil fertility, limit nutrient leaching losses, and significantly reduce the negative impacts of pests, including weeds (Kugbe et al., 2018).

2.3 ADVANTAGES AND DISADVANTAGES OF INTERCROPPING

Smallholder farmers who only have a small plot of land to feed and sustain their family are the most likely to use intercropping. Land and capital are the key limits in this situation, whereas labor may be plentiful. These farmers must maximize their land's overall output by maximizing growth elements such as light, water, and nutrients. Some of the benefits of growing two or more crops at the same time are as follows (Aye & Howeler, 2012):

Increased food variety in different crops as grain and root or tuber crops provide carbs; grain legumes provide protein; and vegetables provide vitamins and fiber.

- Increased yield stability or income and reduced risk of total crop failure.
- Reduced incidence of pests and diseases.

- Reduced weed competition.
- Reduced soil erosion by establishing an early ground cover between the rows of the slow-growing, long duration crop.
- Increased efficient use of land and labor, while the latter can also help with other operations during the year.
- Increased yield and total net income per unit area of land.

However, intercropping also has some disadvantages:

- Eliminates the use of automation for planting, weeding, and harvesting, as well as the use of some pesticides for weed management and fertilizer application.
- Increases in the complications of the management of each crop individually are possible.
- Requires more labor per unit area.
- Diminishes each crop's production due to intercrop competition, although this is usually offset by an increase in the total value of all crops in the system.

Intercropping systems must be structured to optimize the system's total net revenue, as well as maximizing the various advantages while minimizing the downsides described above. This necessitates the careful selection of the various crops to be planted, as well as the most suitable varieties of each crop, the most effective plant densities and planting arrangements, the relative timing of planting each crop, the most effective fertilization, the amounts and balance of nutrients, and the application times, as well as their distribution among the various crops.

2.4 EVALUATION OF THE EFFICIENCY OF INTERCROPPING SYSTEMS

1. Land equivalent ratio (LER)
   Described by Mead and Willey (1980) as follows:

$$\text{LER} = \Sigma_i = 1^n \left(Y_i^I / Y_i^M \right)$$

where:
Y_i^I = yield of crop i in intercropping
Y_i^M = yield of crop i in monocropping
n = number of crops in the intercropping system

2. Area time equivalent ratio (ATER)
   Considers time factor along with land area. Proposed by Hiebsch and McCollum (1987), it is calculated as

$$\text{ATER} = \Sigma i = 1^n \left[\left(Ti^M / Ti^I \right) \left(Yi^I / Yi^M \right) \right]$$

where:

Ti^M = duration of crop i in monocropping

Ti^I = total duration of the intercropping system

3. Relative crowding coefficient (K)

Computed for (km) and (ks) and for the two crops (K) according to Hall (1974) as follows:

$$Kab = \frac{Y_ab}{Y_{aa} \times Z_{ab}} - \frac{Y_{ba}}{Y_{bb} \times Z_{ba}}$$

where:

Z_{ab} = sown proportion of species a (in mixture with b)

Z_{ba} = sown proportion of species b (in mixture with a)

$$Kba = \frac{Y_{ab}}{Y_{aa} \times Z_{ab}} - \frac{Y_{ba}}{Y_{bb} \times Z_{ba}}$$

A species with a coefficient of less than, equal to, or greater than 1 means it has produced fewer yields, the same yield, or more yield than expected, respectively.

The component crop with the higher coefficient is the dominant one. To determine if there is a yield advantage to mixing, the product of the coefficient is formed by multiplying $K_{ab} \times K_{ba}$.

If $K > 1$, there is a yield advantage.

If $K = 1$, there is no difference.

If $K < 1$, there is a yield disadvantage.

4. Aggressivity (A)

This was proposed by McGilchrist (McGilchrist & Trenbath, 1971) and was determined according to the following formula:

$$A_{ab} = \frac{Y_{ab}}{Y_{aa} \times Z_{ab}} - \frac{Y_{ba}}{Y_{bb} \times Z_{ba}}$$

where:

Z_{ab} = sown proportion of species a (in a mixture with b)

Z_{ba} = sown proportion of species b (in a mixture with a)

An aggressivity value of zero indicates that the intercropped crops are equally competitive. For any other situation, both crops will have the same numerical value, but the sign of the dominant crop will be positive and the dominated crop will be negative. The greater the numerical value of Agg, the higher the difference between the actual and expected yields.

5. Competition ratio (CR)

The competition ratio was calculated using the formula proposed by Willey and Rao (1980).

The CR of crop b is the reciprocal of the CR of crop a.

$$CRa = (LERa/LERb)(Sb/Sa)$$

where:

S_a = relative space occupied by crop a

S_b = relative space occupied by crop b

2.5 CROPPING PATTERN IN EGYPT

Egypt is located in the northeastern part of the African continent, in a temperate zone with dry warm climates. Agriculture in Egypt represents the largest sector of the Egyptian economy and contributes about 30% of the national income. Egypt has gained many agricultural developments through agriculture intensification, which is considered the main approach to achieving economic growth. In Egypt, cropping depends on five main crops, namely wheat, maize, clover, cotton, and sugarcane. Wheat and maize are used to make flour (Abo-Doma & Azzam, 2007; Azzam & Mahrous, 2010), clover is used to make feed (Abd El-Naby Zeinab et al., 2014; Azzam et al., 2022), cotton is used to make fiber and edible oil, and sugarcane is used to make sugar (El-Geddawy et al., 2008). Cereals, legumes, fibers, forages, vegetables, and fruit crops are all part of the Egyptian cropping pattern (Ouda & Zohry, 2018). Several intercropping systems have been implemented in Egypt.

2.5.1 INTERCROPPING CEREALS WITH LEGUMES

Egypt suffers from a shortfall in the manufacture of grain and oil crops, which has led to widening food gaps between production and consumption due to population growth. The government aims to improve the efficiency of these crops by introducing new varieties, using modern farming methods, and cultivating new lands, and intercropping plays an important role in solving the problem (Awaad & El-Naggar, 2018).

By adopting intercropping systems on a wide scale, farmers will be able to boost their productivity and profitability from resource-poor agricultural systems while also decreasing risks and enhancing the soil quality in the long run. For cereals such as maize and wheat, intercropping has improved yields and soil fertility. The philosophy of intercropping depends on exploiting available environmental resources and increasing the yield per unit area. Thus, intercropping legumes with cereals provides many benefits for the soil (Sherif et al., 2005).

Maize is one of the most grown cereal crops in Egypt for human consumption and animal feed. Plant density and distribution may play an important role in maize growth and grain yield. It is well known that a good distribution of maize plants creates a canopy to intercept more light energy, thereby increasing vegetative growth and grain yield (de Oliveira et al., 2007; Sherif et al., 2005); controlling plant population density can be achieved through controlling hill spacing. At low population density, the grain yield is limited to the amount of plants per square meter (El-Mekser et al., 2009).

2.5.1.1 Intercropping Maize with Soybean

Maize and soybean intercropping is an important agricultural system for increasing the productivity of Egyptian farms without any additional costs. The morphological and physiological differences between the two crops provide mutual benefits. Furthermore, this system reduces the depletion of elements from the soil more than the solid cultivation of maize (Tsubo et al., 2005). Intercropped soybeans use one of the following series of replacements: 50% maize and 50% soybean, 75% maize and 25% soybean, or 25% maize and 75% soybean.

Ijoyah (2014) found that the intercropping of soybeans and maize boosted maize grain output and total intercrop yield, where the land equivalent coefficient was greater than 0.25, and the land equivalent ratio was greater than 1.0. L. Feng et al. (2021) revealed that all intercropping patterns, (2:4) or (3:4), were beneficial in improving the nitrogen equivalent ratio (NER), the land equivalent ratio, and the nitrogen utilization efficiency (NUE), which were all greater than 1.

In Egypt, Sahar (1994) found that the highest maize yield was obtained with a (2:2) pattern and received the highest nitrogen dose (408 kg N ha⁻¹). The highest yield of soybean was obtained with a (2:4) pattern and received the highest nitrogen dose (408 kg N ha⁻¹). The maximum value of the land equivalent ratio was obtained when a mix of these components received a low N dose (288 kg t ha⁻¹) and was oriented in a (2:4) pattern (Figure 2.1). Abou-Keriasha et al. (2009) found significant benefit when LER, A, and CR were higher, which was expressed with a higher MAI value. Abou-Kerisha et al. (2010) also showed that the yield of maize grain was dominant while the intercropped crops were dominated, where the highest land equivalent ratio yielded 1.15 and monetary 776.96 LE when maize was intercropped with soybean. Addo-Quaye et al. (2011) showed that an arrangement of one row of corn alternating with double rows of soybean recorded the best yield with respect to soybean. Sherif and Gendy (2012) reported that the highest soybean yields (3.140 t ha⁻¹) were obtained when soybean Giza 22 cv. was intercropped with maize, planted at 60 cm between hills, and thinned to one plant/hill (24,000 plants ha⁻¹). However, the highest yields of maize (2649 t ha⁻¹) were obtained from maize planted at a rate of 48,000 plants ha⁻¹ and intercropped with Giza 22 soybean cv. with a maximum land equivalent ratio of 1.80. Net returns of $295.20 were obtained when soybean cultivar Giza 22 was intercropped with maize at the highest population density (48,000 plants ha⁻¹) when hills were spaced 60 cm apart and thinned to two plants/hill.

FIGURE 2.1   Maize intercropping with soybean under Egyptian conditions.

2.5.1.2 Intercropping Maize with Groundnut

C. Feng et al. (2021) reported an LER of 1.3 for the maize/peanut intercropping system. The LER was unaffected by continent, intercrop planting pattern, or temporal niche differentiation; however, sowing maize earlier than peanut boosted the partial LER of maize and decreased the partial LER of peanut. A raised N rate increased the partial LER of maize and lowered the partial LER of peanut, but had no effect on the total LER. This meta-analysis demonstrates that maize/peanut intercropping is more effective in terms of land use than solitary crops, achieving a "win-no-win" yield advantage on average, owing primarily to maize.

Sherif et al. (2005) found that when intercropping maize with peanut, increasing the spacing of maize with one plant per hill increased maize yield and yield components, except for plants that carried two ears. The yield and most of the studied characteristics decreased by delaying the planting date of maize due to unfavorable growth conditions. Furthermore, decreasing the planting density of maize and delaying its planting date led to an increase in the yield and attributes of groundnut due to decreasing interspecific competition. Additionally, delaying the planting date of maize increased the land use efficiency (LER) and decreased the competition ratio. Whereas dense maize plants (50 cm apart at two plants per hill) increased the LER and the relative crowding coefficient (K).

It can be inferred that soybean and groundnut can be intercropped successfully with maize for more efficient land use and higher economic return.

2.5.1.3 Intercropping Maize with Cowpea

Abou-Keriasha et al. (2011) noted that a reduction in intercropped cowpea yield when intercropping with maize (taller plant) might be due to the more shading effect of taller maize plants on shorter cowpea plants and and due to intercepted light, and competition for nutrients, water, and carbon dioxide. Dahmardeh et al. (2009) showed that the intercropping of maize and bean in different planting ratios (100:100, 50:100, 100:50, 25:75, 75:25, 50:50, 0:100, and 100:0) significantly affected the quantitative and qualitative characteristics of the forage. Sowing the crops in a 100:100 ratio yielded the maximum yield of green fodder (65.7 t ha⁻¹). The maximum maize grain output (9.0 t ha⁻¹) resulted from a 75:25 ratio of maize and cowpea, whereas the highest cowpea grain yield (3.9 t ha⁻¹) resulted from a 50:100 ratio of maize and cowpea, respectively. Sowing the crops in a 100:100 ratio yielded the greatest (2.26) and the highest land equivalent ratio (Azim et al., 2000). The impact of intercropping maize (*Zea mays*) and cowpea (*Vigna unguniculata*) on fodder biomass output and silage qualities was investigated. Thirteen maize fodders were grown alone and in combination with cowpea at seed ratios of 85:15 and 70:30, respectively. At the time of heading, the fodder was harvested (at about 35% dry matter). In comparison to maize alone, the findings showed a considerable increase in biomass and crude protein output when maize was intercropped with cowpea at seed ratios of 70:30 and 85:15. (I) Maize alone, (II) maize and cowpea (85:15), (III) maize and cowpea (70:30), and (IV) maize supplemented with 2.5% urea were all used to make silages. According to the findings, intercropping maize and cowpea at a seed ratio of 70:30 improved fodder productions and provided high-quality silage. Adipala et al. (2002) demonstrated that planting cowpea within 15 minutes of maize had a substantial impact on the growth and yield of cowpea. Simultaneous planting yielded higher yields in most cases (land equivalent ratio for cowpea/maize intercropping systems was higher than 1). Ofori and Stern (1987) indicated that delaying maize planting in a cowpea + maize intercropping system increased cowpea yield by 46%. It was also noted that delaying cowpea sowing increased maize yield, but decreased cowpea yield (Figure 2.2).

2.5.1.4 Intercropping Maize or Wheat with Tomato

A study was aimed at determining the growth and yield of corn influenced by different row intercropping patterns with an indeterminate tomato (Pino et al. 1994). The results indicated that two rows of corn intercropped with a single row of tomato were the most vigorous. Corn intercropped with a single row of tomato recorded the highest weight of 1000 seeds while three rows of corn intercropped with two rows of tomato recorded the highest yield and return on investment. The study's conclusion further indicated that row intercropping tomatoes with corn is beneficial to corn farmers as it could significantly reduce insect damage and increase net income. Mohamed et al. (2013) reported that maize intercropped with tomato can modify the microclimate for tomato, where it protects the tomato fruits from sun damage in July and August. On other hand, the authors stated that the reproductive processes in tomato are more sensitive to high temperatures than the vegetative processes. Pino et al. (1994) discovered that maize intercropped with tomato reduced pests and diseases that usually exist in tomato monocultures. Abd El-Aal and Zohry (2003) mentioned that intercropping tomato with maize saved irrigation water by 40% compared with solid treatments. Tomato fruits were significantly affected by intercropping with maize along with application of phosphate source and doses. The

FIGURE 2.2 Intercropping maize with cowpea in Egypt under Egyptian conditions (Medina & Pimentel, 2012).

damage to tomato fruits was decreased and marketable yield increased. These could be traced back to the height of the maize plants that act as a shadow on tomato plants, protecting fruits from the sun's rays and reducing the effect of direct burning of fruits. They added that the greatest advantage of using intercropping is to maximize the use of a land unit and water to achieve maximum production (Figure 2.3).

Wheat can be intercropped with tomato systems in winter. These two systems modify the microclimate for tomato plants, where wheat or faba bean plants protect tomato plants from low temperatures in January and February (Abd El-Zaher et al., 2013). Fernandez-Munoz et al. (1995) indicated that exposing tomato plants to low temperatures reduces pollen production, shed, viability, and tube growth. Furthermore, intercropping wheat with tomato can increase water use efficiency, where the tomato grows tap deep strong root systems, which facilitate the absorption of soil moisture deeper than wheat root systems (Figure 2.4).

2.5.2 INTERCROPPING CEREALS WITH SUGAR CROPS

2.5.2.1 Intercropping with Sugarcane

Sugarcane is the main source of refined sugar and the sole source for the molasses industry in Egypt. In addition, products from industry are used as raw materials in the plywood, paper, and pulp industries. Molasses, a by-product of sugarcane, is used to create ethyl alcohol, active yeast, citric and acetic acid, and dextran a plasma substitute. Sugarcane is planted in wide rows (100 cm) and takes several months to develop its canopy, during which the soil and solar energy go to waste. During its early growth stages, the growth rate of sugarcane is slow, with the leaf canopy providing a sufficient uncovered area for growing another crop (Nazir et al., 2002), which allows intercropping. In this case, the intercropped crop will not need any extra irrigation water as it will use the water applied to sugarcane to fulfill its water

FIGURE 2.3 Intercropping maize with tomato under Egyptian conditions.

FIGURE 2.4 Intercropping wheat with tomato under Egyptian condition.

requirements. Furthermore, intercropping of sugarcane provides extra income for farmers during the early growth stages of sugarcane (Ouda & Zohry, 2018).

In Egypt, sugarcane has two growth cycles: autumn and winter season sugarcane.

Autumn season sugarcane: The percentage of plantations of autumn season sugarcane amounts to 30% of the total sugarcane area. Sugarcane is planted from mid-September to mid-November and is harvested 16 months later (El-Kholi, 2008).

Spring season sugarcane: Sugarcane is planted in February, March, and April, and the percentage of plantations during this season is 70% of the total area under sugarcane planting. The crop remains in the field for 12–13 months till harvesting (El-Kholi, 2008).

When intercropping wheat with sugarcane, the wheat does not require any further irrigation water as it utilizes the water applied to sugarcane to fulfill its water requirements. Wheat intercropping with sugarcane also provides farmers with additional income during the early stages of sugarcane growth. Sugarcane is planted in September, wheat is cultivated in November with 40% of its optimum planting density, and wheat is harvested in April under the intercropping wheat with sugarcane system (Figure 2.5). This technique yields 40% more wheat while reducing sugarcane yields by no more than 5% (Ahmed et al., 2013).

2.5.2.2 Intercropping with Sugar Beet

Sugar beet (*Beta vulgaris* L.) is an important crop not only in Egypt, but also all over the world as a source for the sugar industry. In Egypt, it is the second sugar after sugarcane and its area has increased significantly by approximately 25.6% over the last 35 years. Consequently, the contribution of sugar beet to sugar production in Egypt increased to 35.5% of total sugar production in 2012 (Motagally & Metwally, 2014). Egyptian sugar beet successfully grows in dirt that has recently been reclaimed and in old lands. It gives higher yield, its growth period is about half the sugarcane season (6–7 months), and it has lower water requirements than sugarcane. Sugar beet is a good candidate for intercropping, where wheat intercropping with sugar beet has been successfully implemented in Egypt (Abou-Elela, 2012). In this technique, sugar beet is the main crop, which is planted with its full planting density and obtains its

FIGURE 2.5   Wheat intercropped with sugarcane under Egyptian conditions.

required water and fertilizer, whereas wheat is considered a secondary crop that uses the applied water and fertilizer for the main crop (sugar beet). To reduce intraspecific competition between the main crop (sugar beet) and the secondary crop (wheat or faba bean), the optimum planting density for wheat should be 25%. Abou-Elela (2012) revealed that the quality characteristics of sugar beet were significantly better with relay intercropping of wheat 42 days after planting sugar beet by using 30 kg seed ha^{-1} which represents 25% of pure stand of wheat (120 kg ha^{-1}). The highest LER values (1.28) and total return income ($733.10) were obtained by sowing wheat on the top of the second terrace of sugar beet. Toaima (2006) studied the effect of cropping two and three rows of wheat, plus sugar beet and wheat in pure stand. They found that all the studied characteristics had a significant effect in the two seasons. On the contrary, the sucrose and purity percentage gave the highest values with three rows of wheat with sugar beet. Ibrahim et al. (2008) also studied the effect of intercropping two and three rows of wheat with sugar beet on a bed of sugar beet of 120 cm with a solid culture of both crops. They showed that root length, root diameter, root fresh weight/plant, root fresh yield/fed, and purity% were reduced by the intercropping system when compared to pure stand. Farghaly et al. (2003) and Gadallah et al. (2006) recorded that different intercropping of wheat with sugar beet resulted in higher gross return per unit area than pure stand.

2.5.3   RELAY INTERCROPPING COTTON WITH WHEAT

During the last decades, cotton (*Gossypium barbadense* L.) growers in Egypt have suffered greatly from the rapid increase in the cost of production, which has not been matched by an equal increase in price policy. Moreover, the predominant deterioration of cotton productivity was a cogent reason for farmers to avoid cotton planting. Research showed that Egypt tried to activate cotton rotation by including some long duration winter crops, particularly wheat (*Triticum aestivum* L.) of which Egypt suffers a deficit of approximately 50%, and to achieve maximum utilization with higher gross income (Sherif et al., 2011).

FIGURE 2.6   Relay intercropping cotton with wheat under Egyptian conditions.

In the wheat–cotton relay intercropping system, both wheat and cotton are grown in the same field for one year. In these systems, strips of winter wheat, which are sown in the fall, are sown with cotton in the spring. From March to April, the two crops are grown together in one field, with the seedling phase of cotton and the maturation phase of wheat overlapping in time and space. After the wheat harvest in early summer, the whole space is occupied by cotton (Figure 2.6). Hussein (2005) reported that sowing cotton as a solid on March 25 gave the highest cotton yield and its attributes. The best sowing date for intercropping cotton with wheat was March 25 or April 15, which produced the highest yield and did not differ significantly as compared with sole planting; whereas the early (5 March) or late (15 May) sowing date resulted in a decrease in seed yield. Toaima (2004) and Toaima et al. (2007) revealed that the effect of relay cropping patterns of cotton with wheat yield and their components was significant. The grain yield was 89% of pure stand wheat. Nevertheless, the plant height and yield components of wheat (length of spike, number of spikes/m², grain weight/spike, and 1000 kernel weight) under relay intercropping were higher than those grown as a control crop (Sherif et al., 2011).

Maximum values of 1.93 and 1.13 for LER and ATER, respectively, were recorded when cotton was planted on March 15 through relay intercropping with wheat variety cv. Sakha 93. The highest wheat yield of 2590.5 kg ha^{-1} was obtained when wheat cv. Giza 168 was relayed intercropped with cotton on March 1. The highest cotton yield was produced when wheat cv. Sakha 93 was relayed intercropped with cotton on March 15, reaching 966.34 kg ha^{-1}. All the above studies concluded that relay intercropping of cotton with wheat, as an intensive cropping system, is recommended to increase the wheat cultivated and the productivity of the unit area.

2.5.4 Intercropping Cereal Crops with Cassava

Cassava (*Manihot esculenta*, Grantz) is regarded as one of the most significant tropical root crops (Figure 2.7). It is known as "Africa's food security crop" (Tewe & Egbunike, 1992). It ranks as the fourth food crop in developing countries. It is a key low-cost carbohydrate source, the cheapest caloric source, and

Cassava plant   **Cassava root crops**

FIGURE 2.7   Intercropping cassava with maize under Egyptian conditions.

contains the maximum concentration of starch compared to other crops (Hair, 1995). The comparatively low cost of cassava production is evident in being one of the cheapest foods in most growing areas. This can be accredited to several factors such as low labor requirements, easy cultivation, and high productivity value for at least three outstanding ecological adaptations, e.g., drought tolerance, ability to grow in sub-optimal soils, and aggressiveness toward weeds and insect pests (A. As Saqui, 1984).

Cassava is near the top of the list of crops that convert the most solar energy into soluble carbs per unit of area. Cassava produces roughly 40% more carbohydrates than rice and 25% more than maize among the starchy staples, making it the cheapest source of calories for both human and animal nutrition. Moisture (70%), starch (24%), fiber (2%), protein (1%), and other components (including minerals) make up the average cassava root (3%). Unlike other crops, cassava thrives in less than ideal conditions, allowing farmers to boost their total agricultural output by utilizing marginal land (Cock, 1982). One of the most significant plant products for humans is plant starch. It is a necessary component of food, accounting for a significant amount of daily caloric consumption. Cassava flour and gari (a processed cassava product) are widely consumed in West Africa. Cassava starch is a type of starch that comes from cassava.

2.5.5 INTERCROPPING CASSAVA WITH MAIZE

Cassava is a late-maturing and low canopy–closure crop. Intercropping crops between cassava rows can reduce the amount of solar radiation wasted in the inter-row areas of the cassava for the first three months after planting, which can be tapped by the intercrops. Okigbo (1978) found no reduction in the yield of intercropped cassava and maize when both were planted at the same time, or two weeks after maize was sown. Kang and Wilson (1981) reported that the best yield was obtained at an

FIGURE 2.8 Cassava intercropping with maize under Egyptian conditions

International Institute of Tropical Agriculture (IITA) non-profit organization station when maize cv. TZPB was planted at 30,000 plants ha^{-1} with cv. Tms 30395 cassava cultivar at 10,000 plants ha^{-1}.

For both crops, the most important factors were competition for light and perhaps nutrients (Lawson, 1981). Sherif and Ibrahim (2011) discovered that the greatest estimated cassava yield was 33.1 t ha^{-1} when maize was planted at its lowest plant density (20,160 plants ha^{-1}) on June 15. While the highest maize yield was estimated at 1470 kg ha^{-1} when planted at its highest plant density (16,800 plants ha^{-1}) on June 15. The highest LER value of 1.25 was obtained when maize was sown later on June 15 at its lowest density (20,160 plants ha^{-1}) (Figure 2.8).

The above studies demonstrate that the production of cassava in newly reclaimed land in Egypt seems very promising and could have a tremendous future in Egypt.

2.6 CONCLUSION

Intercropping has many advantages, especially for the developing world. When factors such as climatic conditions, the timing of the intercrop planting, and the crop used for intercropping are right, then intercropping will be very successful.

REFERENCES

A As Saqui, M. (1984). The potential of cassava in optimizing small-farm productivity in Liberia. In *Proceedings, Sixth Symposium of the International Society for Tropical Root Crops*, hosted by CIP in Lima, Peru, 21–26 February 1983.

Abd El-Aal, A., & Zohry, A. (2003). Natural phosphate affecting maize as a protective crop for tomato under environmental stress conditions at Toshky. *Egyptian Journal of Agricultural Research, 81*(3), 937–953.

Abd El-Naby Zeinab, M., Azzam, C. R., & Abd El-Rahman, S. S. (2014). Evaluation of ten alfalfa populations for forage yield, protein content, susceptibility to seedling damping-off disease and associated biochemical markers with levels of resistance. *Journal of American Science*, *10*(7).

Abd El-Zaher, S., Shams, A., & Mergheny, M. (2013). Effect of intercropping pattern and nitrogen fertilization on intercropping wheat with tomato. *Egyptian Journal of Applied Sciences*, *28*(9), 474–489.

Abo-Doma, A., & Azzam, C. R. (2007). Hunting of some differentially expressed genes under salt stress in wheat. Egyptian Journal of Plant Breeding, *11*, 233–244.

Abou-Elela, A. (2012). Effect of intercropping system and sowing dates of wheat intercropped with sugar beet. *Journal of Plant Production*, *3*(12), 3101–3116.

Abou-Keriasha, M., Abd El-Hady, M., & Nawar, F. (2009). Response of some cowpea varieties to intercropping with maize under upper Egypt condition. Egyptian Journal of Applied Sciences, *24*, 495–514.

Abou-Keriasha, M., Gadallah, R., & El-Wakil, N. (2011). The influence of preceding crops and intercropping maize with cowpea on productivity and associated weeds. *Egyptian Journal of Agronomy*, *33*(1), 1–18.

Abou-Kerisha, M. A., Sherif, S. A., & Mohamed, W. K. (2010). Maize grain yield response to intercropping with three legume fodder crops. *Egyptian Journal of Agricultural Research*, *88*(4), 1259–1276.

Addo-Quaye, A., Darkwa, A., & Ocloo, G. (2011). Yield and productivity of component crops in a maize-soybean intercropping system as affected by time of planting and spatial arrangement. *Journal of Agricultural and Biological Science*, *6*(9), 50–57.

Adipala, E., Ocaya, C., & Osiru, D. (2002). Effect of time of planting cowpea (*Vigna unguiculata* (L.) Walp) relative to maize (*Zea mays* L.) on growth and yield of cowpea. *Tropicultura*, *20*(2), 49–57.

Ahmed, A., Ahmed, N., & Khalil, S. (2013). Effect of intercropping wheat on productivity and quality of some promising sugar cane cultivars (Autumn plant). *Minia Journal of Agricultural Research and Development*, *33*, 597–623.

Andrews, D., & Kassam, A. (1976). The importance of multiple cropping in increasing world food supplies. *Multiple Cropping*, *27*, 1–10.

Awaad, H., & El-Naggar, N. (2018). Role of intercropping in increasing sustainable crop production and reducing the food gap in Egypt. In *Sustainability of agricultural environment in Egypt: Part I* (pp. 101–118).

Aye, T. M., & Howeler, R. (2012). Cassava agronomy: Intercropping systems. In *The cassava handbook: A reference manual based on the Asian regional cassava training course, held in Thailand*. Bangkok, Thailand: Centro Internacional de Agricultura Tropical (CIAT), the Department of Agriculture (DOA) and the Thai Tapioca Development Institute (TTDI) of Thailand, 613–625.

Azim, A., Khan, A., Nadeem, M., & Muhammad, D. (2000). Influence of maize and cowpea intercropping on fodder production and characteristics of silage. *Asian-Australasian Journal of Animal Sciences*, *13*(6), 781–784.

Azzam, C. R., Abd El-Naby, Z. M., Abd El-Rahman, S. S., Omar, S. A., Ali, E. F., Majrashi, A., & Rady, M. M. (2022). Association of saponin concentration, molecular markers, and biochemical factors with enhancing resistance to alfalfa seedling damping-off. *Saudi Journal of Biological Sciences*, *29*(4), 2148–2162.

Azzam, C. R., & Mahrous, M. A. (2010). Performance and genetic relationships among ten Egyptian wheat cultivars as revealed by RAPD-PCR analysis. *Egyptian Journal of Plant Breeding*, *14*, 87–102.

Cock, J. H. (1982). Cassava: A basic energy source in the tropics. *Science*, *218*(4574), 755–762.

Dahmardeh, M., Ghanbari, A., Syasar, B., & Ramroudi, M. (2009). Effect of intercropping maize (*Zea mays* L.) with cow pea (*Vigna unguiculata* L.) on green forage yield and quality evaluation. *Asian Journal of Plant Sciences*, *8*(3), 235–239.

Dariush, M., Ahad, M., & Meysam, O. (2006). Assessing the land equivalent ratio (LER) of two corn [*Zea mays* L.] varieties intercropping at various nitrogen levels in Karaj, Iran. *Journal of Central European Agriculture, 7*(2), 359–364.

de Oliveira, T. K., Macedo, R. L. G., Venturin, N., Botelho, S. A., Higashikawa, E. M., & Magalhães, W. M. (2007). Solar radiation in understory of agrosylvopastoral system with eucalypt on different spacings. *Cerne, 13*(1), 40–50.

El-Geddawy, D. I., Azzam, C. R., & Khalil, S. (2008). Somaclonal variation in sugarcane through tissue culture and subsequent screening for molecular polymorphisms. *Egyptian Journal of Genetics And Cytology, 37*(2).

El-Kholi, M. (2008). Sugar Crops Research Institute, Giza (Egypt): A profile. *Sugar Tech, 10*(3), 189–196.

El-Mekser, H. K. A., Mosa, H., Maha, G., & El-Ghonemy, M. (2009). Effect of row orientation, row spacing and plant population density on grain yield and other agronomic traits in maize (*Zea mays* L.). *Alexandria Journal of Agricultural Research, 54*(3), 17–27.

Enyi, B. (1973). Effects of intercropping maize or sorghum with cowpeas, pigeon peas or beans. *Experimental Agriculture, 9*(1), 83–90.

Eskandari, H. (2011). Intercropping of wheat (*Triticum aestivum*) and bean (*Vicia faba*): Effects of complementarity and competition of intercrop components in resource consumption on dry matter production and weed growth. *African Journal of Biotechnology, 10*(77), 17755–17762.

Farghaly, B., Zohry, A., & Bassal, S. (2003). Crops management for intercropping sugar beet with some essential crops to maximize area unit productivity. *Journal of Agricultural Science, Mansoura University, 28*(7), 5183–5199.

Faris, M., Burity, H., Reis, D., & Mafra, R. (1983). Economic analysis of bean and maize system in monoculture vs. associated cropping. *Field Crop Research, 1*, 319–335.

Feng, C., Sun, Z., Zhang, L., Feng, L., Zheng, J., Bai, W., Gu, C., Wang, Q., Xu, Z., & van der Werf, W. (2021). Maize/peanut intercropping increases land productivity: A meta-analysis. *Field Crops Research, 270*, 108208.

Feng, L., Yang, W.-T., Zhou, Q., Tang, H.-Y., Ma, Q.-Y., Huang, G.-Q., & Wang, S.-B. (2021). Effects of interspecific competition on crop yield and nitrogen utilisation in maize-soybean intercropping system. *Plant, Soil and Environment, 67*(8), 460–467.

Fernandez-Munoz, R., Gonzalez-Fernandez, J., & Cuartero, J. (1995). Variability of pollen tolerance to low temperatures in tomato and related wild species. *Journal of Horticultural Science, 70*(1), 41–49.

Gadallah, R., Abdel-Galil, A., & Nawar, F. (2006). Maximizing productivity by intercropping some winter crops on sugar beet. *Journal of Agricultural Science, Mansoura University, 31*(5), 2601–2614.

Hair, S.K.O. (1995). Tropical Research and Education Center, University of Florida. New Crop Fact Sheet: 1–6.

Hall, R. (1974). Analysis of the nature of interference between plants of different species. I. Concepts and extension of the de Wit analysis to examine effects. *Australian Journal of Agricultural Research, 25*(5), 739–747.

Hiebsch, C., & McCollum, R. (1987). Area- ×-time equivalency ratio: A method for evaluating the productivity of intercrops. *Agronomy Journal, 79*(1), 15–22.

Hussein, S. M. (2005). Planting, date, pattern and fertilizers levels for cotton grown in relay intercropping with wheat. *Zagazig Journal of Agricultural Research, 32*, 1403–1425.

Ibrahim, E., Badr, M., & Abd El-Zaher, S. (2008). Response of some intercropping systems of wheat with sugar beet to bio-mineral nitrogenous fertilization. In *Proceeding (The Second Field Crop Conference) FCRI*. ARC, Giza, Egypt.

Ijoyah, M. (2014). Maize-soybean intercropping system: Effects on striga control, grain yields and economic productivity at Tarka, Benue State, Nigeria. *International Letters of Natural Sciences, 14*.

Kang, B., & Wilson, G. (1981). Effect of maize plant population and nitrogen application on maize cassava intercrop. In *Tropical root crops: Research strategies for the 1980s: Proceedings of the First Triennial Root Crops Symposium of the International Society for Tropical Root Crops-Africa Branch*, 8–12 September 1980, Ibadan, Nigeria.

Kugbe, X. J., Yaro, R. N., Soyel, J. K., Kofi, E. S., & Ghaney, P. (2018). Role of intercropping in modern agriculture and sustainability: A review. *British Journal of Science.16*(2).

Lawson, T. (1981). *Agroclimatology section*. Annual Report, International Institute of Tropical Agriculture, Ibadan, Nigeria.

McGilchrist, C., & Trenbath, B. (1971). A revised analysis of plant competition experiments. *Biometrics, 27*, 659–671.

Mead, R., & Willey, R. (1980). The concept of a 'land equivalent ratio'and advantages in yields from intercropping. *Experimental Agriculture, 16*(3), 217–228.

Medina, L., & Pimentel, B. (2012). Growth and yield of corn as affected by different row intercropping patterns with indeterminate tomato. *Philippine Journal of Crop Science (Philippines)*, 60–65.

Mohamed, W., Ahmed, N. R., & Abd El-Hakim, W. (2013). Effect of intercropping dates of sowing and N fertilizers on growth and yield of maize and tomato. *Egyptian Journal of Applied Sciences, 28*(12B), 625–644.

Motagally, F. A., & Metwally, A. (2014). *Maximizing productivity by intercropping onion on sugar beetAsian Journal of Crop Science, 6*(3), 226–235.

Nazir, M. S., Jabbar, A., Ahmad, I., Nawaz, S., & Bhatti, I. H. (2002). Production potential and economics of intercropping in autumn-planted sugarcane. *International Journal of Agriculture and Biology, 4*(1), 140–142.

Ofori, F., & Stern, W. (1987). Relative sowing time and density of component crops in a maize/cowpea intercrop system. *Experimental Agriculture, 23*(1), 41–52.

Okigbo, B. (1978). *Cropping systems and related research in Africa*. AAASA.

Ouda, S. A., & Zohry, A. E.-H. (2018). Cropping pattern to face climate change stress. In *Cropping pattern modification to overcome abiotic stresses* (pp. 89–102). Springer, Cham.

Pino, M., A., De-Los, Bertoh, M., & Espinosa, R. (1994). Maize as a protective crop for tomato in conditions of environmental stress. *Cult Trop, 15*, 60–63.

Reganold, J. P. (1992). *Effects of alternative and conventional farming systems on agricultural sustainability*. Food and Fertilizer Technology Centre, Taipei, Taiwan, 5.

Sherif, S., & Gendy, E. (2012). Growing maize intercropped with soybean on beds. *Egyptian Journal of Applied Sciences, 27*(9), 409–423.

Sherif, S. A., & Ibrahim, S. T. (2011). Intercropping maize with cassava in sandy soil. *Egyptian Journal of Applied Sciences, 26*(11), 744–755.

Sherif, S. A., Ibrahim, S. T., & Mohamed, W. K. (2011). Relay intercropping of cotton with wheat in reclaimed land. *Egyptian Journal of Agronomy, 33*(1), 51–65.

Sherif, S. A., Zohry, A., & Sahar, T. (2005). Intercropped with groundnut on growth, yield and yield components of both crops. *Arab Universities Journal of Agricultural Sciences, 13*(3), 771–791.

Tewe, O., & Egbunike, G. (1992). Utilization of cassava in non-ruminant livestock. In *Cassava as livestock feed in Africa*. IITA, Ibadan, Nigeria.

Toaima, S. (2004). Effect of relay intercropping cotton with sugar beet under different levels of fertilizers on its yield and yield components. *Egyptian Journal of Agricultural Research, 82*(4), 1641.

Toaima, S. (2006). Responce of onion, faba bean, and wheat to intercropping with fodder beet under different fertilizer levels of NPK. *Minufiya Journal of Agricultural Research, 31*(4), 333–345.

Toaima, S., Gadallah, R., & Mohamadin, E. (2007). Effect of relay intercropping systems on yield and its components of wheat and cotton. *Annals of Agricultural Science-Cairo-*, *52*(2), 317.

Tsubo, M., Walker, S., & Ogindo, H. (2005). A simulation model of cereal–legume intercropping systems for semi-arid regions: I. Model development. *Field Crops Research*, *93*(1), 10–22.

Willey, R., & Rao, M. (1980). A competitive ratio for quantifying competition between intercrops. *Experimental Agriculture*, *16*(2), 117–125.

3 Cereal Yield in Dry Environments

Adaptability of Barley vs. Wheat

Waleed Fouad Abobata, Clara R. Azzam, and Sahar A. Sherif

CONTENTS

DOI: 10.1201/9781003250845-3

3.1 INTRODUCTION

According to the International Agriculture Counseling Group, 1.4 billion people rely on agriculture in arid regions where water, not land, is scarce. Dry environments are associated with yield losses in marginal lands, in either early fall or winter (initial dry conditions) or late spring (terminal dry environment) (Ceccarelli & Grando, 1996). Potentially, the Mediterranean Basin's eastern side is also vulnerable to future climate change, which is expected to reduce the length of the rainy season, increase the frequency of dry conditions, and minimize barley (*Hordeum vulgare* L.) and wheat (*Triticum aestivum* L.) production (Dodig et al., 2015; Samarah, 2016).

Agriculture is confronted with a diverse range of challenges in cereal production. Climate change is one of these challenges, as is global water scarcity (Lobell et al., 2011). Significant concerns in sub-Saharan Africa include persistent biotic and abiotic stresses, inherently high climate variability, the threat of higher temperatures, and a harsher dry environment (due to climate change). Africa's rapid population growth is fueling an increase in food demand (Araus & Cairns, 2014).

While agriculture is faced with the challenge of meeting rising food demands, despite the existing and anticipated adverse effects of climatic changes on yields represented by temperatures and dry environments, the necessity of developing sustainable agriculture in dry environment conditions in times of global climate change highlights the difficulty of this task. Climate change has made the sustainability of cereal grain yield a challenge for food security. With a projected population of 9.7 billion people by 2050 and climatic changes, increasing cereal production has become an agricultural challenge.

Several simulation models have previously been proposed to predict climatic changes and the occurrence of dry environments, and it is critical to develop strategic crops to meet the stress of dry environments that limit crop yield production. Wheat (*T. aestivum* L.) and barley (*H. vulgare* L.) are two of the most crucial and widely grown crops in the world due to their social and economic value as food and feed. These crops are susceptible to drought stress. In 2013, dry environment stress affected nearly 65 million hectares of wheat production. Climate change and global warming are anticipated to increase the occurrence of dry environments, resulting in crop productivity losses in agriculture.

In Africa, wheat is cultivated over approximately 10 million hectares. It is a significant staple crop in many countries and an essential commodity throughout Africa. Wheat consumption in African countries has steadily increased over the last two decades due to the increasing population and socioeconomic changes associated with urbanization and changing food preferences. Support is needed to enable the production of improved and sustainable wheat-based technologies and innovations suitable for Africa's various agroecological zones and to improve the sustainable dissemination, scaling-up, and promotion of wheat-based technologies and innovations outside the value chain (Eltaher et al., 2018).

One of the world's five major crop species is barley, commonly used in animal feed, human food, malting, brewing, and distilling. The zone with the best conditions for producing optimal malting quality in barley cultivars has a combination of low rainfall and soil fertility with local crop rotational requirements. While barley users require a consistent raw material, a specific quality, and a reasonable price, barley breeders must develop new barley cultivars. Barley breeders want to improve cultivars for markets that require high-quality parameters, such as a clean and bright grain, low moisture, and disease resistance. When practicing breeding, it is critical to provide barley cultivars with the ability to produce high-quality barley in various environmental conditions (Knežević et al., 2004). Barley has been grown since antiquity. It is the most exhaustive adapted cereal crop, and it can withstand the vicissitudes of climate, such as temperature extremes and water scarcity. Until it was replaced by wheat, barley was the essential staple food for many countries. A shift in people's food preferences toward wheat has resulted in a drop in the demand for barley, resulting in a significant reduction in the area under barley cultivation.

A dry environment is a perennial issue, mainly in the Middle East, Asia, and Africa. Barley is commonly used in various applications, ranging from livestock feed to specialized foodstuffs. It is best known for producing malt, used in brewing and distilling. This plant species is a common crop in dry environments with low inputs and a high risk of crop failure because dry conditions reduce yield, as well as grain quality. The International Center for Agricultural Research in the Dry Areas carried out pragmatic strategies in barley breeding programs for stressed environments. Decentralizing breeding to exploit locally adapted germplasm, assessing correct germplasm in target environments, selecting farmers' fields, and using heterogeneous seed (mixtures and landraces) were all part of their strategy (Ceccarelli et al., 2010).

Crop breeding produces novel barley and wheat genotypes that are highly tolerant to the dry environment by hybridizing promising dry environment–tolerant genotypes and then selecting among their progeny. Previous researches have shown that dry environment tolerance is a polygenic characteristic, and genetic composition helps analyze the gene networks that control dry environment tolerance (Dawson et al., 2015).

Crop breeding research is critical for developing new barley and wheat cultivars with high dry environment tolerance. The grain yield must be increased in tandem with dry environment tolerance. To accomplish this, potential genotypes with genotypic differences for dry environment tolerance should be identified first (Baenziger, 2016). The selected characteristics are recorded on entire elite genotypes to define dry environment tolerance. There must also be some morphological, physiological, or yield-related characteristics that distinguish between dry environment–tolerant and susceptible genotypes, with high heritability estimates and a significant positive correlation with grain yield. The selection should then be made based on dry environment tolerance and yield. Following the identification and selection of tolerant genotypes, the breeding program could begin by hybridizing the selected genotypes as donor parents (Reynolds, 2001).

Single nucleotide polymorphism (SNP) markers are the most widely used method for covering wheat and barley genomes (Hussain et al., 2018; Thabet et al., 2018). Many SNPs are used in genome-wide association studies (GWAS) and quantitative trait loci (QTLs) mapping to explain the genetics of complex characteristics such as dry environment tolerance by identifying genomic regions or genes that may control target characteristics. GWAS or QTL mapping could be used to identify new genes controlling dry environment tolerance (Sukumaran et al., 2018; Zeng et al., 2014).

Improving dry environment stress tolerance is a difficult mission for barley and wheat researchers. The understanding of dry environment tolerance has advanced because of advancements in three master research fields: plant physiology, crop breeding, and plant genetics. Combining data from physiology, breeding, and genetics may aid in identifying the most waterless environment–tolerant genotypes with the genes controlling dry environment tolerance (Sallam et al., 2019). This chapter examines current advances in physiology, breeding, and genetic research to develop adaptability and tolerance to dry environments and the adaptation of barley versus wheat to these conditions.

3.2 PHYSIOLOGICAL AND BIOCHEMICAL REACTIONS TO DRY ENVIRONMENTS

A lack of water can cause dramatic morphological, physiological, biochemical, and molecular changes in a dry environment. Depending on the native environment, these changes can occur at any growth stage and reduce crop growth and production. As a result, genotypes may be studied for dry environment tolerance at various growth stages, as some genotypes may tolerate a dry environment at the seedling or germination stage. The characteristic that can measure the effect of dry environment stress on plants is used to identify dry environment tolerance. As a result, it is essential to determine the proper dry environment–tolerant characteristics (Zeng et al., 2014).

The investigation of various physiological and biochemical characteristics such as photosynthetic and osmotic adjustment, water-use efficiency (WUE), leaf turgor, relative water content (RWC), abscisic acid (ABA) content, transpiration efficiency (TE), stomatal conductance, carbohydrates, soluble stem mobilization, and senescence, as well as developmental, morphological, and ultrastructural characteristics such as leaf emergence, leaf waxiness, leaf area index (LAI), tiller development, stomatal densities, flowering time, cell membrane stability, maturity rate, yield, and yield components, leads to the concept of pyramiding the physiological characteristics and definitions of ideotypes for dry environment conditions (Blum, 1996; Ceccarelli et al., 2010).

According to the current physiological and biochemical performance constraints, water stress is the primary yield-limiting factor. In that case, physiological experience suggests that deep-rooted genotypes will have an advantage over others because they supply moisture from deeper levels in the soil. However, selecting for yield alone will not ensure that promising genotypes have the deepest roots because dry environment tolerance may be conferred through the genetic superiority of other mechanisms, such as superior spike photosynthesis, osmotic adjustment, heat-tolerant metabolism, accumulation and remobilization of stem reserves, and good emergence and establishment under dry environment stress (Reynolds, 2001).

Several researchers have previously investigated the dry environment tolerance of barley and wheat. However, the development of such crops for dry environment tolerance is hampered for various reasons. First, a dry environment can cause dramatic alterations in the physiological characteristics of plants, which must be understood and measured. Second, the interaction of genotypic × environmental (GE) factors will impact on selection. Third, dry environment tolerance is a complicated characteristic regulated by several genes, resulting in a slight genetic contribution; however, these are critical for genetically improving dry environment tolerance. Additional factors are also linked to crops, such as the structure and composition of barley and wheat genomes. Plant physiology and morphology, breeding, genetics, and gene expression research can be used to investigate dry environment stress (Abo-Doma & Azzam, 2007; Zeng et al., 2014).

When plants are subjected to dry environment conditions, they undergo physiological changes to withstand stress (Vinocur & Altman, 2005). Plants that thrive in dry environments try to accumulate proline content, chlorophyll content, amino acids, soluble sugars, and non-enzymatic and enzymatic antioxidant activities (Abid et al., 2016). Because of their importance in adapting to future climate change scenarios, plant physio-morphological characteristics are essential to a plant breeding program for improving dry environment tolerance (Bowne et al., 2012). Furthermore, the genes that control these physiological alterations are critical for breeders and geneticists because they are valuable sources of genetic improvement for dry environment tolerance. The breeding program demonstrates the significant physiological differences between susceptible and tolerant barley and wheat genotypes (Sallam et al., 2019) (Figure 3.1).

3.2.1 PHOTOSYNTHESIS

Plant development and grain yield are primarily driven by photosynthesis. As a result, knowing the physiological base of a plant's response to a dry environment is critical to its function. The photosynthetic rate of cereals is reduced by dry environment stress (Dawood et al., 2019). Factors influencing global plant development (nutrient and water uptake from soil, growth of photosynthetic tissue for carbon fixation and storage, nutrient and carbon relocation during grain filling) significantly impact grain yield (Li et al., 2013). While the number of florets determines the grain's number per inflorescence, the grain weight is determined by grain size and the protein content and starch accumulated during grain filling (Azzam & Abd El-Kader, 2010; Gambín & Borrás, 2010). The genetics of grain architecture is complex, involving paternal and maternal hormonal regulation, developmental signals, and the integration of environmental information such as the photoperiod and abiotic and biotic stresses (Kesavan et al., 2013).

The reproductive stage of barley and wheat development begins by transitioning the vegetative meristem into the reproductive meristem (inflorescence primordial). The stage concludes with physiological grain maturity, as evidenced by grain desiccation and entry into dormancy (Saini & Westgate, 1999). The initiation of floral,

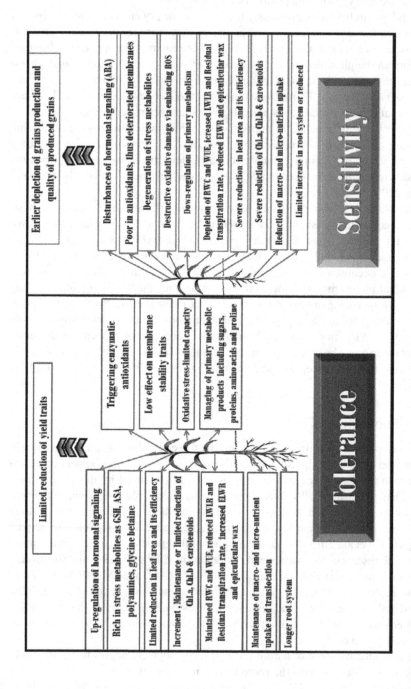

FIGURE 3.1 Significant physiological differences between susceptible and tolerant barley and wheat genotypes (adapted from Sallam et al., 2019).

inflorescence, and florets' differentiation, female and male gametogenesis, pollination, fertilization, and seed improvement are all sub-stages of the generative stage. The overall grain size is influenced by the nuclei's number formed during the syncytial phase, which exceeds 2000 in *Hordeum* and *Triticum* (Saini & Westgate, 1999). The grain is a caryopsis in barley and wheat, a kind of fruit fused with the seed and fruit coats. Husks surround the caryopsis. The caryopsis comprises a diploid embryo and a sizable triploid endosperm held together by maternal-origin tissues (testa and pericarp). The endosperm, outer bran, and inner germ make up each grain. Wheat grains are a good source of folate, copper, phosphorus, selenium, manganese, niacin, calcium, thiamine, and vitamin B6 (Figure 3.2). Barley grains, on the other hand, are

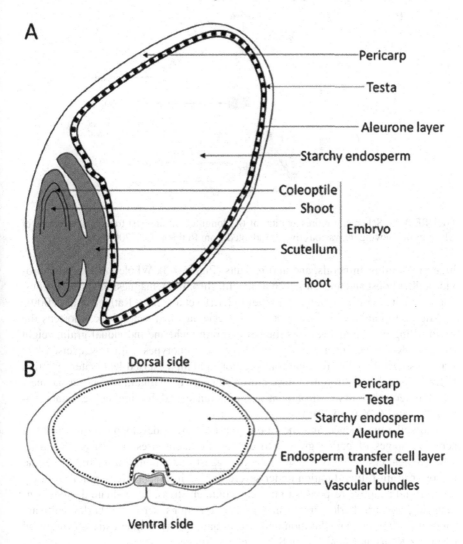

FIGURE 3.2 Schematic representation of (A) longitudinal and (B) transverse sections of barley grain viewing the various tissues (adapted from Li et al., 2013).

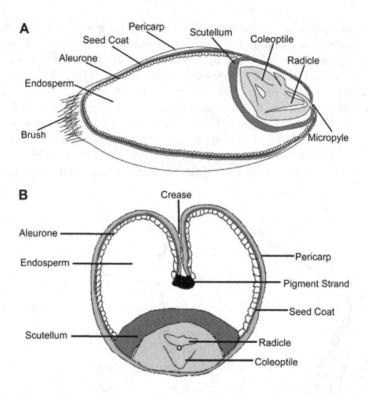

FIGURE 3.3 Schematic representation of (A) longitudinal and (B) transverse sections of wheat grain viewing the various tissues (adapted from Rathjen et al., 2009).

high in vitamins, minerals, and antioxidants (Figure 3.3). Whole barley grains provide a fiber boost and nutrients such as niacin, chromium, magnesium, phosphorous, copper, manganese, selenium, and vitamin B1 (Li et al., 2013; Rathjen et al., 2009).

Middle to late dry environment stress accelerates leaf senescence, shortens the grain-filling period, and reduces the barley grain yield and individual grain weight (Sánchez-Rodríguez et al., 2010). Furthermore, genotypes with low excised-leaf water loss (ELWL), low residual transpiration, and high excised-leaf water retention (ELWR) in a dry environment have a high capacity to maintain leaf water balance, which reflects their dry environment stress tolerance, and thus higher yield stabilization (Akrami & Yousefi, 2015).

The spike emergence and the initial phase of grain development are the most sensitive stages of barley growth to dry environment stress (Saini & Westgate, 1999; Sehgal et al., 2018). These findings suggest that a high grain-filling rate and a shorter grain-filling duration under severe dry environment stress conditions may be essential adaptive responses to dry environment stress. To evaluate dry environment responses in barley grown under such conditions, screening barley cultivars for grain-filling rate and duration under terminal dry environment stress is required (Sánchez-Rodríguez et al., 2010; Sehgal et al., 2018).

Metabolic alterations of photosynthetic activity might be caused by an imbalance in light capture and use (Kingston-Smith & Foyer, 2000), a reduction in Rubisco activity, injury to chloroplast membranes (Amirjani & Mahdiyeh, 2013), the disintegration of the chloroplast structure and photosynthetic system, chlorophyll photooxidation, the devastation of chlorophyll substrates, the suppression of chlorophyll biosynthesis, and an increase in chlorophyllase.

Furthermore, drought-tolerant wheat cultivars increase total chlorophyll at the pre-and post-anthesis stages and have an additional stable photosynthetic level. Thus, for limited grain yield reduction, breeders should select barley and wheat cultivars that can sustain photosynthetic systems and photochemical efficiency in a dry environment (Dawood et al., 2019).

3.2.2 WATER AND NUTRIENT RELATIONS

The important plant–water relations parameters are relative water content, water loss rate, excised-leaf water retention, succulence index, and residual transpiration rate. The relative water content, which measures the plant water status by reflecting the tissue metabolic activity, is the most helpful indicator of dehydration tolerance. Under dry environment stress conditions, the barley's grain yield is adversely correlated with leaf water potential (Samarah, 2016). The alteration in water loss to excised-leaf water loss could be used to estimate the plant's leaf water relations, particularly comparing hydrated leaves to leaves subjected to deficit irrigation. Dry environment stress increases excised-leaf water retention, reflecting the leaf's water retention mechanism under stress, which could be attributed to leaf rolling or a reduction in the exposed leaf surface area (Lonbani & Arzani, 2011). During the generative stages of barley and wheat, there is a significant positive correlation between grain yield and relative water content under dry environment stress.

Plants have evolved epicuticular waxes, which are organic cuticle combinations that cover the outer surface of plant tissues, to control water loss caused by epidermal conductance. Epicuticular wax may be an essential characteristic in dry environment–tolerant genotypes. They produce additional epicuticular wax on the leaves, reducing water loss from the surface of the plant leaf (Mukhtar et al., 2012). Water constraints combined with low N are the primary constraints to wheat yield, affecting chlorophyll fluorescence, leaf–water relations, and photosynthetic processes, resulting in the limited growth rate of the plant, early senescence, reduced duration of grain filling with restricted grain weight, and reduced crop productivity (Mobasser et al., 2014).

Dry environments have been demonstrated to reduce macronutrient uptake and translocation (N, P, and K^+) in numerous plant species, presumably due to reduced root volume (Noman et al., 2018). Nutrients are unavailable in dry soils. Water constraints combined with low N are the primary constraints to wheat yield, affecting chlorophyll fluorescence, leaf–water relations, and photosynthetic processes resulting in early senescence (Mobasser et al., 2014). The radius of water-filled pores decreases when the soil water content decreases, P mobility decreases, and tortuosity increases (Faye et al., 2006). Furthermore, dry environment stress reduces active cation transportation and membrane permeability (K^+, Ca_2^+, and Mg_2^+). These cations are absorbed at a lower rate through the roots (Farooq et al., 2012). Dry environment stress decreases Ca_2^+ concentrations in aboveground biomass, attributed to a decrease in transpiration flux (Kingston-Smith & Foyer, 2000). Noman et al. (2018) reported a similar reduction in calcium, potassium, and phosphorus levels in wheat plant roots and shoots when subjected to water stress. Some micronutrients, such as Fe, Mn, and Mo, can become deficient in a dry environment (Hu & Schmidhalter, 2005).

3.2.3 OXIDATIVE STATUS

3.2.3.1 Reactive Oxygen Species (ROS)

The negative effects of dry environment stress are determined by the timing, duration, and magnitude of the stress (Hasanuzzaman et al., 2018). ROS production is proportional to the degree of water stress, which causes membrane peroxidation, organelle peroxidation, enzyme activation and/or inactivation, and nucleic acid breakdown (Outoukarte et al., 2019). Low malonic dialdehyde (MDA) levels have been linked to the dry environment tolerance of wheat (Zhang et al., 2011). It is worth noting that increased lipoxygenase (LOX) enzyme activity is accountable for polyunsaturated fatty acid oxidation. As a result, it increases lipid peroxidation under stresses (Sánchez-Rodríguez et al., 2010). The diversity of reactive nitrogen species (RNS) is slightly greater than that of ROS. Increased uncontrolled creation of ROS and RNS may cause changes in macromolecules, which can serve as markers for oxidative and nitrosative stress (protein carbonylation and lipid peroxidation; lipid nitration, S-nitrosylation, and protein tyrosine nitration). Peroxynitrite, an excellent oxidant that could mediate the tyrosine nitration of proteins and might be a significant biomarker for nitrosative stress in the higher plants, is produced by the reaction of nitric oxide and superoxide radicals (Corpas et al., 2007).

Methylglyoxal is another stress metabolite that accumulates in plant cells due to normal physiological processes such as photosynthesis, but is dramatically increased in response to various abiotic stresses (Gomez Ojeda et al., 2013). It has been stated that methylglyoxal production increases in dry environments (Hossain et al., 2012; Nahar et al., 2017).

3.2.3.2 Antioxidant System

The production of antioxidant enzymes including catalase (CAT), peroxidase (POD), ascorbate peroxidase (APX), superoxide dismutase (SOD), dehydroascorbate reductase (DHAR), monodehydroascorbate reductase (MDHAR), glutathione reductase, and glutathione peroxidase (GPX) due to drought is the adaptive mechanism in cereals such as barley and wheat. The expression pattern of SOD, APX, and CAT depends on the plant's developmental phase and genotype under dry environment stress in barley. These genes may play a vital role in controlling dry environment stress (Dudziak et al., 2019). Wheat plants exposed to a mild dry environment increase their APX activity, while prolonged water deficit decreases due to increased MDA production (Nikolaeva et al., 2010). The wheat-tolerant genotypes have higher POD activity, higher phenolic content, and a lower damage index, demonstrating greater stomatal closure (Outoukarte et al., 2019). Glutathione reductase (GR) enzyme is essential for maintaining reduced glutathione (GSH), ascorbate (AsA) pools, and an appropriately reduced GSH/oxidized glutathione (GSSG) ratio that is more essential in defining plant tolerance to biotic and abiotic stresses than GSH content (Wang & Frei, 2011). Wheat has been shown to accumulate phenolic compounds in response to abiotic stresses such as drought (Outoukarte et al., 2019). The antioxidant part of phenolics and flavonoids, which minimizes accessibility and the potentials of ROS under dry environment–induced oxidative stress and improves plant protection by reducing lipid peroxidation, results in a significant increase in phenols and flavonoids in wheat plants flag leaves under deficit irrigation. The activities of one or more antioxidant enzymes increase in plants subjected to dry environments, and these enzymes may work in concert to avoid cellular damage. This increased activity is associated with a greater tolerance to dry environments (Sharma et al., 2014).

To summarize, identifying genes that encode these enzymatic activities in wheat and barley under dry environment stress is critical in the breeding program.

3.2.4 OSMOTIC BALANCE AND HORMONAL EFFECT

Plant adaptation to water deficit is divided into three categories: dry environment escape, dehydration tolerance, and dehydration avoidance, or a combination of all three. Osmolyte accumulation is a tolerance mechanism in dry environmental conditions that permits cells to manage the structural integrity of the membrane, and dehydration allows them to tolerate cellular dehydration and the dry environment (Loutfy et al., 2012). In dry climates, plants produce and accumulate compatible solutes such as polyols, sugars, and amino acids to help with osmotic balance, water absorption, and water retention (Hussain et al., 2018). Under rain-fed conditions, however, there is a significant relationship between total proteins and wheat grain yield (Farshadfar et al., 2008). Also, Noman et al. (2018) discovered an enhancement in shoot proteins in wheat plants grown under water stress conditions.

Dry environment stress also changes the endogenous rates of glycine betaine that protects cells from water scarcity by maintaining the osmotic balance between intracellular and extracellular environments, enhancing the quaternary structure of proteins, such as membrane proteins, and antioxidant enzyme protection, as well as the oxygen-releasing compound of photosystem II (Gou et al., 2015). It also controls the cytoplasm's pH, regulates the potential of intracellular osmosis, and stabilizes the structure of the cell membrane of wheat under dry environment stress conditions (Huseynova et al., 2016). The genetic variation of such osmotic changes could help improve barley selection programs for dry environment tolerance in wheat (e.g., selecting genotypes with greater proline content under dry environment stress than under standard conditions) (Sultan et al., 2012). When exposed to a dry environment, ABA accumulation in the stem and leaves or root exudates increases, while leaf cytokinin content decreases (Yang et al., 2004). Reduced 1-aminocyclopropane-1-carboxylic (ACC) acid concentrations and ethylene and increased ABA concentrations increase the grain-filling rate in developing grains of wheat in a mild dry environment. However, in a severe dry environment, ethylene, ABA, and ACC concentrations are excessively high, thereby reducing the grain-filling rate (Yang et al., 2004).

3.3 BREEDING FOR DRY ENVIRONMENT TOLERANCE

Polyploidy is common in wheat. *T. boeoticum* and *T. urartu* are two wild diploid (non-polyploid) wheat. Each diploid wheat cell has two complements of seven chromosomes ($2n = 2x = 14$), tetraploid ($2n = 4x = 28$), or hexaploid ($2n = 6x = 42$) wheat

(Levy & Feldman, 2002). The large wheat genome (1C approximately 17 Gbp) is polyploid and contains a high proportion of repetitive DNA. When combined, these characteristics impede the molecular analysis of the wheat genome. Due to its diploid nature ($2n = 2x = 14$), large chromosome (6–8 m), self-fertility, ease of hybridization, high level of normal and easily inducible variation, ease of doubled haploid (DH) production, and wide adaptation, barley is a beneficial genetic experimental organism. Many traditional genetics studies, such as cytogenetics and mutagenesis, benefit from these characteristics (Knežević et al., 2004).

Although identifying adaptive physiological mechanisms and their genetic markers can be time-consuming and expensive, the information is permanent once the initial investment is made. Depending on the resources available, the information can be used at various stages of the breeding process. In a low-investment scenario, information on the critical physiological characteristics of potential parental lines can be gathered. For example, it may be worthwhile screening an entire crossing block or a subset of commonly used parents to generate a catalog of valuable physiological characteristics or their genetic markers. The data can be used strategically in cross-design, increasing the likelihood of transgressive segregation events that combine desirable characteristics. In a scenario where more resources are available to screen for physiological characteristics, the same selection criteria could be applied to the segregating generations, yield characteristics, or any intermediate phase, depending on where the optimal genetic gains from the selection are obtained (Reynolds, 2001).

Collaboration between physiologists and breeders results in the rapid and effective screening of physiological characteristics for incorporation into breeding programs. Plant architecture can be an essential indicator of water stress tolerance. Root characteristics are also crucial in water exploration in soil. Tiller development and leaf growth are critical morphological characteristics studied in barley that respond to dry environments (Teulat et al., 1997).

The issue in barley breeding is the relationship between the selection and the target environment. Barley breeders divide breeding in two directions: selection for broad and specific adaptation and selection for wide selection has conducted enough breeding cycles under optimal or suboptimal environments and cultivar selected for favorable and unfavorable environments. Breeding for exact adaptation to adverse conditions is frequently associated with undesirable breeding outcomes because it is usually related to reduced quality or yield potential under favorable conditions. Given this, breeders must develop cultivars that maximize yield under favorable conditions (Knežević et al., 2004).

Dry environment tolerance is a highly complex characteristic that can be advanced from various angles. Breeding dry environment–tolerant lines necessitates functional assays for dry environment tolerance, a critical component of a practical plant breeding program. Dry environment–tolerant wheat and barley genotypes can be chosen based on their dry environment–tolerance characteristics. These genotypes are crossed to incorporate multiple tolerance genes for dry environment stress to improve dry environment tolerance. Breeders have traditionally relied on phenotypic selection for the desired characteristics. Because dry environment tolerance has a low heritability, it must be tested over a longer period or in different target environments (Eltaher et al., 2018).

Furthermore, because spatial variation frequently affects measurements of dry environment tolerance, the trials require multiple replications. Because the environments necessitate the selection of different types of dry environment tolerance, high ranks of G × E can lead to no progress in dry environment tolerance. As a result, the G × E interaction is significantly complex in breeding programs (Eltaher et al., 2018). Plant breeders have incorporated DNA molecular markers into the breeding programs to overcome the low heritability of dry environment tolerance, which has resulted in a significant improvement in dry environment tolerance in cereals (Zeng et al., 2014).

3.3.1 GENETICS FOR DRY ENVIRONMENT TOLERANCE AT DIFFERENT GROWTH STAGES

Climatic changes significantly impact the agricultural sector's response to dry environment stress by limiting the productivity and production of critical crops (e.g., wheat and barley). The severity of dry environment stress depends entirely on the environment. As a result, primary meteorological data on dry environments is required prior to designing a breeding program to enhance dry environment tolerance. Dry environment tolerance genetic variation can be studied under controlled conditions in greenhouses, growth chambers, and field conditions (Sallam et al., 2019). Numerous additional factors in dry environments, such as heat stress, erratic weather, and soil mineral nutrition, affect a field experiment on dry environment stress. As a result, testing dry environment tolerance under controlled conditions is beneficial, but primarily to supplement work in the field where numerous variables are uncontrollable (Sallam et al., 2019). The characteristics used to define dry environment tolerance differ depending on the stage of development. To understand the genetic variation in dry environment tolerance, this section focuses on the most critical development and growth phases in barley and wheat: germination, seedling, reproduction, and grain filling (Ashmawy et al., 2016).

3.3.1.1 At the Germination Stage

Seed germination is a sequence of events that begins with water absorption and ends with the radical emerging of the seedling from the seed coat. In barley and wheat, it is a sensitive stage for dry environment stress, reducing germination and seedling emergence. Understanding the normal variation and genetic foundation of germination and related characteristics in dry environments enhances barley and wheat growth and yield. Breeders aim to simulate great osmotic stress to investigate the effect of dry environment stresses on germination. Breeders concentrate on the optimal concentration to differentiate the tested genotypes according to their dry environment tolerance. The germination percentage (G percent), germination rate, and germination pace are the principal characteristics used to assess germination in dry environments. However, some related characteristics could be scored through the germination test, such as the root length, the shoot length, and the shoot:root ratio of germinated seeds (Thabet et al., 2018).

3.3.1.2   At the Seedling Stage

The seedling stage is extremely vulnerable to moisture stress. By assessing the genetic variation at this stage, it is possible to increase the intensity of the selection in breeding for drought-tolerant varieties (Hameed et al., 2010). Genetic variation investigation of dry environment tolerance focuses primarily on leaf and root characteristics during this stage. Understanding the relationship between these characteristics is critical for improving the breeding efficiency of dry environment–tolerant wheat and barley. Dry environment–tolerant plants can tolerate prolonged low relative leaf water content and water deficit (Basu et al., 2016), which can be measured by scoring days to wilting, leaf wilting, and stay green characteristics. These characteristics, known as morphological characteristics, can be visually scored. They can also serve as good predictors of physiological changes. The terms leaf wilting and days to wilting (DTW) refer to the loss of leaf water content, whereas staying green refers to the loss of chlorophyll content. A selection index was formed for every group to overcome dry environment stress: a tolerance index (DTW and sum of leaf wilting [SLW]) and a recovery index (days to regrowth, regrowth biomass, and dry environment survival rate). The two indices were combined to create the dry environment tolerance index (DTI) that strongly correlates with all tolerance and recovery characteristics and tolerance and recovery indices (Sallam et al., 2019). Due to the susceptibility of the seedling stage to a dry environment, wheat and barley breeders should select characteristics that address the essential aspects of dry environment tolerance (Abdel-Ghani et al., 2015).

3.3.1.3   At the Flowering and Grain-Filling Stages

Dry environment stress occurs during the flowering phase and may extend to the grain-filling stage, influencing the number of seeds/spike and the kernel weight, which are essential grain yield components. Breeders frequently use indirect selection and well-correlated, yield-related characteristics to improve the grain yield in dry environments because the grain yield is a complicated characteristic controlled by many genes (Sallam et al., 2019). Breeders use yield characteristics such as seedling vigor, plant height, days to heading, days to maturity, spike length, number of spikelets/spike, root architectural characteristics, number of grains/spike, thousand kernel weight, grain yield/spike, grain yield, biological yield, and the harvest index to assess dry environment stress on wheat and barley plants (Manschadi et al., 2006). Yield has two major components: grain number/unit area and grain weight. The grain number is determined during the pre-anthesis stage, and the grain weight is determined during the post-anthesis stage. Dry environment stress must be investigated during the pre-anthesis and post-anthesis stages. Because dry environment stress reduces the number of fertile spikes/unit area during crop establishment and tillering, as well as the number of grains/spike, it affects yield potential at the sink level (Maccaferri et al., 2011). Dry environment stress at the pre-anthesis phase in wheat can have more significant yield reductions than dry environment stress at the post-anthesis phase of growth. Barley is distinguished by its rapid pre-anthesis growth and ability to produce many tillers covering the soil surface and reducing water evaporation. This characteristic explains why barley outperforms wheat in dry areas (Ceccarelli et al., 2010). The shoot apical meristem in barley goes through three stages before flowering: vegetative, inflorescence, and floral. In each phase, the apical meristem generates different organs, which are influenced by a group of genes that establish and maintain meristem identity or transition (Bäurle & Dean, 2006).

Despite growing interest in individual morphological root characteristics and their functional implications for water capture under water-stressed conditions, there is still no agreement on whether an extensive root system contributes to wheat and barley adaptation to water-stressed environments (Petrarulo et al., 2015; Zeng et al., 2014), because the individual root characteristics, including the depth of rooting, root elongation rate, root distribution at depth, the diameter of the xylem vessel, angle of seminal roots, and root : shoot dry matter ratio, do not adequately describe a root system as large or small.

Breeders frequently choose low rainfall environments with irrigation to stress genotypes and compare them to identical genotypes in well-watered environments. Furthermore, a breeding program for improving dry environment tolerance varies depending on genotype performance, which is measured using G × E (Sallam et al., 2019).

3.3.2 Distinct Genotyping and Phenotyping for Improving Dry Environment Tolerance in Wheat and Barley

For the characteristic of interest, breeders have traditionally relied on phenotypic selection. Tolerance to dry environments can be achieved via direct or indirect selection of a related characteristic(s) that is more heritable or easier to identify. Because above-average moisture occasionally occurs, breeding for improved dry environment tolerance in cereals must be combined with high yield potential. From these crosses, plant breeders select an elite progeny for dry environment tolerance. Due to the low heritability of dry environment tolerance, it must be tested over a longer period or at multiple locations in the target environments. To uncover the genetic basis of dry environment tolerance in wheat and barley, the phenotypic and genetic variation of relevant characteristics in large populations with dense genetic maps is required (Araus & Cairns, 2014).

During this phase, genetic variation investigations of dry environment tolerance focus primarily on leaf and root characteristics. In dry environments, genotypic variation in root system characteristics has significant functional implications for water uptake and crop productivity. Water characteristics such as root architecture and vertical root distribution patterns are critical for improved adaptation in such environments (Manschadi et al., 2006).

Understanding the regular variation and genetic basis of germination and related characteristics in dry environments will aid the growth and yield of barley and wheat. In wheat and barley, high-throughput phenotyping (HTP) was used to evaluate genotypes for dry environment–tolerant characteristics such as seedling vigor, seminal root characteristics, and physiological characteristics (Richard et al., 2015).

High-throughput phenotyping is a novel technique for rapidly screening thousands of genotypes for different traits. In greenhouses or growth chambers, this technology requires a highly automated facility with good environmental controls, accurate sensing techniques, and robotics (Araus & Cairns, 2014).

The main impediment to implementing this technology is the high cost of skilled labor, which many educational institutions cannot afford. Most breeders can only evaluate dry environment tolerance in the field or under controlled conditions by scoring the primary characteristics, such as leaf rolling, staying green, and leaf wilting, which are frequently visually achieved in a reasonable attempt to incorporate physiological assays into plant breeding (Sallam et al., 2019).

3.3.3 THE USE OF NANOTECHNOLOGY TO IMPROVE BREEDING AND DRY ENVIRONMENT TOLERANCE

Plant breeders have recently developed an interest in agricultural nanotechnology, which is defined as the application of nanoparticles (NP) that may have beneficial effects on crops, with tools to improve productivity and tolerance to various biotic and abiotic stresses (Jasrotia et al., 2018). Nanotechnology has three advantages: it is inexpensive, it has a low consumption rate, and it has low phytotoxicity (Taran et al., 2017), though nanoparticles can have positive and harmful biological effects based on their concentration (Olkhovych et al., 2016). For example, dry environment tolerance in barley was investigated using SiO_2 and TiO_2 nanoparticles used in the field during the reproductive stage. Under dry environment stress, SiO_2 improved yield components, but TiO_2 reduced seed yield components at some concentrations (Ghorbanian et al., 2017).

Notably, under dry environment stress, both genotypes responded differently to nanoparticles, which can be explained by a genetic variation used in wheat breeding to improve dry environment tolerance (Taran et al., 2017). Because the primary goal of breeding research is to identify genes that control such genetic variation in order to genetically improve dry environment tolerance in wheat and barley, identifying such genes is critical.

3.4 GENETICS FOR DRY ENVIRONMENT TOLERANCE IN WHEAT AND BARLEY

Genetics is undergoing a significant shift from "one gene at a time" to a genomic approach, which promises to lead to an understanding of all genes that make up an organism, potentially leading to a reshuffling of the genome for new knowledge and new cultivars at a rate that evolution or traditional plant breeding has not been able to achieve. Barley genetics has made significant contributions to the field, both practically and theoretically (Kleinhofs, 2000).

Understanding the genetics of dry environment stress tolerance, such as a quantitative characteristic influenced by many quantitative trait loci and environmental factors, continues to be a source of consternation for plant biologists and geneticists (Fleury et al., 2010).

The genetics of these confounding factors at the molecular, physiological, biochemical, and biological levels and functions is involved in adaptation processes to dry environment stress conditions (Tricker et al., 2018). Tolerance to dry environments is a complicated characteristic. To determine the genes underlying dry environment–tolerant characteristics and the stage, mechanism, or process in which they are involved, intensive and integrative genetic, genomic, and molecular research is required. The genetic and molecular mechanisms underlying the dry environment tolerance of wheat and barley will eventually lead to the development of dry environment–tolerant varieties (Thabet et al., 2018).

3.4.1 The Genetic Basis of Dry Environment Tolerance

In wheat and barley, molecular markers and genome sequencing were used to conduct genetic analyses of dry environment tolerance. Wheat has approximately 800 QTLs for dry environment–tolerant characteristics (agronomic, physiological, root, and yield-related characteristics) discovered through genetic analyses, bi-parental mapping, and GWAS, with 700 and 110 QTLs and marker-trait associations (MTAs) discovered, respectively (Gupta et al., 2017); barley had 500 QTLs (Acuña-Galindo et al., 2015) and 90 MTAs (Thabet et al., 2018). A large-effect, stable QTL that controls many dry environment tolerance-related characteristics at various developmental stages would be ideal for crop improvement, but this has yet to be discovered.

Advances in wheat and barley genome sequencing, combined with cutting-edge bioinformatics tools, are assisting in QTL mapping and linking minor effect QTLs to physical positions on the genome, leading to candidate gene prediction and characterization (Sallam et al., 2019).

Ppd-H1 was discovered to be a photoperiod response gene that controls adaptation to different environments by modulating flowering time in response to long days (LDs). It also incorporates stress signals from the dry environment into floral development in barley (Wiegmann et al., 2019). Ppd-H1 has previously been linked to several shoot and spike-related characteristics in barley. It is an essential gene in coordinating the development of various plant organs with reproductive timing (Pham et al., 2019).

In addition, the circadian clock regulates stress cues and controls stress adaptation. Clock gene transcripts were downregulated when exposed to dry conditions. Ppd-H1 variation, on the other hand, had no consistent effects on clock gene expression, neither under control nor in a dry environment. This research backs up previous findings that the natural mutation in ppd-H1 had no effect on the expression of other barley clock homologs under control, osmotic, or high-temperature stress (Ejaz & von Korff, 2017).

3.4.2 Functional Validation of Dry Environment Tolerance QTLS and Candidate Genes

Recent technological advances, such as microarrays and the identification of gene-rich regions, indicate that genome size is less of an issue than previously thought. Nonetheless, gene functionality identification has been resolved, and transformation technology is now widely and consistently available. It will be possible to insert DNA from any species into economically important crop plants. Functional validation can make an essential contribution to barley characterization and germplasm collection. Although the process of describing genetic variability at the molecular level is nearly trivial, the mechanism for generating and maintaining DNA polymorphism is complex. The variability of barley gene pools provides opportunities to increase the profitability and productivity of barley breeding (Knežević et al., 2004).

QTL tools contribute significantly to barley genetics and breeding by assessing genome location and identifying the numbers of genes that determine complex phenotypes. Breeders will be able to systematically differentiate between structure and function, and partition end-point phenotypes into their regulatory and structural components using tools for sequencing gene-rich regions (Knežević et al., 2004). The functional validation and cloning of predicted candidate genes underlying dry environment tolerance QTLs have encountered challenges. Most QTLs are unstable in different environments. They were discovered using various marker types (simple sequence repeat [SSR], diversity arrays technology [DArT], amplified fragment length polymorphisms [AFLPs], and SNPs) and mapped in populations with other parents (Sallam et al., 2019).

Many transcription factor (TF) family members, such as NAC, dehydration-responsive element-binding protein (DREB), myeloblastosis (MYB), basic leucine zipper (bZIP), WRKY, and TZF, have been recognized as being involved in dry environment tolerance, as have protein kinases, such as mitogen-activated protein kinases (MAPK), calcium-dependent protein kinases (CDPK), and protein phosphatases (Sallam et al., 2019). The HvP5CS gene, which encodes delta-1-pyrroline-5-carboxylate synthase (P5CS), has been identified as the primary dry environment–tolerant gene in wheat and barley (Azzam et al., 2009; Sallam et al., 2019). Cloning may become more efficient and routine using high-throughput and precise phenotyping and genotyping. Many candidates for gene-based association mapping encoding many TFs involved in dry environment tolerance are provided by GWAS and must be validated and cloned. Hundreds of critical genomic regions were identified using the traditional QTL mapping approach, which aided in detecting the loci underlying the variation of dry environment tolerance-related characteristics in barley and wheat and elucidating the genetic factor of this complex characteristic. Multi-environmental field conditions are commonly used to assess genotype performance when using a different type of bi-parental population, such as a recombinant inbred line (RIL) population (Ejaz & von Korff, 2017), and a doubled haploid population (Obsa et al., 2016).

Finally, many QTLs for root-related characteristics under dry environment stress conditions were discovered using barley DH and RIL populations (Naz et al., 2014), which were validated by meta QTL (MQTL) (Li et al., 2013). While dozens of potential QTLs for dry environment tolerance-related characteristics in wheat and barley have been identified, very few have been validated or used in breeding programs to improve yield under dry environment stress. To improve dry environment tolerance in elite wheat Indian cultivars, desired alleles from some QTLs for several dry environment–related characteristics have been incorporated into breeding programs. The common QTLs between barley and wheat show promise in marker-assisted selection (MAS). Their efficacy has been tested in different locations, under different dry environment conditions, and in different genera.

3.4.3 GENOMICS ANALYSES OF DRY ENVIRONMENT TOLERANCE

Genomic selection (GS) is a relatively new approach to predicting genotype performance in wheat and barley. It is now being used to breed for dry environment tolerance. The few investigations that used this methodology to obtain genomic estimated breeding values (GEBVs) for grain yield discovered that they were between 0.4 and 0.50, indicating that synthetic wheat genotypes contributed to improved grain yield under dry environment stress (Jafarzadeh et al., 2016). Assessing the GEBVs for dry environment tolerance-related characteristics could be a valuable resource for genetic improvement and yield enhancement under dry environment stress conditions.

Functional genomics and QTL mapping are the most effective approaches for identifying essential genes and networks that mediate yield response under dry environment stress.

Marker coverage of the plant genome is critical for identifying the most relevant QTL associated with a characteristic of interest. The most common type of marker in genomic studies is single nucleotide polymorphism. Recent advanced technologies based on high-throughput next-generation sequencing (NGS) allow for cheap and quick deep sequencing of the genomes of model and non-model crops, greatly expanding the genetic information available. NGS includes a variety of sequencing technologies and genotyping methods, such as restriction site–associated sequencing (RADseq) (Chutimanitsakun et al. 2011), diversity array technology sequencing (DArTseq) (Cruz et al., 2013), and exome capture. Recent barley genome sequencing has enabled the accurate positioning and location of markers on the chromosome and an effective QTL search in barley germplasm (Mascher et al., 2017).

To date, some QTLs involved in the dry environment stress response in barley have been identified. Jabbari et al. (2018) used association mapping based on LD to find eight markers across the 3H, 5H, and 6H chromosomes significantly associated with grain number/spike under irrigated and water deficit conditions in barley. Honsdorf et al. (2017) discovered a previously unknown wild barley QTL allele on chromosome 4H that increased the thousand grain weight when subjected to terminal dry environment stress. Similarly, a QTL on chromosome 4H related to increased biomass under both dry environment and control conditions was identified by GWAS in a study involving the offspring of a cross between wild barley accessions and an elite barley cultivar (Pham et al., 2019). These findings suggest that in barley breeding programs, wild barley *H. spontaneum* can develop dry environment–tolerant alleles. In doubled haploid barley populations, several QTLs for grain plumpness and yield have been identified, with significant QTL × environment (QE) interaction (Obsa et al., 2017). To date, several dry environment tolerance-associated genes in plant species have been identified (Zhang et al., 2011).

Furthermore, the expression of miRNAs changes in response to dry environment stress. Under dry environment conditions, Hackenberg et al. (2015) discovered an upregulated miRNA, hvu-miR5049b. Furthermore, the authors reported that hvu-miR168-5p was only upregulated in leaves when exposed to a dry environment. At the same time, its expression level in barley roots remained unchanged, implying that some of the dry environment–regulated miRNAs can be expressed differently in barley tissues. Ferdous et al. (2017) determined that Hv-miR827 enhances dry environment tolerance in barley. These miRNAs can regulate many different genes involved in various biological and metabolic processes in plants, such as growth, development, and hormone signaling, thereby influencing plant stress response. Aside from miRNAs, changes in chromatin structure are caused by epigenetic factors such as DNA methylation and histone modifications in response to environmental conditions. Open and closed chromatin states, in response to changing environments, cause gene activation and gene silencing, respectively, and regulate a wide range of developmental processes in plants (Kapazoglou et al., 2013). DNA methylation and chromatin dynamics have been identified as tolerance mechanisms in crops to dry environment stress (Kamal et al., 2021; Kapazoglou et al., 2013). Transcription factors play an essential role in the regulatory networks that underpin plant responses to abiotic stresses (Golldack et al., 2014).

Collin et al. (2020) recently demonstrated that a barley mutant carrying ABA INSENSITIVE 5 (ABI5) genes (HvABI5) is more drought tolerant than its parents. ABI5 is a transcription factor of the basic leucine zipper family that functions in the ABA network. ABA is essential for regulating plant responses to abiotic stresses. ABA-dependent signaling changes the activity of stress-responsive genes in response to dry environment stress, regulating physiological processes such as photosynthesis, stomatal closure, and osmoprotectant biosynthesis (Cantalapiedra et al., 2017; Martignago et al., 2020). Alternative splicing (AS) was also discovered to differ between genotypes as a critical mechanism controlling dry environment–responsive gene expression in barley (Jayakodi et al., 2020). During the transcription of DNA to RNA, first precursor mRNAs (pre-mRNAs) are formed, which contain introns that disrupt the protein-coding regions. Splicing is a required step in the removal of introns from pre-mRNAs. Recent advances in barley genome sequencing have paved the way for new avenues of genetic research, such as QTL studies and current genome-wide association studies (Jayakodi et al., 2020; Mascher et al., 2017; Pham et al., 2019). To maintain homeostasis and functional stability in response to climate change, barley (*H. murinum* subsp. *leporinum*) undergoes epigenetic regulation of gene expression (Chano et al., 2021). To predict the genetic architecture and candidate genes of dry environment tolerance-related characteristics such as leaf, root, and yield, 108 bread wheat genotypes with 9,646 SNPs (Qaseem et al., 2018) and 200 bread wheat genotypes with 20,881 SNPs (Beyer et al., 2019) were used. The most recent wheat genome sequences were used to physically map the most consistent and important genomic regions associated with many agronomic and physiological properties in wheat under dry environmental stress (Bhatta et al., 2018).

3.4.4 GENETIC ENGINEERING OF DRY ENVIRONMENT–TOLERANT GENES IN WHEAT AND BARLEY

Modern crop breeding technology, including transgenic material, can speed up the breeding process and focus on specific quality characteristics. It also allows for the manipulation of quality. One of genetic engineering's primary goals is to produce stable inheritance and expression of dry environment–tolerant plants with single or multiple desired characteristics in subsequent generations (Knežević et al., 2004). Water-use efficiency, root weight, and biomass accumulation were enhanced in transgenic wheat lines under dry environment stress by expressing HVA1, the barley gene (Bahieldin et al., 2005). Overexpression of HvSNAC1 in barley improved dry environment tolerance and tolerance to other biotic stresses, such as *Ramularia cellocygni*, the fungal infection (Al Abdallat et al., 2014).

3.5 ADAPTATION OF BARLEY AND WHEAT TO DRY ENVIRONMENTS

Land allocation to wheat and barley is based on the assumption that these species are more sensitive to stress than others. Barley has a larger embryo than wheat, contributing to characteristics that may play a role in dry environment adaptation, such as having more seminal roots, longer roots, and greater leaf area during crop establishment (Richards, 2006). There have been few comparative studies of wheat and barley in the field where barley consistently outperformed wheat in low-yielding conditions and vice versa. In terms of grain weight, barley was as stable as bread wheat (Savin et al., 2015). Thus, the monoculture of barley in harsh environments does not necessarily improve productivity in the field, tending to reduce the diversification of production, thereby increasing system uncertainty. Wheat and barley of similar phenology must be compared under realistic field conditions to test the hypothesis that barley outyields wheat under severe stress (Savin et al., 2015).

Three main statements that are commonly accepted for cereal crops grown in dry environments and that have been confirmed by many researchers are (1) crops are exposed to a terminal dry environment; (2) barley outyields wheat in dry conditions; and (3) water deficit outweighs the benefit of nitrogen fertilization. Distelfeld et al. (2009) reported that barley is more adaptable to a wide range of environments than wheat due to the allelic diversity in VRN and PPD genes, as also reported by Ceccarelli et al. (2010) and Tambussi et al. (2005). They reported that barley is distinguished by its rapid pre-anthesis growth and ability to form a large number of tillers that cover the soil surface and reduce water evaporation. This characteristic explains why barley outperforms wheat in dry areas. Furthermore, Richards (2006) stated that barley has a larger embryo than wheat, which contributes to characteristics that may play a role in dry environment adaptation, such as barley having more seminal roots, longer roots, and a greater leaf area during crop establishment.

3.5.1 ADAPTATION OF BARLEY AND WHEAT TO DRY ENVIRONMENTS: MORPHOLOGICAL CHARACTERS

Barley plant adaptation to environmental conditions has been studied for many years. Genetic research has focused on selecting cultivars that are more tolerant to environmental stress caused by various environmental conditions and cultivars with improved quality, increased yield, and yield stability. Many genes are primarily responsible for abiotic stress tolerance (Russell et al., 2016). In classical times, wheat became more important as a source of human food, while barley was fed to animals (Nordblom, 1985). The amount and distribution of rainfall vary significantly within and between seasons (Ceccarelli & Grando, 1996).

Furthermore, the timing of abiotic stresses, such as low temperatures, is unpredictable, resulting in a high genotype × year × location interaction variance. As a result, the genetic gains for a grain yield with direct selection are expected to be slow. Using an analytical approach when selecting for yield under dry conditions is justified by the desire to enhance the efficiency of traditional breeding.

Based on the models of Fischer and Turner (1978) and Passioura (1977), morphological characteristics such as prostrate winter growth habit, good early ground cover, cold tolerance, vigorous seedling growth, light plant color at anthesis, and early ear emergence have been found to be positively related to grain yield under dry conditions (Acevedo et al., 1991). However, Ceccarelli and Grando (1996) concluded that indirect selection should be based on a combination of characteristics rather than individual characteristics.

Differences in grain yield are explained in the analytical approach by understanding the characteristics relevant to the plant ideotype (Fischer & Turner, 1978). Such characteristics can only be used as indirect selection criteria by breeders if they meet specific criteria: (1) there must be sufficient genetic variation available; (2) they must have a high heritability; (3) there must be a good correlation with yield in a dry environment; and (4) they must be accessible and inexpensive to screen. Grain yield is determined by transpiration, transpiration efficiency, and the harvest index (Fischer & Turner, 1978; Passioura, 1977). Manschadi et al. (2006) discovered that genetic differences in root architecture between drought-tolerant wheat (SeriM82) and barley (Mackay) genotypes might be linked to two distinct adaptation strategies. In contrast, rather than increasing total water use, the adaptation of barley cv. Mackay is based on developing a large and shallow root system with tremendous potential for water extraction early in the season to optimize soil water extraction timing. As a result, root architecture appears to be an essential target characteristic for the genetic improvement of root systems to increase wheat productivity under water constraints. They proposed that, like barley, selecting drought-tolerant wheat genotypes should improve root system characteristics to allow more water uptake during the grain-filling period. Because the actual Es is primarily determined by the amount of radiation reaching the soil, which is the source of water loss during winter via evaporation, strategies to reduce Es must emphasize early crop establishment and good ground cover in winter. Early growth vigor, tillering capacity, cold tolerance, and prostrate growth habit increase crop water availability (Cooper et al., 1983; Fischer & Turner, 1978; Richards, 2006).

The amount of dry matter produced by the crop per unit of water transpired is transpiration efficiency. Because evaporative demands are lowest during the cool winters in the Middle East, the ability to maintain growth under low temperatures, i.e., good early growth vigor, increases TE (Acevedo et al., 1991; Richards, 2006). However, suppose the vapor pressure deficit is high early in the season. In this case, good early growth vigor may lead to excessive water use, and a trade-off between TE and the harvest index may occur (Passioura, 1977). If this is the case, then a poorer early growth vigor combined with a prostrate growth habit to ensure adequate ground cover may be a better strategy.

Plant color is another characteristic that has been linked to TE (Fischer & Turner, 1978). Acevedo et al. (1991) found that lower chlorophyll a/b ratios resulted in a darker leaf color but not higher chlorophyll content/unit leaf area. Because photosystems in the leaves primarily contain chlorophyll a, darker leaves with a lower chlorophyll a/b ratio may have a higher content of antenna chlorophyll and thus better interception of incoming radiation. The harvest index can be manipulated by earlier heading in terminal dry environment stress (Passioura, 1977). Being early is advantageous because of the lower TE in the spring (Fischer & Turner, 1978).

3.5.2 ADAPTATION OF BARLEY AND WHEAT TO DRY ENVIRONMENTS: APICAL DEVELOPMENT, LEAF, AND TILLER APPEARANCE

Barley originates in the Middle East (*H. vulgare* L.), where it has been cultivated for at least 8000 years (Ejaz & von Korff, 2017; Knežević et al., 2004; Russell et al., 2016). Barley landraces are still widely grown in this region. The evaluation of landraces collected in Jordan and Syria revealed several characteristics that may be adaptive to barley grown in low-rainfall harsh environments (Ceccarelli & Grando, 1996; Mikołajczak et al., 2016). Although there is considerable variation within collection sites, there are significant differences in the average expression of the morphological characteristics between sites. Landraces from the desert and steppe areas of east and northeast Syria, where terminal dry environment stress occurs in conjunction with low winter temperatures, had a prostrate winter growth habit, dark winter plant color, poor early growth vigor, and cold tolerance. Jordanian landraces with mild winters and terminal dry environment stress combined a more erect winter growth habit, light winter plant color, and good early growth vigor with cold sensitivity and the ability to recover from cold damage (Ceccarelli & Grando, 1996; Mikołajczak et al., 2016; Van Oosterom & Acevedo, 1992). This plant ideotype was similar to spring barleys found in Australia and Egypt, where terminal dry environment stress is combined with mild winters. Both ideotypes resulted in inadequate ground cover in spring when combined with an appropriate earliness, which is critical for achieving good yields in low-rainfall harsh environments experiencing terminal dry environment stress (Acevedo et al., 1991; Cooper et al., 1983; Fischer & Turner, 1978; Van Oosterom & Acevedo, 1992). Understanding the significance of these ideotypes for low-rainfall harsh regions requires examining plant development.

There are two critical stages in the development of the barley apex's pre-anthesis; the first stage is the transition from leaf primordia to spikelet primordia initiation. This marks the end of the apex's vegetative phase and the start of the spike initiation phase. The transition is characterized by an increase in the rate of primordia initiation. The second stage occurs when the meristematic dome has stopped forming new primordia, and the maximum number of primordia (MP) has been initiated (Kirby et al., 1985). When the spike initiation phase is over, the spike growth phase, which lasts from the MP stage to anthesis, begins. The observed differences in plant ideotype landraces may be due to differences in the duration of the three phases (vegetative, spike initiation, and spike growth phase). The appearance of cereal leaves and tillers is an expression of cereal growth. Increased leaf and tiller numbers can increase ground cover and, as a result, the fraction of incoming radiation intercepted by the crop. Water losses due to direct soil evaporation can be reduced as a result (Cooper et al., 1983; Fischer & Turner, 1978; Mikołajczak et al., 2016). According to Ceccarelli et al. (2010), barley is distinguished by its rapid pre-anthesis growth and ability to form many tillers that cover the soil surface and reduce water evaporation.

As long as temperatures do not fall below a base temperature of around 0°C, leaf appearance rate is a linear function of temperature (Kirby et al., 1985). The tiller appearance rate is proportional to the rate of leaf appearance. Significant genotypic differences in the leaf appearance rate have been stated for barley and wheat (Kirby et al., 1985).

At various stages of development, barley can be grazed. Grazing at the tillering stage is a common practice in northeast Syria. Such grazing reduces grain and straw yield, but this is more than offset by forage revenue (Nordblom, 1985). The crop is allowed to recover after grazing to produce grain. Low precipitation, low winter temperatures, high temperatures, and vapor pressure deficits limit the barley yield in spring. The majority of rain falls during the cool winter months (Mikołajczak et al., 2016), and little or no residual soil water is available after the hot and dry summer. Increasing the fraction of available water transpired by the crop and optimizing water-use efficiency, defined as the aboveground biomass produced/unit of evapotranspiration, are two mechanisms for increasing dry matter production under water-stressed conditions (Fischer & Turner, 1978).

Direct soil evaporation is a significant source of water loss, with losses of up to 60% of seasonal evapotranspiration reported (Cooper et al., 1983). These losses can be reduced by achieving good early ground cover through a prostrate growth habit, the ability to grow at low temperatures, and cold tolerance (Richards, 2006). In terms of WUE, rapid early biomass accumulation is also advantageous. WUE decreases when temperature, radiation, and vapor pressure deficits rise sharply (Dodig et al., 2015). When WUE is still high, a high crop growth rate (CGR) should produce additional biomass in water-limited environments.

The CGR is proportional to the radiation intercepted by the leaf surfaces early in the season. Suppose the leaf area index, defined as the leaf area/unit ground area, is less than 3 (Poorter & Remkes, 1990), then the intercepted fraction of radiation is primarily determined by the leaf area. A high leaf area early in the season may thus increase the CGR early in the season, assuming that the crop has adequate cold tolerance. For a fast CGR and high biomass, it may be preferable to produce many small and narrow leaves rather than fewer large leaves. The specific leaf area (SLA), defined as the leaf area/unit leaf dry weight, significantly influences the growth rate (m^2 kg^1). SLA variations can be attributed to variations in leaf thickness, density, or both. If water is not a constraint, a high SLA can maximize plant growth; however, in low-productivity environments, a low SLA will likely allow for better survival (Witkowski & Lamont, 1991).

3.5.3 ADAPTATION OF BARLEY AND WHEAT TO DRY ENVIRONMENTS: PLANT IDEOTYPE AND GRAIN YIELD

Early heading in cereals is a critical escape mechanism under terminal stress, and correlations between earliness and grain yield have been extensively reported (e.g., Ceccarelli & Grando, 1996; Fischer & Turner, 1978; Van Oosterom & Acevedo, 1992). Frost occurs in harsh environments, but the timing and severity are highly unpredictable. As a result, the heading date must be early enough to allow for adequate grain filling but late enough to avoid cold damage. Cultivars less affected by environmental fluctuations are preferable in environments with a narrow optimum period for heading dates.

There is debate over whether yield testing in a breeding program for dry areas should be carried out under favorable or unfavorable conditions. Pfeiffer (1988) reported that yield selection in favorable environments improved dry environment tolerance in bread wheat. However, Ceccarelli and Grando (1996) argue that for barley in harsh environments, segregating populations should be screened as soon as possible in the dry target environment. This is done to avoid the loss of tolerant material when selecting some cycles in an ideal environment. However, seed availability is limited in early generations, and yield is challenging to assess conventionally (Acevedo et al., 1991). In dry environments, where all material is at risk in arid years, an early, indirect yield assessment in a more favorable environment would be highly beneficial. Many plant attributes with potential yield benefits in a dry environment have been identified, primarily on theoretical grounds, such as ground cover, winter growth vigor, winter growth habit, plant color, and cold tolerance (Acevedo et al., 1991; Fischer & Turner, 1978; Passioura, 1986; Richards, 2006). Clear evidence of their utility as indirect selection criteria, on the other hand, is scarce. This is not surprising given that Ceccarelli and Grando (1996) and Van Oosterom and Acevedo (1992) demonstrated that different combinations of relatively simple characteristics could produce comparable yields. Individual characteristics must thus be considered a component of the overall plant ideotype.

In the Middle East, water and temperature stress are significant constraints on cereal production. Annual rainfall variation has been reported to explain 75% of the variation in grain yield for wheat across 16 environments in this region, with annual rainfall ranging from 230 to 765 mm (Blum, 1996). Temperature variations are also linked to variations in grain yield. Ceccarelli and Grando (1996) reported average barley grain yields of 1562 and 32 kg ha^{-1} from a single site in northern Syria over two seasons with comparable rainfall and distribution but varying temperatures. Reduced responsiveness to environmental fluctuations can reduce crop failure risk if marginal yields occur due to adverse environmental conditions. Two contrasting plant ideotypes for barley (*H. vulgare* L.), adapted to terminal dry environment–stressed harsh environments, have been identified (Van Oosterom & Acevedo, 1992). The first ideotype is typical of spring-type barleys, with rapid seasonal development that results in excellent early growth vigor and early heading on the one hand, but cold sensitivity on the other. The second ideotype is vernalization dependent, with low early growth vigor, prostrate winter growth, low cold tolerance, and medium early heading. The second ideotype may produce more stable yields in areas prone to dry environments and cold stress due to differences in development and growth. Rainfall has the greatest impact on grain yield, but multiple regression analyses have revealed that temperature can also have a significant impact (Ceccarelli et al., 2010).

3.6 CONCLUSION

When plants are subjected to abiotic stress, such as dry environments, various physiological, metabolic, and defense systems are activated to allow plants to survive and maintain growth and productivity (Valliyodan & Nguyen, 2006). The tolerance/

sensitivity genetics of dry environments is complicated, and the linked characteristics are complicated and polygenic, making the development of dry environment–tolerant cultivars difficult. On the other hand, proteomics, transcriptomics, and gene expression investigations have enabled the identification of the factors involved in the regulation of the synthesis of several proteins that may provide stress tolerance.

Plants have a system for detecting and responding to environmental cues such as dry conditions. The perceived signal is transduced, resulting in genes encoding proteins involved in dry environment tolerance (Figure 3.4) (Gahlaut et al., 2016). Plants adapt to dry environments through two distinct mechanisms, dry environment avoidance and dry environment tolerance, which are not mutually exclusive. Table 3.1 lists the various morphological and physiological characteristics involved in each dry environment's adaptation mechanism.

Barley is an excellent model plant for studying the genetics of drought adaptation because not only is it a valuable crop but it also has a high degree of genetic variability for stress tolerance. Throughout the wheat belt, barley is grown for grain, green forage, and, to a lesser extent, hay. The main grain production areas are distributed similarly to wheat. Since the early 1970s, production has steadily increased, and barley is now second only to wheat crop in size.

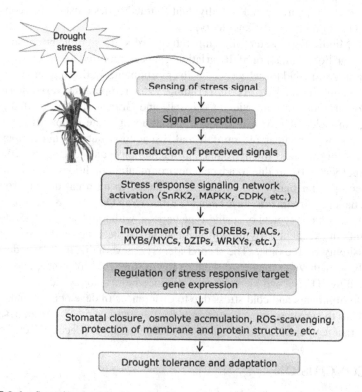

FIGURE 3.4 Steps involved in the expression of drought tolerance, from the perception of drought stress and transducing the signal through transcription factors for the activation of genes involved in adaptation (adapted from Gupta et al., 2017).

TABLE 3.1
A Summary of the Morphological and Physiological Characteristics and Adaptation Mechanisms under Dry Environment Stress Conditions

No.	Adaptation mechanism/trait	Ease of use (+/++/+++)	References
	I. Avoidance		
1.	Leaf rolling	+++	Araus (1996)
2.	Leaf glaucousness	+++	Tsunewaki and Ebana (1999)
3.	Shoot vigor	+++	Richards (1996)
4.	Transpirational cooling (cooler canopy)	++	Pinto et al. (2010)
5.	Stomatal conductance	+	Reynolds et al. (1994)
6.	Early maturation	+++	Tewolde et al. (2006)
7.	Membrane stability	+	Reynolds et al. (1994)
8.	Green flag leaf area (stay green)	+++	Kohli et al. (1991)
9.	Root vigor and architecture	+	Hurd (1974)
	II. Tolerance		
1.	Photosynthetic rate	+	Al-Khatib and Paulsen (1999)
2.	Chlorophyll content	++	Farooq et al. (2009)
3.	ABA accumulation	+	Innes et al. (1984)
4.	Osmoprotectant accumulation	+	Abebe et al. (2003)
5.	Soluble sugar content	+	Rosa et al. (2009)
6.	Generation of reactive oxygen species	+	Gill and Tuteja (2010)

Difficult: +; easy: ++; very easy: +++.

The early maturity of barley, when compared to wheat, allows it to be grown successfully for grain in dry environments where the season is cut short by low-water/high-temperature stress. Its earlier maturity has led to it being regarded as drought tolerant in many countries, but this is more of a reflection of drought escape than of resistance per se. Its drought tolerance, combined with its root system's excellent soil-binding properties and resistance to sand blast, allows for vigorous vegetative growth on dry sand dunes where wheat would barely survive. The responses of plants to drought stress are complex, and they have evolved various strategies to mitigate the negative effects of harsh environments by altering their physiological, molecular, and cellular functions. Applying specific agronomic practices, on the other hand, plays an important role in adaptation to dry environments.

REFERENCES

Abdel-Ghani, A. H., Neumann, K., Wabila, C., Sharma, R., Dhanagond, S., Owais, S. J., Börner, A., Graner, A., & Kilian, B. (2015). Diversity of germination and seedling traits in a spring barley (*Hordeum vulgare* L.) collection under drought simulated conditions. *Genetic Resources and Crop Evolution*, 62(2), 275–292.

Abebe, T., Guenzi, A. C., Martin, B., & Cushman, J. C. (2003). Tolerance of mannitol-accumulating transgenic wheat to water stress and salinity. *Plant Physiology*, 131(4), 1748–1755.

Abid, M., Tian, Z., Ata-Ul-Karim, S. T., Cui, Y., Liu, Y., Zahoor, R., Jiang, D., & Dai, T. (2016). Nitrogen nutrition improves the potential of wheat (*Triticum aestivum* L.) to alleviate the effects of drought stress during vegetative growth periods. *Frontiers in Plant Science, 7*, 981.

Abo-Doma, A., & Azzam, C. R. (2007). Hunting of some differentially expressed genes under salt stress in wheat. Egyptian Journal of Plant Breeding, *11*, 233–244.

Acevedo, E., Craufurd, P., Austin, R., & Perez-Marco, P. (1991). Traits associated with high yield in barley in low-rainfall environments. *The Journal of Agricultural Science, 116*(1), 23–36.

Acuña-Galindo, M. A., Mason, R. E., Subramanian, N. K., & Hays, D. B. (2015). Meta-analysis of wheat QTL regions associated with adaptation to drought and heat stress. *Crop Science, 55*(2), 477–492.

Akrami, M., & Yousefi, Z. (2015). Biological control of fusarium wilt of tomato (*Solanum lycopersicum*) by *Trichoderma* spp. as antagonist fungi. *Biological Forum, 7*(1), 887

Al Abdallat, A., Ayad, J., Abu Elenein, J., Al Ajlouni, Z., & Harwood, W. (2014). Overexpression of the transcription factor HvSNAC1 improves drought tolerance in barley (*Hordeum vulgare* L.). *Molecular Breeding, 33*(2), 401–414.

Al-Khatib, K., & Paulsen, G. M. (1999). High-temperature effects on photosynthetic processes in temperate and tropical cereals. *Crop Science, 39*(1), 119–125.

Amirjani, M. R., & Mahdiyeh, M. (2013). Antioxidative and biochemical responses of wheat to drought stress. *Journal of Agricultural and Biological Sciences, 8*(4), 291–301.

Araus, J. (1996). Integrative physiological criteria associated with yield potential. In *Increasing yield potential in wheat: Breaking the barriers* (pp. 150–167). Mexico, CIMMYT.

Araus, J. L., & Cairns, J. E. (2014). Field high-throughput phenotyping: The new crop breeding frontier. *Trends in Plant Science, 19*(1), 52–61.

Ashmawy, H., Azzam, C. R., & Fateh, H. S. (2016). Variability, heritability and expected genetic advance in barley genotypes irradiated with gamma rays التباينات، كفاءة التوريث و التحسين الوراثي المتوقع في in m3, m4 and m5 generations= التراكيب الوراثية للشعير المعاملة بأشعة جاما في الأجيال الطفورية الثالث و الرابع و الخامس. *Egyptian Journal of Plant Breeding, 203*(3795), 1–17.

Azzam, C. R., & Abd El-Kader, M. H. (2010). Effect of wheat variety on protein patterns, bread and cake quality. *Egyptian Journal of Plant Breeding, 14*(1), 135–157.

Azzam, C. R., Edris, S., & Mansour, A. (2009). Changes in wheat P5CS gene expression in response to salt stress in wheat. *Egyptian Journal of Genetics and Cytology, 38*(2).

Baenziger, P. (2016). Wheat breeding and genetics. In *Reference module in food science* (pp. 1–10).

Bahieldin, A., Mahfouz, H. T., Eissa, H. F., Saleh, O. M., Ramadan, A. M., Ahmed, I. A., Dyer, W. E., El-Itriby, H. A., & Madkour, M. A. (2005). Field evaluation of transgenic wheat plants stably expressing the HVA1 gene for drought tolerance. *Physiologia Plantarum, 123*(4), 421–427.

Basu, S., Ramegowda, V., Kumar, A., & Pereira, A. (2016). Plant adaptation to drought stress. *F1000Research, 5*.

Bäurle, I., & Dean, C. (2006). The timing of developmental transitions in plants. *Cell, 125*(4), 655–664.

Beyer, S., Daba, S., Tyagi, P., Bockelman, H., Brown-Guedira, G., & Mohammadi, M. (2019). Loci and candidate genes controlling root traits in wheat seedlings—A wheat root GWAS. *Functional & Integrative Genomics, 19*(1), 91–107.

Bhatta, M., Morgounov, A., Belamkar, V., Yorgancılar, A., & Baenziger, P. S. (2018). Genome-wide association study reveals favorable alleles associated with common bunt resistance in synthetic hexaploid wheat. *Euphytica, 214*(11), 1–10.

Blum, A. (1996). Crop responses to drought and the interpretation of adaptation. In *Drought tolerance in higher plants: Genetical, physiological and molecular biological analysis* (pp. 57–70). Springer, Dordrecht.

Bowne, J. B., Erwin, T. A., Juttner, J., Schnurbusch, T., Langridge, P., Bacic, A., & Roessner, U. (2012). Drought responses of leaf tissues from wheat cultivars of differing drought tolerance at the metabolite level. *Molecular Plant, 5*(2), 418–429.

Cantalapiedra, C. P., García-Pereira, M. J., Gracia, M. P., Igartua, E., Casas, A. M., & Contreras-Moreira, B. (2017). Large differences in gene expression responses to drought and heat stress between elite barley cultivar Scarlett and a Spanish landrace. *Frontiers in Plant Science, 8*, 647.

Ceccarelli, S., & Grando, S. (1996). Drought as a challenge for the plant breeder. *Plant Growth Regulation, 20*(2), 149–155.

Ceccarelli, S., Grando, S., Maatougui, M., Michael, M., Slash, M., Haghparast, R., Rahmanian, M., Taheri, A., Al-Yassin, A., & Benbelkacem, A. (2010). Plant breeding and climate changes. *The Journal of Agricultural Science, 148*(6), 627–637.

Chano, V., Domínguez-Flores, T., Hidalgo-Galvez, M. D., Rodríguez-Calcerrada, J., & Pérez-Ramos, I. M. (2021). Epigenetic responses of hare barley (*Hordeum murinum* subsp. *leporinum*) to climate change: An experimental, trait-based approach. *Heredity, 126*(5), 748–762.

Chutimanitsakun, Y., Nipper, R., & Cuesta-Marcos, A., Cistu e, L., Corey, A., Filichkina, T., Johnson, E. A., & Hayes, P. M. (2011). Construction and application for QTL analysis of a restriction site associated DNA (RAD) linkage map in barley. *BMC Genomics, 12*(4), 1–13.

Collin, A., Daszkowska-Golec, A., Kurowska, M., & Szarejko, I. (2020). Barley ABI5 (abscisic acid INSENSITIVE 5) is involved in abscisic acid-dependent drought response. *Frontiers in Plant Science, 11*, 1138.

Cooper, P., Keatinge, J., & Hughes, G. (1983). Crop evapotranspiration—A technique for calculation of its components by field measurements. *Field Crops Research, 7*, 299–312.

Corpas, F. J., Luis, A., & Barroso, J. B. (2007). Need of biomarkers of nitrosative stress in plants. *Trends in Plant Science, 12*(10), 436–438.

Cruz, V. M. V., Kilian, A., & Dierig, D. A. (2013). Development of DArT marker platforms and genetic diversity assessment of the US collection of the new oilseed crop lesquerella and related species. *PloS one, 8*(5), e64062.

Dawood, M. F., Abeed, A. H., & Aldaby, E. E. (2019). Titanium dioxide nanoparticles model growth kinetic traits of some wheat cultivars under different water regimes. *Plant Physiology Reports, 24*(1), 129–140.

Dawson, I. K., Russell, J., Powell, W., Steffenson, B., Thomas, W. T., & Waugh, R. (2015). Barley: A translational model for adaptation to climate change. *New Phytologist, 206*(3), 913–931.

Distelfeld, A., Li, C., & Dubcovsky, J. (2009). Regulation of flowering in temperate cereals. *Current Opinion in Plant Biology, 12*(2), 178–184.

Dodig, D., Zorić, M., Jović, M., Kandić, V., Stanisavljević, R., & Šurlan-Momirović, G. (2015). Wheat seedlings growth response to water deficiency and how it correlates with adult plant tolerance to drought. *The Journal of Agricultural Science, 153*(3), 466–480.

Dudziak, K., Zapalska, M., Börner, A., Szczerba, H., Kowalczyk, K., & Nowak, M. (2019). Analysis of wheat gene expression related to the oxidative stress response and signal transduction under short-term osmotic stress. *Scientific Reports, 9*(1), 1–14.

Ejaz, M., & von Korff, M. (2017). The genetic control of reproductive development under high ambient temperature. *Plant Physiology, 173*(1), 294–306.

Eltaher, S., Sallam, A., Belamkar, V., Emara, H. A., Nower, A. A., Salem, K. F., Poland, J., & Baenziger, P. S. (2018). Genetic diversity and population structure of F3: 6 Nebraska winter wheat genotypes using genotyping-by-sequencing. *Frontiers in Genetics, 9*, 76.

Farooq, M., Hussain, M., Wahid, A., & Siddique, K. (2012). Drought stress in plants: An overview. In *Plant responses to drought stress* (pp. 1–33).

Farooq, M., Wahid, A., Kobayashi, N., Fujita, D., & Basra, S. (2009). Plant drought stress: Effects, mechanisms and management. In *Sustainable agriculture* (pp. 153–188). Springer, Dordrecht.

Farshadfar, E., Ghasempour, H., & Vaezi, H. (2008). Molecular aspects of drought tolerance in bread wheat (*T. aestivum*). *Pakistan Journal of Biological Sciences: PJBS, 11*(1), 118–122.

Faye, I., Diouf, O., Guisse, A., Sene, M., & Diallo, N. (2006). Characterizing root responses to low phosphorus in pearl millet [*Pennisetum glaucum* (L.) R. Br.]. *Agronomy Journal, 98*(5), 1187–1194.

Ferdous, J., Whitford, R., Nguyen, M., Brien, C., Langridge, P., & Tricker, P. J. (2017). Drought-inducible expression of Hv-miR827 enhances drought tolerance in transgenic barley. *Functional & Integrative Genomics, 17*(2), 279–292.

Fischer, R., & Turner, N. C. (1978). Plant productivity in the arid and semiarid zones. *Annual Review of Plant Physiology, 29*(1), 277–317.

Fleury, D., Jefferies, S., Kuchel, H., & Langridge, P. (2010). Genetic and genomic tools to improve drought tolerance in wheat. *Journal of Experimental Botany, 61*(12), 3211–3222.

Gahlaut, V., Jaiswal, V., Kumar, A., & Gupta, P. K. (2016). Transcription factors involved in drought tolerance and their possible role in developing drought tolerant cultivars with emphasis on wheat (*Triticum aestivum* L.). *Theoretical and Applied Genetics, 129*(11), 2019–2042.

Gambín, B., & Borrás, L. (2010). Resource distribution and the trade-off between seed number and seed weight: A comparison across crop species. *Annals of Applied Biology, 156*(1), 91–102.

Ghorbanian, H., Janmohammadi, M., Ebadi-Segherloo, A., & Sabaghnia, N. (2017). Genotypic response of barley to exogenous application of nanoparticles under water stress condition. *Annales Universitatis Mariae Curie-Sklodowska, sectio C–Biologia, 72*(2).

Gill, S. S., & Tuteja, N. (2010). Reactive oxygen species and antioxidant machinery in abiotic stress tolerance in crop plants. *Plant Physiology and Biochemistry, 48*(12), 909–930.

Golldack, D., Li, C., Mohan, H., & Probst, N. (2014). Tolerance to drought and salt stress in plants: Unraveling the signaling networks [Review]. *Frontiers in Plant Science, 5.* https://doi.org/10.3389/fpls.2014.00151

Gomez Ojeda, A., Corrales Escobosa, A. R., Wrobel, K., Yanez Barrientos, E., & Wrobel, K. (2013). Effect of Cd (II) and Se (IV) exposure on cellular distribution of both elements and concentration levels of glyoxal and methylglyoxal in Lepidium sativum. *Metallomics, 5*(9), 1254–1261.

Gou, W., Tian, L., Ruan, Z., Zheng, P., Chen, F., Zhang, L., Cui, Z., Zheng, P., Li, Z., & Gao, M. (2015). Accumulation of choline and glycinebetaine and drought stress tolerance induced in maize (*Zea mays*) by three plant growth promoting rhizobacteria (PGPR) strains. *Pakistan Journal of Botany, 47*(2), 581–586.

Gupta, P. K., Balyan, H. S., & Gahlaut, V. (2017). QTL analysis for drought tolerance in wheat: Present status and future possibilities. *Agronomy, 7*(1), 5.

Hackenberg, M., Gustafson, P., Langridge, P., & Shi, B. J. (2015). Differential expression of micro RNAs and other small RNAs in barley between water and drought conditions. *Plant Biotechnology Journal, 13*(1), 2–13.

Hameed, A., Goher, M., & Iqbal, N. (2010). Evaluation of seedling survivability and growth response as selection criteria for breeding drought tolerance in wheat. *Cereal Research Communications, 38*(2), 193–202.

Hammad, S. A., & Ali, O. A. (2014). Physiological and biochemical studies on drought tolerance of wheat plants by application of amino acids and yeast extract. *Annals of Agricultural Sciences, 59*(1), 133–145.

Hasanuzzaman, M., Nahar, K., Anee, T., Khan, M., & Fujita, M. (2018). Silicon-mediated regulation of antioxidant defense and glyoxalase systems confers drought stress tolerance in *Brassica napus* L. *South African Journal of Botany, 115*, 50–57.

Honsdorf, N., March, T. J., & Pillen, K. (2017). QTL controlling grain filling under terminal drought stress in a set of wild barley introgression lines. *PloS one, 12*(10), e0185983.

Hossain, M. A., Piyatida, P., da Silva, J. A. T., & Fujita, M. (2012). Molecular mechanism of heavy metal toxicity and tolerance in plants: Central role of glutathione in detoxification of reactive oxygen species and methylglyoxal and in heavy metal chelation. *Journal of Botany, 2012*.

Hu, Y., & Schmidhalter, U. (2005). Drought and salinity: A comparison of their effects on mineral nutrition of plants. *Journal of Plant Nutrition and Soil Science, 168*(4), 541–549.

Hurd, E. (1974). Phenotype and drought tolerance in wheat. *Agricultural Meteorology, 14*(1–2), 39–55.

Huseynova, I. M., Rustamova, S. M., Suleymanov, S. Y., Aliyeva, D. R., Mammadov, A. C., & Aliyev, J. A. (2016). Drought-induced changes in photosynthetic apparatus and antioxidant components of wheat (*Triticum durum* Desf.) varieties. *Photosynthesis Research, 130*(1), 215–223.

Hussain, H. A., Hussain, S., Khaliq, A., Ashraf, U., Anjum, S. A., Men, S., & Wang, L. (2018). Chilling and drought stresses in crop plants: Implications, cross talk, and potential management opportunities. *Frontiers in Plant Science, 9*, 393.

Innes, P., Blackwell, R., & Quarrie, S. (1984). Some effects of genetic variation in drought-induced abscisic acid accumulation on the yield and water use of spring wheat. *The Journal of Agricultural Science, 102*(2), 341–351.

Jabbari, M., Fakheri, B. A., Aghnoum, R., Mahdi Nezhad, N., & Ataei, R. (2018). GWAS analysis in spring barley (*Hordeum vulgare* L.) for morphological traits exposed to drought. *PloS One, 13*(9), e0204952.

Jafarzadeh, J., Bonnett, D., Jannink, J.-L., Akdemir, D., Dreisigacker, S., & Sorrells, M. E. (2016). Breeding value of primary synthetic wheat genotypes for grain yield. *PloS One, 11*(9), e0162860.

Jasrotia, P., Kashyap, P., Bhardwaj, A., Kumar, S., & Singh, G. (2018). Nanotechnology scope and applications for wheat production: A review of recent advances.

Jayakodi, M., Padmarasu, S., Haberer, G., Bonthala, V. S., Gundlach, H., Monat, C., Lux, T., Kamal, N., Lang, D., & Himmelbach, A. (2020). The barley pan-genome reveals the hidden legacy of mutation breeding. *Nature, 588*(7837), 284–289.

Kamal, K. Y., Khodaeiaminjan, M., Yahya, G., El-Tantawy, A. A., Abdel El-Moneim, D., El-Esawi, M. A., Abd-Elaziz, M. A., & Nassrallah, A. A. (2021). Modulation of cell cycle progression and chromatin dynamic as tolerance mechanisms to salinity and drought stress in maize. *Physiologia Plantarum, 172*(2), 684–695.

Kapazoglou, A., Drosou, V., Argiriou, A., & Tsaftaris, A. S. (2013). The study of a barley epigenetic regulator, HvDME, in seed development and under drought. *BMC Plant Biology, 13*(1), 1–16.

Kesavan, M., Song, J. T., & Seo, H. S. (2013). Seed size: A priority trait in cereal crops. *Physiologia Plantarum, 147*(2), 113–120.

Kingston-Smith, A., & Foyer, C. (2000). Bundle sheath proteins are more sensitive to oxidative damage than those of the mesophyll in maize leaves exposed to paraquat or low temperatures. *Journal of Experimental Botany, 51*(342), 123–130.

Kirby, E., Appleyard, M., & Fellowes, G. (1985). Effect of sowing date and variety on main shoot leaf emergence and number of leaves of barley and wheat. *Agronomie, 5*(2), 117–126.

Kleinhofs, A. (2000). The future of barley genetics. *Barley Genet, 1,* 6–10.

Knežević, D., Pržulj, N., Zečević, V., Đukić, N., Momčilović, V., Maksimović, D., Mićanović, D., & Dimitrijević, B. (2004). Breeding strategies for barley quality improvement and wide adaptation. *Kragujevac Journal of Science, 26,* 75–84.

Kohli, M., Mann, C., & Rajaram, S. (1991). Global status and recent progress in breeding wheat for the warmer areas. Wheat for the Nontraditional Warm Areas. *A Proceedings of the International Conference,* Foz do Iguacu (Brazil); 29 Jul–3 Aug 1990.

Levy, A. A., & Feldman, M. (2002). The impact of polyploidy on grass genome evolution. *Plant Physiology, 130*(4), 1587–1593.

Li, M., Lopato, S., Kovalchuk, N., & Langridge, P. (2013). Functional genomics of seed development in cereals. In *Cereal genomics II* (pp. 215–245). Springer, Dordrecht.

Lobell, D. B., Schlenker, W., & Costa-Roberts, J. (2011). Climate trends and global crop production since 1980. *Science, 333*(6042), 616–620.

Lonbani, M., & Arzani, A. (2011). Morpho-physiological traits associated with terminal drought stress tolerance in triticale and wheat. *Agronomy Research, 9*(1–2), 315–329.

Loutfy, N., El-Tayeb, M. A., Hassanen, A. M., Moustafa, M. F., Sakuma, Y., & Inouhe, M. (2012). Changes in the water status and osmotic solute contents in response to drought and salicylic acid treatments in four different cultivars of wheat (*Triticum aestivum*). *Journal of Plant Research, 125*(1), 173–184.

Maccaferri, M., Sanguineti, M. C., Demontis, A., El-Ahmed, A., Garcia del Moral, L., Maalouf, F., Nachit, M., Nserallah, N., Ouabbou, H., & Rhouma, S. (2011). Association mapping in durum wheat grown across a broad range of water regimes. *Journal of Experimental Botany, 62*(2), 409–438.

Manschadi, A. M., Christopher, J., deVoil, P., & Hammer, G. L. (2006). The role of root architectural traits in adaptation of wheat to water-limited environments. *Functional Plant Biology, 33*(9), 823–837.

Martignago, D., Rico-Medina, A., Blasco-Escámez, D., Fontanet-Manzaneque, J. B., & Caño-Delgado, A. I. (2020). Drought resistance by engineering plant tissue-specific responses. *Frontiers in Plant Science, 10,* 1676.

Mascher, M., Gundlach, H., Himmelbach, A., Beier, S., Twardziok, S. O., Wicker, T., Radchuk, V., Dockter, C., Hedley, P. E., & Russell, J. (2017). A chromosome conformation capture ordered sequence of the barley genome. *Nature, 544*(7651), 427–433.

Mikołajczak, K., Ogrodowicz, P., Gudyś, K., Krystkowiak, K., Sawikowska, A., Frohmberg, W., Górny, A., Kędziora, A., Jankowiak, J., Józefczyk, D., Karg, G., Andrusiak, J., Krajewski, P., Szarejko, I., Surma, M., Adamski, T., Guzy-Wróbelska, J., & Kuczyńska, A. (2016). Quantitative trait loci for yield and yield-related traits in spring barley populations derived from crosses between European and Syrian cultivars. *PloS One, 11*(5), e0155938. https://doi.org/10.1371/journal.pone.0155938

Mobasser, H. R., Mohammadi, G. N., Abad, H. H. S., & Rigi, K. (2014). Effect of application elements, water stress and variety on nutrients of grain wheat in Zahak region, Iran. *JBES, 5,* 105–110.

Mukhtar, A., Asif, M., & Goyal, A. (2012). Silicon the non-essential beneficial plant nutrient to enhanced drought tolerance in wheat. In *Crop plant.* CBS Publishers & Distributors Pvt. Ltd. https://doi.org/10.5772/4564

Nahar, K., Hasanuzzaman, M., Alam, M., Rahman, A., Mahmud, J.-A., Suzuki, T., & Fujita, M. (2017). Insights into spermine-induced combined high temperature and drought tolerance in mung bean: Osmoregulation and roles of antioxidant and glyoxalase system. *Protoplasma, 254*(1), 445–460.

Naz, A. A., Arifuzzaman, M., Muzammil, S., Pillen, K., & Léon, J. (2014). Wild barley introgression lines revealed novel QTL alleles for root and related shoot traits in the cultivated barley (*Hordeum vulgare* L.). *BMC genetics*, *15*(1), 1–12.

Nikolaeva, M., Maevskaya, S., Shugaev, A., & Bukhov, N. (2010). Effect of drought on chlorophyll content and antioxidant enzyme activities in leaves of three wheat cultivars varying in productivity. *Russian Journal of Plant Physiology*, *57*(1), 87–95.

Noman, A., Ali, Q., Naseem, J., Javed, M. T., Kanwal, H., Islam, W., Aqeel, M., Khalid, N., Zafar, S., & Tayyeb, M. (2018). Sugar beet extract acts as a natural bio-stimulant for physio-biochemical attributes in water stressed wheat (*Triticum aestivum* L.). *Acta Physiologiae Plantarum*, *40*(6), 1–17.

Nordblom, T. (1985). Livestock-crop interactions: The decision to harvest or to graze mature grain crops. Discussion Paper No. 10. *Discussion Paper (ICARDA)*.

Obsa, B. T., Eglinton, J., Coventry, S., March, T., Guillaume, M., Le, T. P., Hayden, M., Langridge, P., & Fleury, D. (2017). Quantitative trait loci for yield and grain plumpness relative to maturity in three populations of barley (*Hordeum vulgare* L.) grown in a low rain-fall environment. *PloS One*, *12*(5), e0178111.

Obsa, B. T., Eglinton, J., Coventry, S., March, T., Langridge, P., & Fleury, D. (2016). Genetic analysis of developmental and adaptive traits in three doubled haploid populations of barley (*Hordeum vulgare* L.). *Theoretical and Applied Genetics*, *129*(6), 1139–1151.

Olkhovych, O., Volkogon, M., Taran, N., Batsmanova, L., & Kravchenko, I. (2016). The effect of copper and zinc nanoparticles on the growth parameters, contents of ascorbic acid, and qualitative composition of amino acids and acylcarnitines in *Pistia stratiotes* L. (Araceae). *Nanoscale Research Letters*, *11*(1), 1–9.

Outoukarte, I., El Keroumi, A., Dihazi, A., & Naamani, K. (2019). Use of morpho-physiological parameters and biochemical markers to select drought tolerant genotypes of durum wheat. *Journal of Plant Stress Physiology*, 1–7.

Passioura, J. (1986). Resistance to drought and salinity: Avenues for improvement. *Functional Plant Biology*, *13*(1), 191–201.

Passioura, J. B. (1977). Grain yield, harvest index, and water use of wheat. *Journal of the Australian Institue of Agricultural Science 43*, 117–120.

Petrarulo, M., Marone, D., Ferragonio, P., Cattivelli, L., Rubiales, D., De Vita, P., & Mastrangelo, A. M. (2015). Genetic analysis of root morphological traits in wheat. *Molecular Genetics and Genomics*, *290*(3), 785–806.

Pfeiffer, W. (1988). Drought tolerance in bread wheat: Analysis of yield improvement over the years in CIMMYT germplasm. In *Wheat production constraints in tropical environments*. Chiang Mai, Thailand. 19–23 January 1987.

Pham, A.-T., Maurer, A., Pillen, K., Brien, C., Dowling, K., Berger, B., Eglinton, J. K., & March, T. J. (2019). Genome-wide association of barley plant growth under drought stress using a nested association mapping population. *BMC Plant Biology*, *19*(1), 1–16.

Pinto, R. S., Reynolds, M. P., Mathews, K. L., McIntyre, C. L., Olivares-Villegas, J.-J., & Chapman, S. C. (2010). Heat and drought adaptive QTL in a wheat population designed to minimize confounding agronomic effects. *Theoretical and Applied Genetics*, *121*(6), 1001–1021.

Poorter, H., & Remkes, C. (1990). Leaf area ratio and net assimilation rate of 24 wild species differing in relative growth rate. *Oecologia*, *83*(4), 553–559.

Qaseem, M. F., Qureshi, R., Muqaddasi, Q. H., Shaheen, H., Kousar, R., & Röder, M. S. (2018). Genome-wide association mapping in bread wheat subjected to independent and combined high temperature and drought stress. *PLoS One*, *13*(6), e0199121.

Rathjen, J. R., Strounina, E. V., & Mares, D. J. (2009). Water movement into dormant and non-dormant wheat (*Triticum aestivum* L.) grains. *Journal of Experimental Botany*, *60*(6), 1619–1631.

Reynolds, M. (2001). *Application of physiology in wheat breeding.* Cimmyt.

Reynolds, M., Balota, M., Delgado, M., Amani, I., & Fischer, R. (1994). Physiological and morphological traits associated with spring wheat yield under hot, irrigated conditions. *Functional Plant Biology, 21*(6), 717–730.

Richard, C., Hickey, L., Fletcher, S., Chenu, K., Borrell, A., & Christopher, J. (2015). High-throughput phenotyping of wheat seminal root traits in a breeding context. *Procedia Environmental Sciences, 29*, 102–103.

Richards, R. (1996). Defining selection criteria to improve yield under drought. *Plant Growth Regulation, 20*(2), 157–166.

Richards, R. A. (2006). Physiological traits used in the breeding of new cultivars for water-scarce environments. *Agricultural Water Management, 80*(1–3), 197–211.

Rosa, M., Prado, C., Podazza, G., Interdonato, R., González, J. A., Hilal, M., & Prado, F. E. (2009). Soluble sugars--metabolism, sensing and abiotic stress: A complex network in the life of plants. *Plant Signaling & Behavior, 4*(5), 388–393. https://doi.org/10.4161/psb.4.5.8294

Russell, J., Mascher, M., Dawson, I. K., Kyriakidis, S., Calixto, C., Freund, F., Bayer, M., Milne, I., Marshall-Griffiths, T., & Heinen, S. (2016). Exome sequencing of geographically diverse barley landraces and wild relatives gives insights into environmental adaptation. *Nature Genetics, 48*(9), 1024–1030.

Saini, H. S., & Westgate, M. E. (1999). Reproductive development in grain crops during drought. *Advances in Agronomy, 68*, 59–96.

Sallam, A., Alqudah, A. M., Dawood, M. F., Baenziger, P. S., & Börner, A. (2019). Drought stress tolerance in wheat and barley: Advances in physiology, breeding and genetics research. *International Journal of Molecular Sciences, 20*(13), 3137.

Samarah, N. H. (2016). Understanding how plants respond to drought stress at the molecular and whole plant levels. In *Drought stress tolerance in plants, vol 2* (pp. 1–37). Springer, Dordrecht.

Sánchez-Rodríguez, E., Rubio-Wilhelmi, M. M., Cervilla, L. M., Blasco, B., Rios, J. J., Rosales, M. A., Romero, L., & Ruiz, J. M. (2010). Genotypic differences in some physiological parameters symptomatic for oxidative stress under moderate drought in tomato plants. *Plant Science, 178*(1), 30–40.

Savin, R., Slafer, G. A., Cossani, C. M., Abeledo, L. G., & Sadras, V. O. (2015). Cereal yield in Mediterranean-type environments: Challenging the paradigms on terminal drought, the adaptability of barley vs wheat and the role of nitrogen fertilization. In *Crop physiology* (pp. 141–158). Elsevier, Academic Press.

Sehgal, A., Sita, K., Siddique, K. H., Kumar, R., Bhogireddy, S., Varshney, R. K., HanumanthaRao, B., Nair, R. M., Prasad, P., & Nayyar, H. (2018). Drought or/and heat-stress effects on seed filling in food crops: Impacts on functional biochemistry, seed yields, and nutritional quality. *Frontiers in Plant Science, 9*, 1705.

Sharma, A. D., Dhuria, N., Rakhra, G., & Mamik, S. (2014). Accumulation of water stress-responsive class-IIl type of boiling stable peroxidases (BsPOD) in different cultivars of wheat [Triticum aestivum. *Acta Biologica Szegediensis, 58*(2), 115–122.

Sukumaran, S., Reynolds, M. P., & Sansaloni, C. (2018). Genome-wide association analyses identify QTL hotspots for yield and component traits in durum wheat grown under yield potential, drought, and heat stress environments. *Frontiers in Plant Science, 9*, 81.

Sultan, M. A. R. F., Hui, L., Yang, L. J., & Xian, Z. H. (2012). Assessment of drought tolerance of some Triticum L. species through physiological indices. *Czech Journal of Genetics and Plant Breeding, 48*(4), 178–184.

Tambussi, E., Nogues, S., Ferrio, P., Voltas, J., & Araus, J. (2005). Does higher yield potential improve barley performance in Mediterranean conditions?: A case study. *Field Crops Research, 91*(2–3), 149–160.

Taran, N., Storozhenko, V., Svietlova, N., Batsmanova, L., Shvartau, V., & Kovalenko, M. (2017). Effect of zinc and copper nanoparticles on drought resistance of wheat seedlings. *Nanoscale Research Letters, 12*(1), 1–6.

Teulat, B., Monneveux, P., Wery, J., Borries, C., Souyris, I., Charrier, A., & This, D. (1997). Relationships between relative water content and growth parameters under water stress in barley: A QTL study. *New Phytologist, 137*(1), 99–107.

Tewolde, H., Fernandez, C., & Erickson, C. (2006). Wheat cultivars adapted to post-heading high temperature stress. *Journal of Agronomy and Crop Science, 192*(2), 111–120.

Thabet, S. G., Moursi, Y. S., Karam, M. A., Graner, A., & Alqudah, A. M. (2018). Genetic basis of drought tolerance during seed germination in barley. *PloS One, 13*(11), e0206682.

Tricker, P. J., ElHabti, A., Schmidt, J., & Fleury, D. (2018). The physiological and genetic basis of combined drought and heat tolerance in wheat. *Journal of Experimental Botany, 69*(13), 3195–3210.

Tsunewaki, K., & Ebana, K. (1999). Production of near-isogenic lines of common wheat for glaucousness and genetic basis of this trait clarified by their use. *Genes & Genetic Systems, 74*(2), 33–41.

Valliyodan, B., & Nguyen, H. T. (2006). Understanding regulatory networks and engineering for enhanced drought tolerance in plants. *Current Opinion in Plant Biology, 9*(2), 189–195.

Van Oosterom, E., & Acevedo, E. (1992). Adaptation of barley (*Hordeum vulgare* L.) to harsh Mediterranean environments. *Euphytica, 62*(1), 1–14.

Vinocur, B., & Altman, A. (2005). Recent advances in engineering plant tolerance to abiotic stress: Achievements and limitations. *Current Opinion in Biotechnology, 16*(2), 123–132.

Wang, Y., & Frei, M. (2011). Stressed food–The impact of abiotic environmental stresses on crop quality. *Agriculture, Ecosystems & Environment, 141*(3–4), 271–286.

Wiegmann, M., Maurer, A., Pham, A., March, T. J., Al-Abdallat, A., Thomas, W. T., Bull, H. J., Shahid, M., Eglinton, J., & Baum, M. (2019). Barley yield formation under abiotic stress depends on the interplay between flowering time genes and environmental cues. *Scientific Reports, 9*(1), 1–16.

Witkowski, E., & Lamont, B. B. (1991). Leaf specific mass confounds leaf density and thickness. *Oecologia, 88*(4), 486–493.

Yang, J., Zhang, J., Wang, Z., Xu, G., & Zhu, Q. (2004). Activities of key enzymes in sucrose-to-starch conversion in wheat grains subjected to water deficit during grain filling. *Plant Physiology, 135*(3), 1621–1629.

Zeng, Z., Teulat, B., Merah, O., Sirault, X., Borries, C., Waugh, R., This, D., Abebe, T., Guenzi, A., & Martin, B. (2014). Detection and validation of novel QTL for shoot and root traits in barley (*Hordeum vulgare* L.). *Journal of Experimental Botany, 9*, 171–180.

Zhang, Y. J., Yang, J. S., Guo, S. J., Meng, J. J., Zhang, Y. L., Wan, S. B., He, Q. W., & Li, X. G. (2011). Over-expression of the Arabidopsis CBF1 gene improves resistance of tomato leaves to low temperature under low irradiance. *Plant Biology, 13*(2), 362–367.

4 Cereal Performance and Senescence

Misbah Naz, Sajid Hussain,
Sania Zaib, Akmal Zubair, Muhammad Tariq,
Zhicong Dai, and Daolin Du

CONTENTS

4.1 INTRODUCTION

Cereal crops such as maize, barley, rice, and wheat are essential to feed the world's growing population and safeguard food security. Field and laboratory trials have enhanced the understanding and importance of the senescence of organs, especially the grain quality and yield of wheat and barley (Sehgal et al., 2018). For this purpose, a balance is required in the timing of the senescence, nutrient contents in grain, nutrient use efficiency (NUE), and yield in order to improve grain cultivars under specific growth conditions and environments. Senescence is the final developmental stage of plant cells, tissues, and organs (Distelfeld et al., 2014). Some cereals species are single-fruit species having only the final development stage for the entire plant. Most people (intentionally or unintentionally) are well aware of plant senescence by visual appearance (Agrawal and Rakwal, 2006). Compelling examples of the senescence of cereal (single-fruit) crops such as the green leaf surface turning yellow within a few weeks, are similar to large-scale plant senescence processes that consider the entire landscape including the autumn coloration of deciduous trees. Although chlorophyll

DOI: 10.1201/9781003250845-4

degradation and (when red color is observed) anthocyanin biosynthesis are responsible for this brilliant performance, many additional molecular and biochemical processes can also lead to aging syndrome. Once senescence begins, it usually mobilizes the phloem's movable nutrients from the senescent plant parts to the developing sinks, such as the seeds and grains of single-fruit crops (Tadeo et al., 2008). In this regard, nitrogen has a special status. It is the most important plant mineral nutrient in terms of quantity. Macromolecules having nitrogen, such as proteins and nucleic acids, are hydrolyzed and changed into glutamine and glutamate before being loaded and transported to the developing sink in the phloem, and other amino acids (Shah and Dubey, 2003).

While striving to understand the goals of plant senescence, it is key to comprehend and explain in a broad context about new achievements or findings, e.g., regarding the source to sink relationship in the senescence process; the cross-talk among senescence regulation; flowering time control (Waldie et al., 2010); and the link between senescence control and abiotic stress, i.e., drought and nutrient availability, and biotic stress, i.e., pathogen factors response to plants. These methods will contribute to the research on aging in basic and applied plant biology (Großkinsky et al., 2018). Single-fruit plants need to initiate the senescence of the whole plant to remobilize and transfer the assimilation stored in the nutrient tissues to the grain (Lester, 2000). The lodging-resistant cultivars and high nitrogen use that keep the plant green even at maturity leads to delayed plant senescence, resulting in a low harvest index (HI) and a large amount of non-structural carbohydrates (NSC) in straw (Thomas, 2013). In general, water stress or osmotic stress at the grain-filling stage leads to premature senescence, low photosynthesis rate, and reduced grain-filling rate. However, it enhances the NSC from the remobilization of nutrients in tissues and grains (Schippers et al., 2007). With proper control of the mild dryness of soil at a later stage in rice and wheat filling, the senescence of the whole plant can be promoted, resulting in higher and better mobilization of carbon from nutrient tissues to the grain at a speedy filling rate. In some cases, senescence is unfavorably delayed due to high nitrogen use or the introduction of hybrid cultivars with strong heterosis (Yang et al., 2003). The benefits of increased NSC remobilization and the augmented grouting speed can exceed the loss period of the low photosynthesis rate and reduced grouting time. This leads to a boost in grain production, a better harvest index, and higher water use efficiency (WUE). Consequently, the productivity of grain per unit area of wheat and barley is based on the number of spikelets per ear, the number of ears per plant, the number of grains per spikelet, and the grain weight per plant (Whaley et al., 2000). These components are controlled by biotic and abiotic factors, which are important in plant developmental stages.

4.2 PHYSIOLOGICAL CHARACTERISTICS TO IMPROVE YIELD STABILITY

High yields of crops under abiotic and biotic stresses at the reproductive stage are a high priority for breeding programs in the age of climate change globally (Chapman et al., 2012). However, in cereal breeding programs, options have been put forward to explore the maintenance of the green or senescent phenotype, which represents the opposite physiological characteristics to improve yield stability to varying degrees (Farooq et al., 2011). Therefore, whether to consider maintaining the green or the aging phenotype is still an ongoing debate and has not yet been fully resolved. The aim is to design target phenotypes to reduce abiotic stress before and after grain flowering, with a focus on regulating the hormone balance that maintains remobilization and the green phenotype in plants. In this regard, to maintain a high grain yield, it is necessary to (i) maintain green traits to increase the number of grains in a stressful environment before and during anthesis and (ii) fine-tune the regulation and molecular mechanism to optimize the quality and weight of the grain under stress after anthesis (Foulkes et al., 2011). These are an ideal choice for stress-resistant strains before flowering and during fertilization to increase development and yield. However, a small amount of effective remobilization behavior during the filling process may optimize obesity, thickness, and efficiency. The Fear of Stress and Forced Breeding Program provides focused directions to ensure targeted results and effective grouting under terminal drought and stress (Anderson and Becker, 2017).

Physiologists have defined "keep green" as prolonged photosynthetic activity duration under unfavorable growth conditions, and its antagonist term is "aging" which is defined as the degradation of chlorophyll contents (Ali et al., 2018). Crop yield under osmotic (water) stress or heat stress largely depends on the photosynthetic products provided by current photosynthesis or by remobilizing stored carbohydrates from the stem. Another source of assimilates that reduce photosynthesis under drought stress includes carbohydrates (sugars, starches, and fructans) that accumulate and are stored in the stems before flowering (Schnyder, 1993). These reserves can be used during the grain-filling process, specifically under a low photosynthesis rate due to drought faced by the grain. This phenomenon is only useful during the seed-filling process, and is not necessarily beneficial during key stages including gametogenesis, flowering, or before (10 days) and after (8 days) fertilization (Barnabás et al., 2008). The relative importance of these two plant strategies is to maintain and remobilize green phenotypes for an effective seed setting and seed-filling mechanism under drought conditions, which depend on the plasticity of the genotype and its ability to cope with severe stress. Therefore, factors that ultimately support crop productivity under abiotic stress (drought stress) depend on how the crop (i) assimilates carbon compounds to protect fertilization and maintain pollen viability and (ii) effectively assimilates distribution to the absorbing organs (Subbarao et al., 1995) (Figure 4.1).

4.3 EVALUATING THE YIELD STABILITY OF THE POPULATION UNDER CHANGING ENVIRONMENT

Several studies suggest using the starch content of each seed as a measure of the drought tolerance index, and they have also shown that an increase in the number of seeds has little effect on seed weight (Cao et al., 2019). Therefore, it is necessary to analyze how maintaining the green and remobilizing phenotypes affects the stability of the yield by affecting the number of grains per plant and the grain weight, which are the stability indicators of yield. Further, drought stress during the critical growth and developmental stages, such as the gametogenesis process, flowering, and at early seed development, which can impair the grain setting rate, hinder the grain

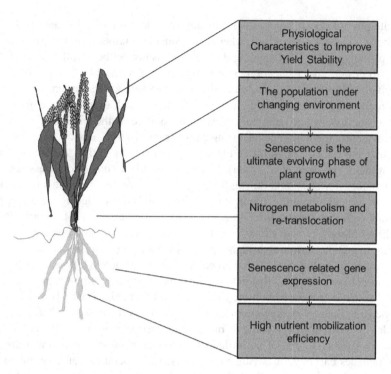

FIGURE 4.1 Morphological and physiological characteristics of annual crops relevant for their responses.

weight and incur losses in yield and its attributes (Ciampitti and Prasad, 2020). This is believed to be due, at least in part, to the reduction in photosynthesis efficacy and changes in the sucrose cleavage process in the reproductive organs. Although inverses and their activities are passively influenced by abiotic and biotic stresses and worsen when abscisic (ABA) levels increase, the source of inverse elasticity is studied very rare (Voegele and Mendgen, 2011). The recently identified elastic inverses and the synthase of sucrose in wheat support the development of molecular markers and are expected to minimize the known conversion of the sugar contents bottleneck to promote an intact sugar source for sensitive reproductive organs. Evaluating the stability of a yield for the increasing population under high temperatures, drought, and heat stresses, and determining the quantitative trait loci (QTLs) stability for the wheat yield mapping population will help to develop climate-based wheat varieties using breeding programs at the field level (Bodner et al., 2015).

4.3.1 Senescence is the Ultimate Evolving Phase of Plant Growth

The transition between maintaining senescence and greenness which impact stability of grain yield is still unclear. Using normalized difference vegetation index (NDVI) technology, many researchers have shown that under terminal pressure, controlling or sustaining the senescence rates and maintaining the greenness at physiological maturity correlate negatively as well as positively with yields (Jagadish et al., 2015).

Therefore, there is a need to use high-output chlorophyll fluorescence or a similar technique with a specific spectral indices for assessing the aging patterns and determining the elusive transition from staying green to aging. With an understanding of the molecular mechanism, there is the potential to achieve yield stability by optimizing assimilation, modifying photorespiration, and enhancing pool intensity, and by defining different strategies under abiotic stress after flowering, so as to functionally cultivate and maintain green in grains (Dubey et al., 2020). In addition, drought-resistant or drought-tolerant wheat germplasm, which can conserve the source strength subjected to stress at the young spore stage, fails to sustain the sink strength under stress at the flowering and grain-filling stages (Shrawat and Armstrong, 2018). This provides a good example of the "maintain green" phenotype before anthesis and the independent operation of the phenotype with enhanced remobilization efficiency (Figure 4.1). Therefore, we have proposed a future grain stress breeding program to take advantage of the said distinctive mechanism by identifying strains or germplasm with these characteristics. Such contrast threads can then be used to develop a phenotype for staying green before flowering and mobilizing after flowering (de Ribou et al., 2013). In cereal crops, the regulation of senescence is at the individual leaf level. Therefore, mineral nutrients are transferred to new leaves from older ones, and finally to the flag leaves, which provide most of the nutrients and photosensitizers used in grain development (Feller and Fischer, 1994). In contrast to dicots, the leaves of cereals plants have basal tissues; their tip is composed of the oldest cells, while the youngest cells are located at the base of the leaf (Fricke, 2002). Because of their systematic arrangement, the leaves of cereal plants are ideal for studying the aging or senescence process. A cross section of a leaf provides a sample of cells at the development stage, which can be studied under in vitro conditions and under various stress levels. Therefore, using the in vitro system, various aspects of senescence can be resolved, eliminating their potential effects on other parts of the plant (Long and Hällgren, 1993).

4.3.2 METABOLISM OF NITROGEN AND RE-TRANSLOCATION

Numerous proteolytic processes that are linked with the senescence mechanism in the entire leaf ensure that the protein of the leaf cells is completely converted to amino acids using the degradation process, ammonium ions, and amides during the process of senescence (Gregersen et al., 2008). The main portion of ammonium ions is integrated into amino acids to be transferred from the senescent leaves, while a small portion evaporates from the leaf crown as ammonia (Huang et al., 2014). Amino acids are transported through the phloem and directly or indirectly through the roots or glumes to the developing grain. Glutamic acid is the main amino acid exported to the phloem in barley, wheat, and other cereals. Therefore, glutamate derived from 2-ketoglutarate plays a key role in the remobilization of nitrogen (Gregersen et al., 2008). The relative availability of glutamine in the phloem at the later stages of wheat senescence shows its importance. Therefore, glutamine synthetase (GS) activity is very important for the re-assimilation of ammonium (NH_4^+) ions into amino acids for transport during aging, especially the cytoplasmic GS form (Lopes et al., 2006).

4.3.3 SENESCENCE-RELATED GENE EXPRESSION

Over time, extensive biological changes occur during grain leaf senescence which involve gene transcription changes (Becker and Apel, 1993). In the past few decades, many specific aging-related genes have been identified and isolated (Table 4.1). Using several screening techniques, these aging-related genes have been characterized, especially in barley. Many clones of cDNA formed from barley leaves that have been subjected to dark treatment to induce senescence and to characterize genes are also upregulated during normal senescence (Kleber-Janke and Krupinska, 1997). Among these, genes encoding protease inhibitors and 4-hydroxyphenylpyruvate dioxygenase have been identified. It is suggested that for barley senescence, one gene is used as a biomarker gene encoded for the small nuclear targeting protein HvS40, which is upregulated during a pathological attack and necrotic leaf senescence. Krupinska et al. (2002) described three other barley senescence-related genes encoding cysteine protease, glycosyltransferase, and GRAB2 homolog (NAC domain transcription factor). Many genes in barley are induced during leaf senescence and are stimulated by abiotic stress (heavy metal). These include proteins encoding heavy metal–associated (HMA) domains, Ca^{2+}-dependent C_2 domain proteins, receptor-like protein kinases, and metal sulfur, the protein gene (Gregersen et al., 2008).

4.3.4 GENETIC REGULATION OF AGING PHENOTYPE

From a large number of research findings, it is obvious that there are many genetic bases for the variation in the senescence time and rate of wheat and barley leaves. Therefore, the regulation of senescence has now become the main objective of breeding programs, which have been passively involved in breeding at the stage of seed maturity (Gregersen et al., 2008). Varieties are stable within a geographical area, indicating that the breeder has inadvertently selected the best maturity date for a specific area. However, if a targeted breeding variety or genetic manipulation can extend the time to maturity, this may increase productivity (Thomas and Howarth, 2000). Many attempts have been made and some have fruitfully delayed the senescence of cereals; however, as a result of other trait selections, accelerated senescence has also occurred (Figure 4.2).

TABLE 4.1
Senescence-Regulated Genes Protein Degradation Function

Pattern	Plant part	Plant species	References
Cysteine proteases	Petals	Alstroemeria pelegrina	Wagstaff et al. (2002)
Cysteine proteases	Leaves	Arabidopsis thaliana	Guo et al. (2004)
Serine proteases	Petals	Narcissus pseudonarcissus	Hunter et al. (2002)
Aspartic proteases	Petals	Alstroemeria pelegrina	Breeze et al. (2004)
Aspartic proteases	Leaves	Brassica napus	Buchanan-Wollaston and Ainsworth (1997)
Serine proteases	Leaves	Petroselinum crispum	Jiang et al. (1999)

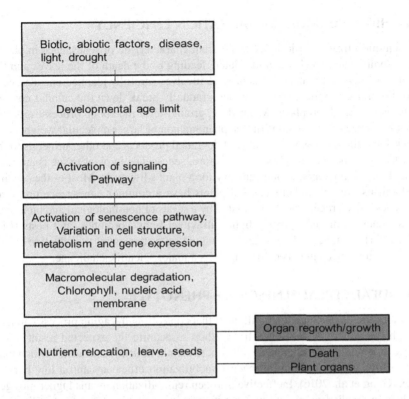

FIGURE 4.2　Whole plant leaf senescence flow chart.

4.4　THE SENESCENCE PROCESS OF BARLEY AND WHEAT

Extensive research has been conducted on the senescence process of barley, wheat, and maize with respect to nitrogen use efficiency (NUE). In the aging mechanism, protein disintegrates and other mineral nutrients are remobilized from old leaves to the remaining plant tissues, specifically the grain. The majority of the degraded protein exists in the chloroplast; Rubisco constitutes the most important part of the protein. Despite extensive research, the proteolytic degradation of Rubisco has not yet been determined. Evidence of matrix protein degradation in chloroplasts is summarized: Rubisco is released from the chloroplasts of the mesophyll cells into small vesicles containing matrix material (RCB = Rubisco-containing corpuscles). These vesicles then assume different degradation pathways. Transcriptome analysis of aging in barley and wheat plants has recognized genes involved in the degradation, metabolism, and regulation processes, which can be used in upcoming approaches designed to alter the aging process. Cultivating crops with traits related to the senescence process, such as larger yields and improved nutrient use efficiency, is complicated. This kind of breeding must solve the dilemma related to delayed senescence and reduced nutrient use efficiency, which may lead to higher yields. Pinpointing the regulatory genes involved in aging may produce results that can efficiently minimize this problem.

4.5   HIGH NUTRIENT MOBILIZATION EFFICIENCY

Nutrient mobilization efficacy is an inherent characteristic of senescence in plants. It is a complicated mechanism ultimately leading to the death of the leaves and the vegetative parts of the plant. This process involves carefully planned encoded gene activation for catabolic enzymes, which gradually break down the cellular components such as in chloroplasts. Meanwhile, ground-level metabolic processes remain intact before cell death to confirm the phenomenon of larger molecular weight components and the successive transport of degraded products and other minerals to the conducting tissue such as phloem. Therefore, senescence in plant tissue is an interesting biological phenomenon necessary for a plant's life cycle. Most of the specific mechanisms are still indefinable, but plants have a common senescence program. The senescence process can be regulated by a series of non-biological and biological factors, such as drought, ozone, light, ultraviolet light, hormones, and pests (Lim et al., 2021). Consequently, further understanding of the signal transduction events and regulatory mechanisms behind aging is a major scientific challenge.

4.6   IDEAL CEREAL SENESCENCE PHENOTYPE

The literature has repeatedly highlighted that controlling the aging process can lead to higher productivity; however, the method to acquire the expected result is not simple. It is vital to understand the key phenomenon of delay senescence which may lead to high yields, but low nitrogen remobilization efficiency and a low harvest index (Gong et al., 2016). Ineffective nitrogen remobilization means higher nitrogen residues in residual crops and higher nitrogen input requirements for fertilization, which may cause environmental problems. Moreover, a delay in senescence is also associated with a slow grain-filling rate. The variety may be susceptible to environmental stresses such as drought and heat in the later stages of crop growth (Yang et al., 2003). Late senescence can also be caused by different abiotic stresses, especially excessive N_2 input into the crops and the supply of water, which can cause serious nutrient and lodging issues (Yang et al., 2003). Whereas accelerated senescence leads to effective nitrogen remobilization and high protein residues, it also leads to lower grain yield, which may be due to the photosynthesis-specific period (Condon et al., 2002). The detailed attention is unlikely to be disrupted in breeding programs (Symonds et al., 1995). It is necessary to avoid the environmental factors that influence aging as they may cause unfavorable deterioration.

4.7   HOW TO ACHIEVE THIS IN PRACTICE

One aspect related to reducing aging that needs to be addressed is the speed at which aging develops (Hafsi et al., 2000). This means that large and high-yielding green stems and leaves should be retained for as long as possible, and they should have the ability to respond to environmental stress (such as high temperatures or dryness) as quickly as possible and mobilize nutrients efficiently. This can be described as a transition in aging, that is, Type A is the opposite of the green phenotype specified by Thomas and Howarth (2000) for Type B. The difficulty with this method is to ensure effective aging and other nutrient remobilization under moderate stress conditions to avoid unfavorable delays in aging, lodging of stems, and harvesting issues. This is entirely reasonable in the plant breeding of wheat and barley but may need to wait for further clarification of genetically determined senescence regulation. Therefore, elucidating the role of aging-related transcription factors, such as the transcription factors of NAC, can open up new ways to manipulate the aging phenotype. Further, manipulating the vital indicators in N_2 remobilization such as cell death mechanisms, proteases, and glutamine synthetase, seems to be a long-lasting strategy for regulating aging factors.

4.8 CONCLUSION AND REMARKS

The product obtained from grain is determined by a synergetic collaboration between source and storage capability. However, the source–sink interaction is far from being fully understood. Therefore, field trials have been conducted on wheat crops to study the response of flag leaves and grains to sink/source operations. The yield of grain can be calculated by the synergistic interaction between the source activity and the storage capacity. Sources recover and supply reserves to sinks (growing grains). The pool strength depends on the number and probable size of the grains per stem, which depend on the plant's ability to actively acquire photochemicals and reserves and accumulate these compounds in its vegetative organs.

In the current literature, the reduced sink-to-source ratio is due to semi-threshing that delays the senescence of wheat leaves, while the higher sink-to-source ratio is due to falling leaves that results in higher reactive oxygen species (ROS) production and endorses the degradation of protein complexes and carbohydrates in the chlorophyll of plants. The interaction between cytokinins (CTKs), jasmonate (JA), and indole-3-acetic acid (IAA) plays a major role in regulating the source capacity and sink strength, thereby affecting the leaf senescence process. Sink and source operations induce many differentially expressed proteins, which are mainly involved in the removal of reactive oxygen species, leaf photosynthesis, carbon and nitrogen metabolism, the production of precursor metabolites, and energy and grain development. The flag leaves of deciduous plants promote the degradation of carbohydrates and proteins by hydrolase activity, which is delayed in the flag leaves of semi-deciduous plants, affecting the C_2 and N_2 metabolism in grains. The removal of plants' ears was increased, but fallen leaves inhibited single-grain growth which indicates that wheat yield potential is limited by storage capacity and the availability of resources. In this chapter, we highlighted that strengthening the source and sink capacity during the breeding program improves the wheat yield potential, making it possible to achieve future yield increases.

REFERENCES

Agrawal GK, Rakwal R. 2006. Rice proteomics: a cornerstone for cereal food crop proteomes. *Mass Spectrometry Reviews.* 25(1):1–53.

Ali A, Gao X, Guo Y. 2018. Initiation, progression, and genetic manipulation of leaf senescence. *Methods in Molecular Biology.* 1744:9–31.

Anderson RI, Becker HC. 2017. Role of the dynorphin/kappa opioid receptor system in the motivational effects of ethanol. *Alcoholism: Clinical and Experimental Research.* 41(8):1402–1418.

Barnabás B, Jäger K, Fehér A. 2008. The effect of drought and heat stress on reproductive processes in cereals. *Plant, Cell & Environment.* 31(1):11–38.

Becker W, Apel K. 1993. Differences in gene expression between natural and artificially induced leaf senescence. *Planta.* 189(1):74–79.

Bodner G, Nakhforoosh A, Kaul H-P. 2015. Management of crop water under drought: a review. *Agronomy for Sustainable Development.* 35(2):401–442.

Breeze E, Wagstaff C, Harrison E, Bramke I, Rogers H, Stead A, Thomas B, Buchanan-Wollaston V. 2004. Gene expression patterns to define stages of post-harvest senescence in Alstroemeria petals. *Plant Biotechnology Journal.* 2(2):155–168.

Buchanan-Wollaston V, Ainsworth C. 1997. Leaf senescence in Brassica napus: cloning of senescence related genes by subtractive hybridisation. *Plant Molecular Biology.* 33(5):821–834.

Cao Q, Li G, Cui Z, Yang F, Jiang X, Diallo L, Kong F. 2019. Seed priming with melatonin improves the seed germination of waxy maize under chilling stress via promoting the antioxidant system and starch metabolism. *Scientific Reports.* 9(1):1–12.

Chapman SC, Chakraborty S, Dreccer MF, Howden SM. 2012. Plant adaptation to climate change opportunities and priorities in breeding. *Crop and Pasture Science.* 63(3):251–268.

Ciampitti IA, Prasad PV. 2020. *Sorghum: state of the art and future perspectives.* John Wiley & Sons.

Condon AG, Richards R, Rebetzke G, Farquhar G. 2002. Improving intrinsic water-use efficiency and crop yield. *Crop Science.* 42(1):122–131.

de Ribou SdB, Douam F, Hamant O, Frohlich MW, Negrutiu I. 2013. Plant science and agricultural productivity: why are we hitting the yield ceiling? *Plant Science.* 210:159–176.

Distelfeld A, Avni R, Fischer AM. 2014. Senescence, nutrient remobilization, and yield in wheat and barley. *Journal of Experimental Botany.* 65(14):3783–3798.

Dubey RK, Tripathi V, Prabha R, Chaurasia R, Singh DP, Rao CS, El-Keblawy A, Abhilash PC. 2020. *Unravelling the soil microbiome: perspectives for environmental sustainability.* Springer.

Farooq M, Bramley H, Palta JA, Siddique KH. 2011. Heat stress in wheat during reproductive and grain-filling phases. *Critical Reviews in Plant Sciences.* 30(6):491–507.

Feller U, Fischer A. 1994. Nitrogen metabolism in senescing leaves. *Critical Reviews in Plant Sciences.* 13(3):241–273.

Foulkes MJ, Slafer GA, Davies WJ, Berry PM, Sylvester-Bradley R, Martre P, Calderini DF, Griffiths S, Reynolds MP. 2011. Raising yield potential of wheat. III. Optimizing partitioning to grain while maintaining lodging resistance. *Journal of Experimental Botany.* 62(2):469–486.

Fricke W. 2002. Biophysical limitation of cell elongation in cereal leaves. *Annals of Botany.* 90(2):157–167.

Gong X, Wheeler R, Bovill WD, McDonald GK. 2016. QTL mapping of grain yield and phosphorus efficiency in barley in a Mediterranean-like environment. *Theoretical and Applied Genetics.* 129(9):1657–1672.

Gregersen P, Holm P, Krupinska K. 2008. Leaf senescence and nutrient remobilisation in barley and wheat. *Plant Biology.* 10 Supplement 1:37–49.

Großkinsky DK, Syaifullah SJ, Roitsch T. 2018. Integration of multi-omics techniques and physiological phenotyping within a holistic phenomics approach to study senescence in model and crop plants. *Journal of Experimental Botany.* 69(4):825–844.

Guo Y, Cai Z, Gan S. 2004. Transcriptome of Arabidopsis leaf senescence. *Plant, Cell & Environment.* 27(5):521–549.

Hafsi M, Mechmeche W, Bouamama L, Djekoune A, Zaharieva M, Monneveux P. 2000. Flag leaf senescence, as evaluated by numerical image analysis, and its relationship with yield under drought in durum wheat. *Journal of Agronomy and Crop Science.* 185(4):275–280.

Huang B, DaCosta M, Jiang Y. 2014. Research advances in mechanisms of turfgrass tolerance to abiotic stresses: from physiology to molecular biology. *Critical Reviews in Plant Sciences.* 33(2–3):141–189.

Hunter DA, Steele BC, Reid MS. 2002. Identification of genes associated with perianth senescence in daffodil (*Narcissus pseudonarcissus* L. 'Dutch Master'). *Plant Science.* 163(1):13–21.

Jagadish KS, Kavi Kishor PB, Bahuguna RN, von Wirén N, Sreenivasulu N. 2015. Staying alive or going to die during terminal senescence—an enigma surrounding yield stability. *Frontiers in Plant Science.* 6:1070.

Jiang WB, Lers A, Lomaniec E, Aharoni N. 1999. Senescence-related serine protease in parsley. *Phytochemistry.* 50(3):377–382.

Kleber-Janke T, Krupinska K. 1997. Isolation of cDNA clones for genes showing enhanced expression in barley leaves during dark-induced senescence as well as during senescence under field conditions. *Planta.* 203(3):332–340.

Krupinska K, Haussuhl K, Schafer A, van der Kooij TA, Leckband G, Lörz H, Falk J. 2002. A novel nucleus-targeted protein is expressed in barley leaves during senescence and pathogen infection. *Plant Physiology.* 130(3):1172–1180.

Lester G. 2000. Polyamines and their cellular anti-senescence properties in honey dew muskmelon fruit. *Plant Science.* 160(1):105–112.

Lim K-J, Paasela T, Harju A, Venäläinen M, Paulin L, Auvinen P, Kärkkäinen K, Teeri TH. 2021. A transcriptomic view to wounding response in young Scots pine stems. *Scientific Reports.* 11(1):1–13.

Long S, Hällgren J-E. 1993. *Measurement of CO_2 assimilation by plants in the field and the laboratory. Photosynthesis and production in a changing environment.* Springer; p. 129–167.

Lopes MS, Cortadellas N, Kichey T, Dubois F, Habash DZ, Araus JL. 2006. Wheat nitrogen metabolism during grain filling: comparative role of glumes and the flag leaf. *Planta.* 225(1):165–181.

Schippers JH, Jing H-C, Hille J, Dijkwel PP. 2007. Developmental and hormonal control of leaf senescence. *Senescence Processes in Plants.* 26:145–170.

Schnyder H. 1993. The role of carbohydrate storage and redistribution in the source-sink relations of wheat and barley during grain filling a review. *New Phytologist.* 123(2):233–245.

Sehgal A, Sita K, Siddique KH, Kumar R, Bhogireddy S, Varshney RK, HanumanthaRao B, Nair RM, Prasad P, Nayyar H. 2018. Drought or/and heat-stress effects on seed filling in food crops: impacts on functional biochemistry, seed yields, and nutritional quality. *Frontiers in Plant Science.* 9:1705.

Shah K, Dubey R. 2003. Environmental stresses and their impact on nitrogen assimilation in higher plants. *Advances in Plant Physiology.* 5:397–431.

Shrawat AK, Armstrong CL. 2018. Development and application of genetic engineering for wheat improvement. *Critical Reviews in Plant Sciences.* 37(5):335–421.

Subbarao G, Johansen C, Slinkard A, Nageswara Rao R, Saxena N, Chauhan Y, Lawn R. 1995. Strategies for improving drought resistance in grain legumes. *Critical Reviews in Plant Sciences.* 14(6):469–523.

Symonds JE, Gibson JB, Wilks AV, Wilanowski TM. 1995. Molecular analysis of a Drosophila melanogaster sn-glycerol-3-phosphate dehydrogenase allozyme variant that has cold labile activity. *Insect Biochemistry and Molecular Biology.* 25(7):789–798.

Tadeo FR, Cercos M, Colmenero-Flores JM, Iglesias DJ, Naranjo MA, Rios G, Carrera E, Ruiz-Rivero O, Lliso I, Morillon R. 2008. Molecular physiology of development and quality of citrus. *Advances in Botanical Research.* 47:147–223.

Thomas H. 2013. Senescence, ageing and death of the whole plant. *New Phytologist.* 197(3):696–711.

Thomas H, Howarth CJ. 2000. Five ways to stay green. *Journal of Experimental Botany.* 51 Spec No(suppl_1):329–337.

Voegele RT, Mendgen KW. 2011. Nutrient uptake in rust fungi: how sweet is parasitic life? *Euphytica.* 179(1):41–55.

Wagstaff C, Leverentz MK, Griffiths G, Thomas B, Chanasut U, Stead AD, Rogers HJ. 2002. Cysteine protease gene expression and proteolytic activity during senescence of Alstroemeria petals. *Journal of Experimental Botany.* 53(367):233–240.

Waldie T, Hayward A, Beveridge CA. 2010. Axillary bud outgrowth in herbaceous shoots: how do strigolactones fit into the picture? *Plant Molecular Biology.* 73(1–2):27–36.

Whaley J, Sparkes D, Foulkes M, Spink J, Semere T, Scott R. 2000. The physiological response of winter wheat to reductions in plant density. *Annals of Applied Biology.* 137(2):165–177.

Yang J, Zhang J, Wang Z, Liu L, Zhu Q. 2003. Postanthesis water deficits enhance grain filling in two-line hybrid rice. *Crop Science.* 43(6):2099–2108.

5 Cereal Responses to Nutrients and Avenues for Improving Nutrient Use Efficiency

Ghulam Haider, Muhammad Ansar Farooq,
Tariq Shah, Saadatullah Malghani,
Masood Iqbal Awan,
Muhammad Habib-ur-Rahman,
Attia Rubab Khalid, and Abdul Ghaffar

CONTENTS

DOI: 10.1201/9781003250845-5

5.1   INTRODUCTION

Food and nutrition insecurities are a major concern for the world's population. About 3 million people are affected in under-developing and developing countries. Of the current world population, 11% are undernourished and 17% suffer from micronutrient deficiencies (FAO et al., 2019). It has been suggested that there will only be enough food for everyone if agricultural production increases. To achieve three sustainable development goals (SDGs): ending hunger and improving nutrition, eliminating poverty, and attaining good well-being by 2030, it will be essential to increase agricultural productivity by 60%–110% (FAO, 2019). Thus, it is necessary to focus on enhancing the yield of cereal crops because they are highly cultivated compared to other crops to meet the world's food requirements (Nitika Sandhu et al., 2021). For years, cereal crops have been an important part of the human diet. A major source of calories in the human diet comes from cereal crops such as maize, rice, and wheat and some are derived from sorghum and millets. About 42% of the calories consumed globally are obtained from rice, wheat, and maize (Matres et al., 2021). Rice is the staple food for more than half of the world's population due to its nutritional value (Fukagawa and Ziska 2019), whereas wheat is the third most important crop regarding its production and is beneficial for human health as it provides essential vitamins, phytochemicals, proteins, and dietary fibers (Shewry and Hey, 2015).

Cereal crops require large amounts of nutrients for better yield. The essential nutrients that plants obtain from the soil are nitrogen (N), phosphorus (P), potassium (K), magnesium (Mg), sulfur (S), and calcium (Ca). Nutrient uptake in cereal plants varies: the uptake of winter wheat is about 200 kg of nitrogen and potassium per hectare while the uptake of micronutrients are a few grams. More and more nutrients are removed from the soil as the crop yield increases. To accomplish a good level of soil nutrients, the application of fertilizers is necessary. Major nutrients for cereal crops, such as nitrogen, potassium, and phosphorus, are provided in the form of chemical fertilizers. The consumption of NPK fertilizer increased by about 5.2% in 2020–2021 and will increase further in the future (Zhang et al., 2020). Applied fertilizers cannot be utilized completely by plants and are lost in the environment. Fertilizers applied in the form of ammonium and urea transform into nitrates. Plants absorb nitrates from the soil for their metabolic processes; however, any remaining nitrates in the soil are lost due to low nitrogen fertilizer recovery and nitrogen use efficiency. The loss of nutrients from the soil has a negative impact on the economy as well as on the environment. Nitrate loss from soil occurs in two ways: leaching and runoff from the surface. The leaching of nitrates increases in the case of soil acidi-fication, and the accumulation of nitrogen and phosphorus in water bodies causes adverse effects on marine biodiversity and human health. Moreover, the emission of nitrogenous oxides, ammonia volatilization, and di-nitrogen due to nitrogen fertil-izers is the main concern for the environment (Jones et al., 2014). Consequently, for food security and sustainable agriculture it is necessary to improve the nutrient use efficiency (NUE) of cereal crops. Crop yield can be increased by higher nutrient use efficiency when applied with lower amounts of fertilizers.

Nutrient management plays a significant role in cereal production and increases farm yield. Sometimes the amount of fertilizers recommended by organizations or governments ignores other factors such as soil type, crop management practices, and weather, leading to the loss of nutrients (Rurinda et al., 2020). However, the con-cept of 4R (right source, right rate, right time, and right place) nutrient stewardship helps in the management of nutrients by guidance that applies the right concentra-tion of nutrients, at the right time, and using the right method which can increase yield and preserve resources. Several genetic approaches have been introduced to improve the NUE in plants. Knowledge about NUE-related traits can aid in improv-ing the genotype with high nutrient efficiency. Several genes are involved in NUE which regulates various mechanisms involved in the absorption and accumulation of nitrogen in plants. The overexpression of nutrient efficiency–related genes in cereal crops effectively enhances nutrient uptake. Various transporters, transcription factors (TFs), and signaling molecules have been identified to increase the NUE. Different transporters were identified to be involved in the transportation and uptake of both macronutrients and micronutrients. These genetic approaches will lead to better cereal crops with high nutrient use efficiency and will contribute to resolving the global issue of food and nutritional insecurities.

This chapter begins with a discussion of the global issue of food and nutrition insecurity in the context of cereal production. In the following section, the basics of plant nutrition for food security is discussed, including essential nutrients for soil with reference to plants, recent trends in NUE, the impact on human health, and the consequences of loss of nutrients. The subsequent section discusses the nutrient management of cereals, including chemical fertilizers and 4R nutrient stewardship, and management for improving NUE in cereal crops. The concluding section is dedi-cated to genetic approaches to improve the nutrient use efficiency in cereal crops.

5.2   CEREAL PRODUCTION AND FOOD SECURITY

Cereals are mostly produced and consumed in Asian countries because almost half of the world's population live in Asian countries. Cereals are a major contributor to world food security particularly in developing countries. There are many threats to cereal production such as climate change, increasing populations, and the distribu-tion of land and other input resources.

5.2.1   WORLD POPULATION, FOOD SECURITY, AND CEREALS DEMAND IN DEVELOPING COUNTRIES

The term "food security" was first used in the mid-1970s, but concerns related to feeding the increasing population were expressed even earlier (FAO, 1975). One of the basic human needs is the supply of food in sufficient quantities, with good nutritional quality at a reasonable price. The definition of food security has been modified many times. According to the newly updated FAO definition:

> food security is a situation that exists when all people, at all the times, have physical, social and economic access to sufficient, safe and nutritious food that meets their dietary needs and food preferences for an active and healthy life.

(FAO, 2002)

The United Nations has made food security a priority and has included it in their sustainable development goals: end hunger, achieve food security and improved nutrition, and promote sustainable agriculture. For food security, the main concern is the increasing world population ultimately raising food demands, forcing consumers to compromise on food quality. In order to feed the increasing population, yield must increase up to 70% by 2050 (FAO, 2021). To feed the entire world's population according to their nutritional demands as well as to resolve the issues related to food security are becoming strategic concerns in developing countries because of the unequal distribution of food resources around the globe. On a global scale, it is not possible for all individuals to have equal access to food of sufficient quantity and quality. In terms of the population, it is anticipated that between 720 and 811 million people around the world endured hunger in 2020 (FAO, 2021). More than 161 million people were added to this hunger pool in 2020 as compared to 2019. The influence of climate change on agriculture and the COVID-19 epidemic both shook the global economy, and are among the causes of food insecurity in the current scenario. It is critical to improve cereal yields in order to alleviate this issue. Developing countries must focus on their production because they are already behind developed countries in cereals production.

5.2.2 THE STATE OF CEREALS PRODUCTION AND CHALLENGES IN DEVELOPING COUNTRIES

Cereals are an important source of human nutrition around the globe. The major cereals are rice, wheat, and maize, which are considered staple food crops for an estimated 50%, 34%, and 13% of the world's population, respectively (IDRC, 2010; Grote et al., 2021). Cereal production statistics showed that the production of wheat, maize, rice, and other food grains was roughly 776.20, 1183.4, 525.1, and 305.4 million tons, respectively (FAO, 2021). Problems in cereal production are mostly associated with the increasing population and climate change. The unavailability of timely inputs and the use of uncertified seeds are also serious issues in improving yield. Sales of improved seed in developing countries in the last few decades accounted for 12% of global commercial seed sales, worth nearly US$4 billion annually (Groosman et al., 1988). In Pakistan today, farmers are still using only 12% certified seed for wheat crops. Rising populations have put pressure on land and food resources. The expansion of cities and huge investments and profits in real estate businesses have resulted in a decrease in cultivated land area. In Africa and East Asia, some land area is available for the expansion of agricultural lands but the shortage of water is the main concern for crop productivity (Bruinsma, 2009). On the one hand, turning marginal land into farmland will result in increased land degradation, carbon dioxide emissions, and biodiversity loss, all of which will have an environmental cost (Grote et al., 2021). On the other hand, increasing cereals production can cover up marginal land and lighten the load on natural ecosystems from being converted to agricultural land. Climate change due to rising CO_2 levels in the atmosphere is a major challenge to future agricultural productivity. The climate influences the distribution and proliferation of weeds, fungal diseases, and insects to a great extent. Changes in the temperature and CO_2 levels will have a direct impact on weeds; however, insects and diseases are unlikely to be directly affected. CO_2 levels may change, but host plant metabolism, development, reproduction, and morphology may be indirectly affected. The overall significance of such developments is not clear at this time, although crop losses owing to weeds is a major concern (Valasai et al., 2005). Asian Development Bank studies showed that increased concentrations of CO_2 appeared to enhance the productivity of agroecosystems for C_3 crops (rice, wheat, and barley) by about 25% and for C_4 crops (millet and maize) by about 10%; however, at the same time, rising temperatures had a negative impact on crop productivity (Valasai et al., 2005). High carbon dioxide concentrations enhance crop yields in various crops, but the nutritional quality of the grain is generally compromised, due to the low availability of Fe and Zn. High carbon dioxide concentrations also decrease the N and protein content of grains of cereal crops. If carbon dioxide continues to rise at the current pace, an increase in the atmospheric temperature by about 1.4°C–5.8°C is expected by the end of this century. If this happens, a decrease in the C_3 plant performance is expected, ultimately lowering plant yield (wheat and rice) by up to 60% owing to high-temperature stress (Valasai et al., 2005). Although greenhouse gases are the consequences of industrialization in developed countries, developing countries are also facing problems. So, technologically advanced countries must help developing countries in taking measures to control climate change.

5.3 BASICS OF PLANT NUTRITION FOR FOOD SECURITY

5.3.1 Recent Trends of Global Nutrient Utilization and NUE in Cereals

Nutrient utilization refers to variations in the quality and quantity of nutrients during nutritional shifts other than simple changes in carbon (C) source, nitrogen source, phosphate source, sulfur source, or amino acids. Plants, for example, may receive inorganic nutrients, water, and carbon dioxide from the environment through their root system. The energy that enables plants to develop is produced by a mixture of organic substances, water, carbon dioxide, and sunshine. Both nutrient depletion in the root environment and nutrient content in the plant tissues can be used to determine plant nutrient absorption. This approach is based on calculating the difference in the amount of a certain ion in the root environment over time.

The NUE concept comprises three primary plant processes: nutrient absorption, assimilation, and use. The capability of crops to absorb and use nutrients for best yields is measured by nutrient use efficiency. Nutrients are divided into two groups based on their concentration in plant tissues (macronutrients and micronutrients), and the functions they perform in plant metabolism are as diverse as their physico-chemical qualities. Solar energy is used to drive selective chemical exchange with the environment, which is actively maintained. Plants, like all living creatures, are principally chemical compartments in thermodynamic disequilibrium with their surroundings. Plants require the intake of at least 13 other vital nutrients from the soil in addition to carbon dioxide and water, which give the structural and metabolic backbone components C, O, and H.

The term plant NUE refers to a multigenic feature that involves several interrelated physiological processes that are influenced by a variety of circumstances. As a result, there are a variety of techniques for defining, analyzing, and possibly improving NUE. The capacity of plants to use nutrients varies significantly between species and cultivars, and this has long been recognized as a potential source of future improve-ment through breeding. Scientists began formulating ideas and terminology to serve as a foundation for study comparisons and discussions in order to build a unified framework for NUE. Since then, numerous studies in various scientific disciplines dealing with various plant species, in various contexts, under various conditions, and focusing on various nutrients have failed to find a single definition for NUE that adequately describes all cases, revealing instead that the issue is far too complex to do so. The efficiency of nutrient acquisition (i.e., the number of nutrients taken up by plants in relation to nutrient supply) and the efficiency of nutrient usage (i.e., the bio-mass generated by the unit of nutrient absorbed by plants) are the two main compo-nents of nutrient use efficiency. It has been observed that there are important limits to increasing these components while maintaining food quality. More understanding of the genetic underpinnings, as well as the physiological and molecular mecha-nisms that govern these efficiencies in plants, might lead to the development of novel solutions to these problems. Nutrient conservation mechanisms used by seagrasses improve nutrient usage efficiency and lower the need for nutrient intake from the environment. Internal recycling, leaf lifespan, and storage are significant processes that demonstrate a high efficiency for keeping nutrients inside plant tissues, which is beneficial in nutrient-poor coastal environments. Best management techniques are the most effective external option for increasing NUE. Plant genetics and physiologi-cal processes, as well as their interactions with best management practices (BMPs), can be employed to improve cropping system efficiency.

5.3.2 TRENDS IN NUTRIENT UTILIZATION AND NUTRIENT USE EFFICIENCY IN CEREALS

The idea of nutrient use efficiency is significant in the evaluation of agricultural production systems. Fertilizer management, as well as soil and plant-water manage-ment, can have a significant influence. Nutrient utilization aims to improve the over-all performance of cropping systems by delivering economically optimal nutrition to the crop while reducing nutrient losses in the field. NUE addresses certain elements of that performance, but not all. As a result, system optimization goals must encom-pass both overall productivity and NUE. The question being addressed, as well as the geographical or temporal scale of interest for which trustworthy data are avail-able, decides the most appropriate formulation of NUE. Global temporal patterns of

NUEs vary by region. Partial nutrient balance (ratio of nutrients removed by crop harvest to fertilizer nutrients applied) and partial factor productivity (crop production per unit of nutrient applied) for N, P, and K are trending upward in Africa, North America, Europe, and the EU-15, while they are trending downward in Latin America, India, and China. Although these worldwide areas may be classified into two categories based on historical trends, the reasons that drive those trends vary greatly within each category. NUE is influenced by a variety of managerial and environmental factors, including plant-water status. Similarly, plant nutritional status has a significant impact on water consumption efficiency.

Cereals (wheat, rice, and maize) are the most important food crops, having a wide range of adaptability to diverse temperatures and inputs. They are in high demand for consumption, accounting for 55% of total planted land worldwide. They are also the most important components of the human diet, accounting for 55% of total calorie and protein requirements. Chemical fertilizer usage has grown practically exponentially over time, even though the nutrient use efficiency of applied nutrients as fertilizers has remained remarkably low. In response, nutrients are lost from soils through multiple pathways or converted into slow cycling within the soil. According to some estimates, up to 70%–80% of additional nitrogen could be lost from rain-fed agroecosystems and 60%–70% from irrigated agroecosystems if they are not effectively maintained. Furthermore, the partial factor productivity of fertilizers used in food grain production has been rapidly decreasing at the national level in recent years. The response of cereal crops to applied fertilizers, for example, has reduced. This has become a major source of concern and a difficulty that must be addressed as a priority.

Maize is renowned as the "queen of cereals" across the world because it has the highest genetic yield potential of all cereals and ranks third globally in term of production after wheat and rice. In many nations, especially in the tropics and subtropics, it is the main source of nutrition. However, abiotic challenges such as moisture deficit due to insufficient and unpredictable winter rainfalls, lack of irrigation infrastructure, and nitrogen stress due to suboptimal N-fertilizer use limit maize yield in the winter. Nutrient shortages reduce crop biomass and grain output, affecting both intercepted photosynthetically active radiation (IPAR), radiation utilization efficiency (RUE), and water productivity. Thus, greater knowledge of how high yield potential maize hybrids use soil moisture, nutrients, and radiation under optimal management might aid in increasing crop output.

Fertilizer rates set by national or district governmental organizations are often comprehensive references for a region or district, disregarding site-specific soils, weather conditions, crop management approaches, and crop types. Low nutrient use efficiency and profitability are common outcomes of such proposals, enhancing the dangers to the environment associated with the loss of unutilized nutrients through emissions or leaching. Nutrient management can increase farm yields, especially in smallholder farming systems, and has a considerable impact on cereal production.

5.3.3 Impact of Plant Nutrition on Food Quality and Human Health

Plants are an integral part of the human diet and provide a sufficient amount of minerals necessary for human nutrition (Graham et al., 2001). They are also a good source of macronutrients that provide energy: carbohydrates, proteins, and lipids. Micronutrients for human health include vitamins A, B complex, C, E, and K; essential lipids; and essential amino acids, which are present in plants in variable quantities. Plant-based foods also contain a range of bioactive compounds that can play an important role in the prevention of chronic diseases. Plants need nutrients for good growth and quality production. Several factors affect their growth including temperature, water, light, and nutrient availability in soil. The availability of adequate amounts of minerals is necessary for plant growth and fertility. Improved crop nutrient management can improve the bioavailability of macro and micronutrients in crops and grains, a prerequisite for successful crop growth and optimal yields in quantity and quality. In addition to water, sunlight, and favorable soil conditions, essential nutrients are crucial to optimizing crop production. Agronomic strategies are most effective in areas with adequate infrastructure for the production, distribution, and application of inorganic fertilizers, and are the only viable options in areas where soils lack sufficient concentrations of mineral elements, required for human nutrition, to support mineral-dense crops. Common strategies are "agronomic" biofortification and genetic biofortification which involve the use of fertilizers containing mineral elements that are deficient in the human diet, primarily Zn, Cu, Fe, I, Se, Mg, and Ca, in combination with

1. appropriate soil amendments, such as composts and manures, to increase the soil concentration of essential elements;
2. acidifying fertilizers, such as urea, ammonium nitrate, ammonium sulfate, and ammonium; and
3. suitable crop rotations, intercropping, or the advent of useful soil microorganisms to increase the phytoavailability of mineral elements.

Crop genotypes with improved abilities to attain mineral elements and store them in edible tissues employ genetic modification. There is sufficient natural genetic variation in the concentrations of mineral elements – commonly lacking in the human diet – in the edible tissues of most crop species to breed for increased concentrations of mineral elements in edible tissues and also scope for targeted genetic modification of crops. The goal of biofortification is to solve some of these problems by increasing the concentration of micronutrients in the edible parts of crops and improving their bioavailability and absorption in the human body after digestion.

5.3.4 IMPACT OF PLANT NUTRITION ON FOOD QUALITY

Plants nutrients are the chemical elements and compounds necessary for plant growth, metabolism, and their external supply. Of more than 100 chemical elements, 17 elements are considered essential elements of plants. Of the 17 basic elements, carbon, hydrogen (H), and oxygen (O) are non-mineral nutrients because they are derived from air and water. The remaining 14 nutrients obtained from the soil through the plant roots include nitrogen, phosphorus, potassium, calcium, magnesium, and sulfur and eight trace elements: boron (B), chlorine (Cl), copper (Cu), iron

(Fe), manganese (Mn), molybdenum (Mo), nickel (Ni), and zinc (Zn). In the absence of these nutrients, plants cannot complete a normal life cycle.

Deficiency in any of these factors limits the plant growth and reduces yield and quality. Each nutrient is needed in different quantities and performs specific functions in plants. For example, seed foods are sources of macronutrients and lipid-soluble vitamins but have low concentrations of Fe and Ca. Some crop quality attributes influenced by nutrients include sugar and protein content, seed size, kernel size, fruit color, flavor, vitamin levels, and kernel hardness. Nutrients acquired from crops and crop products are components of several types of proteins, enzymes, nitrogen, sulfur, and phosphorus. These are important for forming/generating plant tissues and for activating several metabolic processes.

Plants need large amounts of nitrogen, phosphorus, and potassium. Nitrogen performs many functions, primarily the rapid growth of plants, an increase in leaf size and quality, and the enhancement of fruit and seed development. It is an integral component of many important plant components such as amino acids and enzymes which are involved in catalyzing many biochemical processes (Njira and Nabwami, 2015). Thus, it has a role in all metabolic processes. By its function, nitrogen influences crop quality as well as crop growth. The protein content and physical grain quality of maize and rice plants increase with the application of nitrogen, as reported in many studies. Excessive amounts of nitrogen can also degrade crop quality. An oversupply of nitrogen impairs calcium absorption, especially when the $NH_4-N : NO_3-N$ ratio is high. Calcium, on the other hand, is required for the formation of robust cell walls.

Sulfur is the most plentiful element on earth. Plants absorb it as sulfates. It plays an important role in photosynthesis, respiration, and the formation of cell membrane structures in plants. It is required for the synthesis of sulfur-containing amino acids such as cysteine and methionine, which are the building blocks of protein and important constituents of vitamins and hormones. The content of sulfur-containing amino acids in plants is an important index for evaluating crop quality. It is also responsible for the disulfide bond formation between cysteine residues that help to stabilize the tertiary structure of protein. It is also required in the synthesis of co-enzyme A and chlorophyll. It has been reported that a deficiency of sulfur leads to the accumulation of non-protein N such as NO_3^- and amine (NH_2) in crops, which is dangerous. Sulfur deficiency reduces the proportion of sulfur-containing amino acids in crop grains, while sulfur application increases the content of sulfur-containing proteins, which increases the nutritional value of grains. Additionally, S supply levels affect wheat flour extraction rates, gluten quality, and baking quality. A deficiency in sulfur stress decreases the content of sulfur and sulfur-containing amino acids, leading to a reduction in the metabolic activities of plants. An S deficiency leads to impaired synthesis of key enzymes involved in the process of carbon metabolism and reduces the rate of photosynthesis, resulting in the accumulation of most reactive oxygen species in plants (Lunde et al., 2008).

Boron is the fifty-first most common element on earth. It is known as an essential element for plants. It is proposed that boron is passively taken up by plant roots in the form of borate or boric acid. Several factors affect the availability of boron for plant root uptake such as the soil pH, texture, soil moisture, temperature, soil calcium carbonate content, and soil organic matter. Boron deficiency directly affects plant growth and reduces yield. It severely affects plant root growth, inhibiting root elongation and cell division in the growing zone of the root tips which results in the death of the root tips. Boron deficiency is reported to cause male sterility in wheat. However, excessive boron concentrations may prove harmful for some plant species and cause toxicity in plants.

It is found that using inorganic micronutrient fertilizers increases the mineral tiers in safe-to-eat quantities of crop plant life with the effect of the fertilizer used depending on the fertilization practice. For example, the soil and foliar utility of Zn fertilizers increases Zn concentrations in a few plants and soils. Zinc is a major metal component and activator of various enzymes involved in metabolic processes and biochemical pathways. Zn is a structural, functional, or regulatory co-factor of a large number of enzymes. It plays an essential role in DNA transcription. Its other functions are catalyzing the oxidation process in plant cells and has vitality for transformation of carbohydrates, formation of chlorophyll and auxins for the plant growth.

Iron is a micronutrient constituent of the enzyme system that contributes oxidation-reduction reactions in plants, regulates respiration and photosynthesis, and reduces nitrates and sulfates. In developing countries, more people depend on grains and legumes as their main food sources of Zn and Fe and more emphasis is being given to these elements. It is found that insufficient and excessive supplies of nutrients can lead to reduced crop quality. For example, a low content of N (as a raw material) will lead to a reduced amount of protein, while a low content of K will lead to a reduced amount of protein due to a decrease in the activation of enzymes that metabolize carbohydrates for amino acid and protein synthesis. Too much NH_4–N suppresses Ca uptake and its functions. On the other hand, low levels of Mg and K reduce the distribution of carbohydrates. It should be noted that nutrients do not work in isolation; therefore, balanced nutrition is needed to optimize crop quality. Along with crop yields, crop quality is another area that needs to be carefully considered as it affects human nutrition and the profitability of agricultural products (Njira and Nabwami, 2015).

5.4   IMPACT OF PLANT NUTRITION ON HUMAN HEALTH

The dietary source of most human minerals, macronutrients (calories), and micronutrients is plants. The nutritional quality of cereals and legumes has a direct effect on human nutrition. Unfortunately, mineral malnutrition is prevalent in both developed and developing countries, and it is estimated that up to two-thirds of the world's population may be at risk of deficiency in one or more essential mineral elements (Stein, 2010). Plant foods also contain a range of bioactive compounds that can play an important role in the prevention of chronic diseases, including heart disease, cancer, stroke, diabetes, Alzheimer's, cataracts, and age-related functional decline. Phytonutrients show promise in protecting against many diseases. There are correlations between the increase in phytonutrient intake and the reduction in chronic diseases. Micronutrients are vital for maintaining health; they are important in cognitive development, in reproductive capabilities and cellular metabolism, and additionally in immune machine responses in humans. Micronutrients for human health include vitamins A, B complex, C, E, and K; essential lipids; and essential amino acids, which are present in plants in variable amounts and, in some instances, are insufficient to provide a nutritional dose. For the maintenance of good human health, a complex diet consisting of a high dose of various plant foods is necessary. However, plants and plant parts vary in their quantity and quality of these vitamins and minerals, and all major food crops lack certain essential micronutrients (Martin and Li, 2017). For example, rice, wheat, and corn, the world's main food crops, are high in calories but contain insufficient levels of several micronutrients to meet the minimum human daily requirements. If food systems do not provide sufficient quantities and enough diversity of foods to continuously meet these requirement, malnutrition will ensue among certain population groups. Because agriculture is the primary source of all nutrients required for human life, those national agricultural systems that do not provide sufficient nutrient output to meet these nutritional needs will ultimately fail, as well as the food systems dependent on them. Due to the low availability of essential elements in soil and the low bioavailability of mineral elements in the edible parts of crops, it is estimated that about 2 billion individuals are deficient in micronutrients such as iron and zinc. The most important factor in determining a crop's nutritional content is the crop's genotype. Manipulation of the crop's genotype through traditional breeding approaches as well as advanced biotech techniques has been used to increase the macro and micronutrient content in crops. Molecular genetic approaches need to be applied more vigorously to develop nutrient-efficient crop species or genotypes/cultivars within species. Improved mineral nutrition traits in plants will help reduce crop production costs and environmental pollution and also benefit animal and human nutrition.

5.4.1 Removal, Loss Mechanism, and Environmental Consequences of Plant Nutrients

5.4.1.1 Removal and Loss of Nutrients

Cereal crops take up large quantities of nutrients from the soil. These nutrients become part of the root, shoot, leaf, and grain biomass and are removed with the harvesting of the crops; however, the fertility of the soil can be compromised if no fertilizer is applied. Nutrient removal from the soil varies from crop to crop. More and more nutrients are removed as the crop yield increases. To replenish the soil nutrients, fertilizers are necessary; however, extensive use of chemical fertilizers beyond the threshold level increases the loss of nitrogen and phosphorus into surrounding lakes, river, and the marine ecosystem (Cui et al., 2020). Nutrients are lost in various ways, such as nitrogen and potassium runoff, and phosphorus is lost due to its solubility in eroding soil. Soil nutrients can be lost due to erosion because eroded soil has high amounts of nutrients that are washed away from the agricultural land. Soil erosion is a threat as it causes nutrient depletion and degrades the soil (Bashagaluke et al., 2018).

About 50% of nitrogen is applied to cereal crops – maize rice 16% and wheat 18% – to accomplish food requirements (Ladha et al., 2016). Nitrogen loss occurs in three ways: leaching, soil N runoff, and gaseous emission. Most of the nitrogen loss occurs through nitrogen leaching, about 6.7%–19%. Plants utilize nitrogen in the form of nitrates (Noguero et al., 2016). Fertilizers that are applied in the form of ammonium and urea transform into nitrates especially in dryland soils. Plants take up nitrates for their metabolic processes; however, any remaining nitrate after take up is lost from the soil due to low nitrogen fertilizer recovery (33% in cereals) and nitrogen use efficiency (Spiertz, 2010). Nitrate is lost from the soil in two ways: leaching and runoff from the surface (Daryanto et al., 2017). Leaching is a physical process in which nitrates are lost when water is drained due to irrigation or rainfall. Ammonia volatilization is another way of losing nitrogen from the soil. It is a chemical process in which ammonium from nitrogen fertilizer at high pH is converted into ammonia gas. Emissions of ammonia increase with the superficial application of the fertilizer and loss can be minimized by absorbing the fertilizers. The ammonia volatilization process in rice plants occurs due to the application of increasing levels of nitrogen, at a pH of about 8.55, and in highly saline soil (Li et al., 2017). However, nitrogen use efficiency is not associated with ammonia volatilization.

Continuous use of phosphorus fertilizer leads to the accumulation of P in the soil and, ultimately, loss of phosphorus occurs in water bodies after its saturation in the soil (Yan et al., 2017). Phosphorus loss can occur in the form of dissolved P or as particulate P that is bound to the soil (Reid et al., 2018). The removal of particulate P from the field soil depends on the erosion of the soil and the quantity of P present in that sediment, whereas dissolved P is lost from the soil due to runoff during rainfall. The amount of P lost depends on the available concentration of P in the soil and the soil mineralogy. Loss of potassium also takes place in the form of runoff, leaching, and erosion.

5.4.1.2   Consequences of Loss of Nutrients in Environment

Water is one of the most important resources not only for human beings but also for all living organisms and ecosystems. The wide application of fertilizers in agricultural fields causes eutrophication of rivers, groundwater, lakes, and marine ecosystems. Eutrophication occurs due to the runoff or leaching of nutrients (nitrogen and phosphorus) in water (Sharabian et al., 2018). The leaching of nitrates increases the acidification of the soil and ultimately affects aquatic organisms. The accumulation of nitrogen and phosphorus in water bodies causes adverse effects on marine biodiversity and human health (Ngatia et al., 2018). Eutrophication causes excessive growth of microbial species and algae. Excessive growth of algae causes algal bloom which is a threat to water quality. The deterioration of water quality not only leads to loss of marine biodiversity but is also harmful to terrestrial organisms. Algal blooms give rise to various water issues, such as the depletion of oxygen levels, and they also secrete different toxins. The high mortality rate of marine life is associated with harmful algal blooms. Harmful cyanobacteria release the neurotoxic chemical beta methylamino alanine (BMAA) and humans can be affected by this toxin through inhalation and ingestion (Cheung et al., 2013). Hepatoxicity in humans is caused by the accumulation of microcystin, a potent toxin produced by cyanobacteria. Acute poisoning with microcystins damages the liver and results in hemorrhagic shock (Zhang et al., 2015). Moreover, consumption of nitrate-rich water causes methemoglobinemia (blue baby syndrome) which reduces the oxygen in the blood and can lead to death in babies (Figure 5.1).

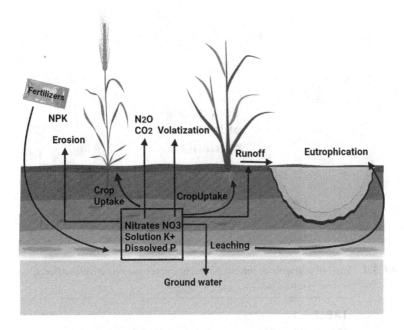

FIGURE 5.1 Representation of the removal and loss of the nutrient from cereal crops.

The use of nitrogen fertilizers has been increased to meet the needs of global food requirements. However, excessive application of nitrogen increases reactive nitrogen species that are harmful to human health. The emission of nitrogenous oxides, ammonia volatilization, and di-nitrogen due to nitrogen fertilizers are the main environmental concerns (Jones et al., 2014). The emission of ammonia and nitrogen dioxide causes toxicity in plants and animals. Synthetic nitrogen fertilizers are important sources of greenhouse gas emission. Cereal crops contribute about 68% to the production of greenhouse gases (Tongwane et al., 2016), as 50% of fertilizers are consumed by cereal crops. Most N_2O is produced by nitrification and denitrification that depend on the concentration of N in the soil. The emission of CO_2 and N_2O increases with increasing applications of nitrogen fertilizers in cereal crops (Figure 5.2).

5.5 OPTIMIZING PLANT NUTRITION FOR ENHANCED NUE AND CEREAL PRODUCTION

Sustainable cereal production is a function of multiple factors viz. crop genetics, an efficient supply of soil and water resources, agronomic practices, climate change impacts, and the application of technologies. Over the last two decades, the production of cereals in Pakistan has increased significantly from 30.5 metric tons (MT) in 2000 to 46 MT in 2020 (20). This increase in cereals yield is mainly attributed to more fertilizer usage coupled with other factors; however, the nutrient use efficiency of crops is 21% lower when compared to 2000 based on data from the FAO (Table 5.1). In 2020, the mineral fertilizer offtake in Pakistan was 4549 MT nutrients

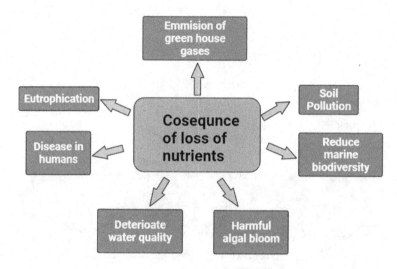

FIGURE 5.2   Loss of nutrients in the environment causes various adverse effects.

TABLE 5.1

Overview of Cereal Production, Nutrient Use, and NUE Trend in Pakistan over the Last Two Decades

Pakistan	2000	2020	% Change
Cropland area	31.7 Mha	32.0 Mha	+1
Cereal production	30.5 Mt	46.0 Mt	+51
Nutrient use	3.0 Mt	4.7 Mt	+57
Crop NUE	35.8%	28.3%	−21

(nitrogen: 3415; phosphorous pentoxide [P_2O_5]: 1084; potassium oxide [K_2O]: 50) (NFDC, 2019–2020). In other words, the ratio of applied N, P, and K in Pakistan during 2019–2020 was about 3.1 : 1 : 0.04 compared with the recommended/desired ratio of 2 : 1 : 0.5 (NFDC, 2019–2020; Vitousek et al., 2009). This imbalance seriously reduces the NUE. In practice, intensive cultivation practices with no proper crop rotation plans and excessive use of nitrogenous fertilizers have rendered soils non-responsive to additional inputs. With the passage of time, farmers have learned to apply phosphatic fertilizers; however, soil test–based balanced use of fertilizers – particularly potash and other micronutrients (e.g., zinc and boron) – are gradually depleting Pakistan's soils (Figure 5.1). Unfortunately, farmers are unaware of the economic and environmental consequences of excessive use of a single nutrient. For example, excessive use of nitrogenous fertilizers causes loss of nitrogen in the form of ammonia (NH_3) and nitrous oxide (N_2O) to the atmosphere and deteriorates air quality, biodiversity, and human well-being (Mahmud et al., 2021). In addition to livestock waste, volatilization of NH_3 from chemical fertilizers contributes ~90% of global NH_3 emissions (Schlesinger and Hartley, 1992; Ferm, 1998). To reduce N losses and increase NUE for sustainable cereals production, it is imperative that

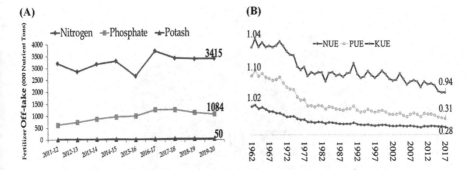

FIGURE 5.3 (A) Comparison of fertilizer offtake pattern in Pakistan over the last 10 years – indicating that the use of nitrogenous fertilizers is much higher across the country, whereas the use of potassic and other micronutrients is minimal. (B) The resultant decrease in nitrogen use efficiency (NUE), phosphate use efficiency (PUE), and potash use efficiency (KUE) are shown.

nutrients removed from the soil are reapplied in the correct balance to minimize waste, economic losses, and environmental impacts. Furthermore, to meet the food demands of the growing world population, better management of essential plant nutrients is necessary for enhanced nutrient uptake and utilization efficiencies (Figure 5.3).

5.5.1 CHEMICAL FERTILIZERS AND THE CONCEPT OF 4R NUTRIENT STEWARDSHIP FOR SUSTAINABLE CEREALS PRODUCTIONS

Chemical fertilizers play a central role in improving the health and productivity of crop plants. According to different estimates, the balanced use of fertilizers contributes approximately 40%–60% to achieving crop yield and the associated food security targets. In practice, an imbalance in the use of fertilizers is common, reducing the NUE of crops with the ultimate consequence of environmental footprints due to chemical fertilizers. To counter these challenges, fertilizer industry stakeholders introduced the concept of 4Rs nutrient stewardship to minimize nutrient losses for agricultural and environmental sustainability. The term 4Rs encompasses the four rights of fertilizer best management practices (FBMPs): right source of nutrients, at the right rate, at the right time, and in the right place. The performance indicators of 4Rs nutrient stewardship are broadly expressed in terms of enhanced per acre yield and the consequent net farm returns with long-term environmental and socioeconomic benefits (Johnston and Bruulsema, 2014). Therefore, the benefits of implementing the 4Rs nutrient stewardship approach are not limited to enhanced cereal productivity but are also a win-win strategy because of its contribution to the accomplishment of SDG13 (climate change) due to its commitment to climate-smart agriculture (Figure 5.4).

5.5.1.1 Right Source

Choosing the right source of fertilizer means selecting a product according to the physical (e.g., texture) and chemical properties (e.g., pH) of the soil and that its chemical

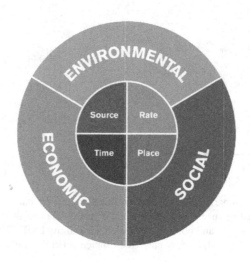

FIGURE 5.4 4Rs nutrient stewardship (right source, right rate, right time, and right place) for enhanced cereal production with multiple environmental and socioeconomic benefits.

composition also matches the nutrient requirements of the crop. Importantly, the right fertilizer source should ensure a supply of essential plant nutrients – mainly N, P, and K as well as other micronutrients (e.g., Zn and B). In this context, a soil test–based assessment of crop nutrient requirements should generally be regarded as the first step in identifying the right fertilizer source, making it easier to decide the best possible combination of fertilizers to use in a crop and site-specific manner. It is pertinent to mention here that commercial fertilizers are available in different forms either as liquid or dry granular products and in different grades such as straight (e.g., urea and calcium ammonium nitrate [CAN]) or compound (e.g., Nitrophos and NPK) fertilizers. Considering both product and soil characteristics, it is extremely important to understand the dynamics of the nutrient interactions with each other and the soil particles. For example, the combined application of phosphorous (e.g., diammonium phosphate [DAP]) and zinc (e.g., zinc sulfate [$ZnSO_4$]) should be discouraged due to the formation of zinc phosphate that is not easily water soluble. Similarly, nitrate-based nitrogenous fertilizers such as CAN should not be used in flooded soils. On the other hand, the choice of liquid or dry granular fertilizer products is largely dependent on the crop conditions and the accurate diagnosis of deficiency symptoms through soil and plant analysis at different growth stages. Generally, farmers prefer dry granular fertilizer products in Pakistan; however, depending on the farmer's finances and other field conditions, the solubility of commercially available fertilizers in Pakistan is debated between the fertilizer industry and the farming community. Based on the least-expensive source of nutrients and the aforementioned factors, the capacity building of farmers to make the right choice of products is a major challenge to improving NUE and the subsequent increase in cereal production.

5.5.1.2 Right Rate

The right rate means the application of fertilizer in the right amount based on site-specific soil analyses results and crop nutrient requirements. In Pakistan, imbalanced fertilizer application is regarded as the main reason for the substantial gap between the average and the potential yield of cereal crops. Imbalanced fertilizer application means both over- and under-application of nutrients. Although adding too much fertilizer seems an unrealistic approach, it is routinely practiced by many farmers due to lack of knowledge or access to soil testing facilities. Over-application of fertilizers poses serious challenges to sustainable cereal production due to losses to the environment and residual buildup in the soils, thereby disturbing the soil health and ultimately reducing plant growth. The under-application of fertilizers induces serious nutrient deficiencies, thus plants cannot complete their growth cycle and demonstrate low productivity. Hence, optimum nutrient application is extremely important to increase cereal productivity, which can be achieved through selecting the right rate of fertilizer. It requires careful attention to numerous soil characteristics (e.g., available nutrient resources in the soil, previous crop residue, and fertilizer management history) as well as other parameters such as crop yield targets and environmental parameters. Nevertheless, soil testing is the main strategy in establishing the right rate of fertilizer application, which can effectively break the stagnation in cereal yields. Soil tests provide an appropriate estimate of soil health (e.g., pH, EC, and exchangeable Na percentage) as well as the soil's fertility status. But this is not routinely practiced in Pakistan, ultimately translating into low NUE and subsequent low cereal productivity. Generally, the soils in Pakistan are deficient in organic matter (96%), phosphorous (92%), potash (50%), zinc (85%), and boron (60%). To promote the balanced use of fertilizers, particularly for farmers with poor access to soil testing facilities, fertilizer companies have devised crop fertilizer recommendations for cereal crops grown in poor, medium, and fertile soils. However, due to horizontal soil variability and diverse agroecological zones, the cropping systems are not responsive to uniform fertilizer application regime. Therefore, striking a balance between the soil nutrient supply capacity and crop nutrient requirements and the environmental conditions is necessary to bridge the gap between the average and potential yield of cereal crops grown in Pakistan.

5.5.1.3 Right Time

The right time means the application of fertilizers in accordance with crop needs at critical growth stages. Optimizing nutrient application timings according to crop demands increases NUE due to maximum crop nutrient uptake and utilization efficiencies and reduced losses to the environment. In general, fertilizer timings have to be synchronized in a site-specific manner, keeping in view the management practices adapted by the farmers and the local environmental conditions influencing the release and availability of nutrients from the fertilizer products. Considering the low cereal productivity and the poor NUE, in particular for nitrogenous fertilizers, there is increasing interest in the use of controlled and slow-release fertilizer technology that provides an opportunity to change the traditional fertilizer timing recommendations. For example, split application of nitrogenous fertilizers is generally practiced to avoid losses of NH_3–N and NO_3–N to the environment. Therefore, the use of stabilizers and inhibitors that can slow down the release of N from the fertilizer product has the potential to enhance NUE coupled with low nutrient losses to the environment, hence better timed for efficient uptake and ultimately increased crop productivity. Controlled-release fertilizers also provide an opportunity to mimic the split fertilizer application due to delayed release and hence the possibility to apply at planting. Thus, the consequences of potential nutrient losses from the traditional split application technology under the influence of soil moisture and weather conditions and also if mismatched with the crop's nutrient requirement can be avoided by the use of slow/controlled-release fertilizer technology. On the other hand, the predominance of calcareous soils in Pakistan demonstrates high P fixation potential. Therefore, it is common to apply phosphatic fertilizers as a basal dose to ensure P availability to young seedlings.

5.5.1.4 Right Place

Right place is the application of fertilizers at the place where crop roots can easily reach and use them. Appropriate fertilizer placement is extremely challenging in a country like Pakistan where most of the landholdings are small; therefore, uniform or broadcast application of fertilizers is a routine practice. In such cases, fertilizers become useless when plants cannot access them. It has been experimentally tested that the right fertilizer placement increases the fertilizer use efficiency of crop plants, enabling the gap between the average and potential yield of cereal crops to be bridged. Adaptation to mechanized farming plays a major role in correct fertilizer placement and thus improved NUE by lowering nutrient application rates – especially in soils that have the potential to fix certain nutrients. This is particularly true for phosphatic fertilizers fixed in soils rich in calcium, iron, and aluminum contents. Thus, the band placement of phosphatic fertilizers as a basal dose near the seed row is critical for nutrient uptake by the cereal crops during critical growth stages. Phosphorous is immobile in nature – meaning it moves only a few centimeters from its place of application. To tackle this issue, the band placement of phosphatic fertilizers via seed drills is practiced by progressive growers in Pakistan. However, extreme care is required in band placement cases that the negative impacts of fertilizer are minimized. For smallholding farmers with poor access to seed drills, farm management practices are extremely important in determining fertilizer placement through broadcast. Keeping in view the calcareous nature of Pakistan's soils, P fixation issues, and broadcast fertilizer application trends by smallholding farmers, they need to be educated in mixing phosphatic fertilizers with farmyard manure in amounts double the amount of phosphatic fertilizer to minimize the P fixation capacity of calcareous soils. This also promotes the concept of integrated nutrient management through the application of organic and chemical fertilizer sources which significantly enhances fertilizer use efficiency and per acre cereal production.

5.6 GENETIC APPROACHES TO IMPROVE NUTRIENT USE EFFICIENCY IN CEREAL CROPS

Nutrient use efficiency is crucial for increasing crop yield and quality while reducing fertilizer inputs and minimizing environmental damage. Major nutrients are a substantial factor in crop yield, and are provided through the use of chemical fertilizers. According to the International Fertilizer Association (IFA), NPK fertilizer consumption in 2020–2021 was estimated at about 198.2 million, 5.2% higher than in 2019–2020. Moreover, this consumption will increase to about 150 million for nitrogen and 120 million for phosphorus by 2040 (Zhang et al., 2020). The increasing demand for food is directly proportional to the availability of plant nutrients that come from chemical fertilizers. The production of fertilizer is heavily dependent on non-renewable fossil fuels. Furthermore, fertilizer that is applied to agricultural land cannot be completely utilized by plants and leaches into the water bodies causing pollution. Consequently, there is an urgent need to improve the nutrient use efficiency of plants for food security and sustainable agriculture. Crop yield can be increased by higher nutrient use efficiency when applied with lower amounts of fertilizer. Recent studies in molecular biology have provided a possible means to address the complex nutritional problems of plants. Using genetic approaches together with phenotypic analysis can elucidate the functions and interactions of plant nutrients at the molecular, cellular, organ, and whole-plant levels. Several genetic approaches have been introduced to improve the NUE in plants (Qaim et al., 2019)

5.6.1 Genotypes with Improved NUE

Different genotypes behave differently in low concentrations of nitrogen. Knowledge of NUE-related traits can improve breeding programs by introducing new genotypes with nutrient use efficiency. Various traits for NUE variation have been identified in cereal crops such as wheat, rice, maize, and barley. A study of wheat indicates that the nitrogen uptake in older genotypes is higher than in new varieties as they were cultivated in the time when no chemical fertilizers were applied. Some modern cultivars of wheat *Triticum aestivum* performed well at very low nitrogen concentrations (Gouis et al., 2000). Various genotypes were identified for NUE-related traits in wheat that perform well in less favorable environmental conditions (Mahjourimajid et al., 2016).

5.6.2 Genes Involved in the NUE

Several genes are involved in the NUE that regulates various mechanisms for the absorption and accumulation of nitrogen in plants. The genes associated with NUE in different cereal crops are divided into six categories (Sandhu et al., 2021): transporter genes, nitrate assimilation genes, signaling molecule, amino acid biosynthesis, transcription factor, and other genes. Among these categories, nitrate assimilation genes and transporter genes are mainly involved in nitrogen uptake; the biosynthesis gene family is involved in nitrogen utilization; and transcription factors and signaling molecules are passively involved or have a passive role in nitrogen uptake and utilization.

5.6.2.1 Nitrogen Transporter Genes

Nitrogen is present in the form of nitrate, the most common form of nitrogen present in the soil. It is actively transported in plants with the help of nitrate transporter (NRTs) genes. These nitrate transporters are encoded by NRT gene families, which are classified into three subfamilies: the NRT1 family, whose members are low-affinity transporters, and the NRT2/NRT family, which mainly encodes high-affinity transporters (Plett et al., 2010).

Approximately 16 low-affinity nitrate transporter 1/peptide transporter family (NPF) genes are expressed in common wheat and are homologous to *Arabidopsis* NPF genes. The expression of a particular transporter in wheat depends on the nitrogen status of the plant and the soil. It has been reported that wheat NPF genes are regulated by the plant nitrogen status, suggesting that nitrogen metabolism is the main regulator of genes involved in nitrate transport. There are four high-affinity NRT2 transporters in rice, among which two genes (OsNRT2.1 and Os-NRT2.2) have strong similarities to NRTs in monocots, while OsNRT2.3 and OsNRT2.4 are more closely related to *Arabidopsis* NRT2. OsNRT2.3 has two subtypes: OsNRT2.3a and OsNRT2.3b. The OsNRT2 gene family was found to play an important role during N uptake and translocation, requiring its partner protein NAR2 to perform this function, in addition to OsNRT2.3b (Chen J.G. et al., 2017). The overexpression of OsNRT2.3b is known to play an important role in high grain yield and NUE in rice (Sandhu et al., 2021; Fen et al., 2016). The high-affinity transport system (HATS) belongs to NRT2 and its protein often requires NAR2 for NRT2 functioning. OsNRT2.1 and OsNRT2.3a interact with OsNAR2.1 helping to stabilize the NRT2 protein and uptake activities. Nevertheless, the OsNRT2.3b transporter does not require NAR2 (Lee S., 2021).

The transport of ammonium ion is mediated by the transporter genes that belong to the ammonium transporter/methylammonium per-Mease (AMT/MEP/Rh) family. Genes related to the AMT family are located on the plasma membrane and can capture 30% of total NH_4^+ uptake in *Arabidopsis thaliana*. AMT family members have also been identified in other cereal crops such as *Oryza sativa*, *Zea mays*, and *Sorghum bicolor*, whereas there is less information about NH_4^+ transporters in wheat. Only the TaAMT1 gene is reported in wheat, which is expressed under arbuscular mycorrhizal fungi. Rice is well studied for the AMT genes. Approximately 12 genes were found in rice and are divided into four subcategories, from OsAMT1 to OsAMT4. OsAMT1;1 transporter is present in the plasma membrane and is constitutively expressed in roots and shoots. The disruption of OsAMT1;1 reduces the uptake of ammonium and seedlings growth as N transport from roots to shoots decreases. The overexpression of OsAMT1;1 in plants enhances N uptake and plants show higher amounts of N assimilation, improved plant growth, and increased grain yield, especially under suboptimal conditions NH_4^+. The expression of OsAMT1;2 and OsAMT1;3 is root specific and induced by N starvation. Only two OsAMT2 transporters, OsAMT2;1 and OsAMT2;2, show expression in both roots and shoots, while OsAMT3;1 shows very weak expression in roots and shoots in rice (Lee S., 2021).

5.6.2.2   Nitrogen Assimilation and Amino Acid Biosynthesis Genes

Nitrogen assimilation is a very important metabolic step followed by nitrogen uptake. This process is regulated by a number of enzymes such as glutamate synthase (GOGAT), glutamine synthetase (GS), nitrate reductase (NR), and nitrite reductase (NiR). Glutamine synthetase and glutamate synthase are involved in nitrogen/nitrate assimilation in the form of amino acids. Initially, NO_3^- is converted to NO_2^- by NR and transported in the plastid where it is reduced by nitrite reductase and converted to NH_4^+. Subsequently, NH_4^+ is assimilated in the form of nitrogen through glutamine synthetase and glutamate synthase (Thomsen et al., 2014). NR requires NADH/NADPH to reduce NO_3^- which is encoded by NADH-NR genes in wheat. The introduction of the NR gene from the tobacco plant into the wheat plant showed an increase in the seed protein content in the usual nitrogen concentration, whereas the variation in the NR gene in sorghum and rice showed no difference in nitrogen assimilation. Genes encoding the GS/GOGAT are responsible for the growth and yield of cereal crops (Table 5.2). Overexpression of OsGS1 increases grain yield whereas an extra copy of the HuGS1 gene in barley increases the grain yield and nitrogen use efficiency as compared to the wild plant. Rice is known to have a small family of GS and GOGAT enzymes present at different locations in a cell. These enzymes have isoform. Among the variable isoforms of GS and GOGAT, the cytosolic GS1;2 isoform and the plastid NADH-GOGAT1 are involved in the assimilation of ammonium ions in the roots. It is reported that under conditions with a high N content, overexpression of the GS1 gene improves the nitrogen scavenging index and NUE, but no change in NUE was observed in a nitrogen-deficient environment.

In addition to these enzymes, glutamate dehydrogenase (GDH), alanine aminotransferase (AlaAT), and asparagine synthetase (AsnS) contribute to nitrogen assimilation and amino acid synthesis. GDH is considered important as it plays a vital role in the nitrogen use efficiency of rice and wheat, as well as catalyzing the interconversion of 2-oxoglutarate and glutamate. AlaAT plays an important role in the synthesis of alanine. HuAlaAT from the barley species was introduced into rice and showed significant nitrogen uptake in the case of nitrogen deficiency (Shrawat et al., 2008). AsnS produces asparagine, a major form of nitrogen in phloem. AsnS catalyzes the transfer of an amide group from glutamine to aspartate to form asparagine. AsnS is encoded by a small gene family having isoform four in wheat and five in barley. The TaASN1 and TaASN2 genes in wheat showed significant increases in grain (Figure 5.5).

5.6.2.3 Signaling Molecule and Transcription Factor in NUE

Transcription factors are the major switches in plant regulatory networks. Various transcription factors related to nitrogen assimilation were found in cereal crops. For cereal crops such as maize, wheat, and rice, DOF, NAC, NAP, GRF, MAD, MYB, NLP, NAM, CUC, and NFY are major transcription factors (Islam et al., 2020).

TABLE 5.2

Transcription Factors for Nitrogen Uptake and Their Functions

Gene Name	Function	References
OsDOf2	Enhance N transportation and grain number	Iwamoto et al. (2016)
OsMYB305	Enhance N uptake under lower application of nitrogen	Wang et al. (2020)
OsNLP1	Improve nitrogen use efficiency, grain yield, and plant growth	Alfatih et al. (2020)
OsMADS57	Increase xylem loading of nitrate and induce nitrate transporter genes	Huang et al. (2020)
TaNAC2	Increase nitrate uptake, biomass, and grain yield	Ha et al. (2015)
NAM-B1	Transfer nitrogen in grains	Islam et al. (2020)

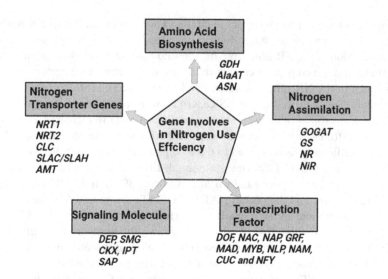

FIGURE 5.5 Representation of the genes family involved in the nitrogen uptake and nitrogen utilization in cereal crops.

Genes from these families contribute to nitrogen use efficiency. The overexpression of the ZmDOF1 gene in the maize plant enhanced nitrogen assimilation in low concentrations of nitrogen. In wheat, the overexpression of DOF1.3 transgenic lines showed enhanced nitrogen assimilation in low nitrogen–stressed conditions (Curci et al., 2017). Nitrogen uptake, tiller number, and growth under low nitrogen concentrations were increased by the overexpression of the OsMYB30 gene in rice. Moreover, the TaNAC2 gene in the wheat plant is associated with nitrate uptake, biomass, and grain yield. Overexpression of NAR2 enhances all these traits in wheat (He et al., 2015). Another transcription factor in wheat, NAM_B1, is reported to transfer nitrogen in the grains. MAD-box TF genes regulate lateral roots growth and N signaling and enhance the expression of the nitrate transporter genes (Chen et al., 2018; Sun et al., 2018). Os_MADS27 transgenic lines show enhanced growth and nitrate accumulation in rice. Various signaling molecules contribute to the transportation of nitrogen (see Figure 5.6).

5.6.3 PHOSPHATE TRANSPORTER AND TRANSCRIPTION FACTORS FOR NUE

Plants take in phosphorus in the form of inorganic phosphate through their roots. Various transporters of Pi were found in cereal crops such as rice and wheat. The PHT1 family Pi transporter, the SPX domain-containing Pi transporter, the vascular phosphate efflux (VPE) transporter, and sulfate transporters (SULTR) play a pivotal role in phosphate transportation in rice plants (Zhang et al., 2020). OsVPE1 and OsVPE2 are vascular phosphate efflux transporters that maintain the homeostasis of phosphate. Two major sulfate transporters, OsSULTR3;3 and SPDT/OsSULTR3;4, are involved in the P distribution. The SPDT/OsSULTR3;4 transporter can directly transport phosphate instead of sulfate. The SPDT mutation changes the distribution

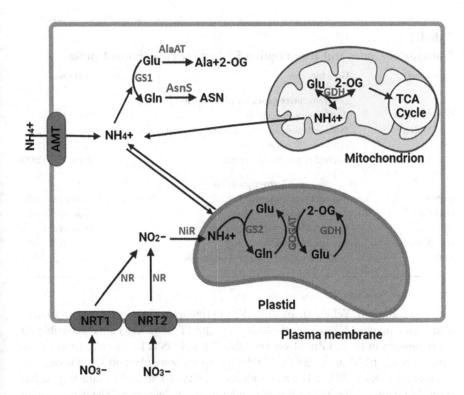

FIGURE 5.6 Schematic diagram representing the transporter and enzymes involved in the nitrogen metabolism process.

of P with reduced P in grains by 20%, while increasing P levels in straw without penalizing other agronomic traits (Table 5.3). The decrease in the concentration of P in grains in SPDT mutants is due to the reduction in phytate content (Yamaji et al., 2017). Therefore, the retention of P in shoots through SPDT manipulation in rice breeding programs will improve nutritional quality and maintain grain yield with less P fertilizer input and environmental pollution (Kopriva and Chu, 2018). Few genes are involved in the inorganic phosphate starvation signaling in hexaploid wheat; for example, TaPHR1-A1 increases the level of other regulatory genes and Pi uptake by increasing the root tips number in wheat.

5.6.4 TRANSPORTER GENE FOR THE UPTAKE OF MICRONUTRIENT

Micronutrients play a vital role in plant growth and balanced crop nutrition. Various transporter genes for the regulation and uptake of micronutrients have been identified in cereal crops. The uptake of zinc and iron Zn-regulated transporter and iron-regulated transporter-like protein (ZIP) has been identified in rice. The identified ZIP members, such as OsZIP1–OsZIP8 except OsZIP6, are involved in the transportation of not only zinc but also cadmium (Cd). OsIRT1 and OSIRT2 have been identified not only as iron transporters but are also involved in the transportation of both

TABLE 5.3

Transporter Genes and Transcription Factors for Phosphorus Uptake

Gene name	Gene function	References
	Transporter genes for phosphorus uptake	
OSPht1/OsPT9	Regulate phosphate uptake	Zhang et al. (2020)
OsSPX-MFS1	Maintain Pi homeostats and transport phosphate in vacuoles	Zhang et al. (2020)
OsVPE1	Phosphate efflux transporter	Zhang et al. (2020)
OsSULTR3;3/SPDT	Involved in the allocation of P in grain	Zhang et al. (2020)
	Transcription factors	
OsPHR3	Activate Pi starvation genes	Zhang et al. (2020)
OsMYB5P	Activate Pi homeostasis genes	Zhang et al. (2020)
OsPTF1	Involved in improving Pi starvation	Zhang et al. (2020)
OsARF16	Regulate Pi starvation	Zhang et al. (2020)
TaPHR-A1	Increase Pi uptake, increase root tip number	Wang et al. (2013)

zinc and cadmium. Yellow stripe–like (YSL) protein mediates the transportation of iron in rice plants. Members of the heavy metal ATPase (HMA) are responsible for the regulation of Fe and Zn. Moreover, OSVIT1 and OSVIT2 vascular transporters have been identified in rice and aid in the transportation of Fe and Zn to leaves and seeds (Bashir et al., 2013). The zinc-induced facilitator-like (ZIFL) family gene has been characterized for its expression in micronutrient deficiency. The expression of ZIFL genes enhances the uptake of zinc and iron in wheat (Sharma et al., 2019). The transporter for phosphorus OsPT2 also mediates the uptake of selenium in rice plants. Together with OsPT2, the transporter NRT1.1B for nitrogen uptake is also responsible for the long-distance transportation of Se (Zhang et al., 2020).

Plants have an enormous number of transporters that are involved in nutrient uptake, transport, and re-translocation, and they play an important role in both grain yield and nutritional quality. A comprehensive understanding of nutrient transporters and the cross-talk among different nutrients can facilitate the enhancement of macronutrient and micronutrient absorption, transport, and re-translocation in plants. This knowledge also constitutes the genetic toolbox for designing and cultivating smart crops with improved grain yield and nutritional quality, but with reduced fertilizer input.

REFERENCES

Alfatih, A., Wu, J., Zhang, Z. S., Xia, J. Q., Jan, S. U., Yu, L. H., & Xiang, C. B. (2020). Rice NIN-LIKE PROTEIN 1 rapidly responds to nitrogen deficiency and improves yield and nitrogen use efficiency. *Journal of Experimental Botany*, *71*(19), 6032–6042.

Bashagaluke, J. B., Logah, V., Opoku, A., Sarkodie-Addo, J., & Quansah, C. (2018). Soil nutrient loss through erosion: Impact of different cropping systems and soil amendments in Ghana. *PLoS One* 13, 1–17.

Bashir, K., Takahashi, R., Akhtar, S., Ishimaru, Y., Nakanishi, H., & Nishizawa, N. K. (2013). The knockdown of OsVIT2 and MIT affects iron localization in rice seed. *Rice*, *6*(1), 1–7.

Bruinsma, J. (2009). *The resource outlook to 2050: By how much do land, water use and crop yields need to increase by 2050?* Paper presented at the Export Meeting on How to Feed the World in 2050. Rome: FAO.

Chen, J. G., Fan, X. R., Qian, K. Y., Zhang, Y., Song, M. Q., Liu, Y., et al. (2017). pOsNAR2.1:OsNAR2.1 expression enhances nitrogen uptake efficiency and grain yield in transgenic rice plants. *Plant Biotechnology Journal* 15, 1273–1283.

Chen, H., Li, D., Zhao, J., Zhang, W., Xiao, K., & Wang, K. (2018). Nitrogen addition aggravates microbial carbon limitation: Evidence from ecoenzymatic stoichiometry. *Geoderma, 329*, 61–64.

Cheung, M. Y., Liang, S., & Lee, J. (2013). Toxin-producing cyanobacteria in freshwater: A review of the problems, impact on drinking water safety, and efforts for protecting public health. *Journal of Microbiology* 51, 1–10.

Cui, N., et al. (2020). Runoff loss of nitrogen and phosphorus from a rice paddy field in the east of China: Effects of long-term chemical N fertilizer and organic manure applications. *Global Ecology and Conservation* 22, e01011.

Curci, P. L., et al. (2017). Transcriptomic response of durum wheat to nitrogen starvation. *Scientific Reports* 7, 1–14.

Daryanto, S., Wang, L., & Jacinthe, P. A. (2017). Impacts of no-tillage management on nitrate loss from corn, soybean and wheat cultivation: A meta-analysis. *Scientific Reports* 7, 1–9.

Facts, I. D. R. C. (2010). *Figures on Food and Biodiversity.* Canada: IDRC Communications, International Development Research Centre.

FAO. (1975). *Soil map of the world 1:5,000,000. Legend.* Paris: UNESCO.

FAO. (2002). *The state of food insecurity in the world 2001.* Rome: FAO.

FAO. (2021). *The state of food security and nutrition in the world 2021.* Rome: FAO.

FAO, IFAD, UNICEF, WFP, and WHO. (2019). *The state of food security and nutrition in the world 2019. Safeguarding against economic slowdowns and downturns.* Rome: FAO.

Ferm, M. (1998). Atmospheric ammonia and ammonium transport in Europe and critical loads: A review. *Nutrient Cycling in Agroecosystems* 51, 5–17.

Fukagawa, N. K., & Ziska, L. H. (2019). Rice: Importance for global nutrition. *Journal of Nutritional Science and Vitaminology (Tokyo)* 65, S2–S3.

Graham, R. D., Welch, R. M., & Bouis, H. E. (2001). Addressing micronutrient malnutrition through enhancing the nutritional quality of staple foods: Principles, perspectives and knowledge gaps. *Advances in Agronomy* 70, 77–142.

Groosman, A., Linnemann, A., & Wierema, H. (1988). Technology development and changing seed supply systems: Seminar proceedings. *Research Report* No. 27. Tilburg, Netherlands: IVO.

Grote, U., Fasse, A., Nguyen, T. T., & Erenstein, O. (2021). Food security and the dynamics of wheat and maize value chains in Africa and Asia. *Frontiers in Sustainable Food Systems* 4, 317.

Ha, N., Feike, T., Back, H., Xiao, H., & Bahrs, E. (2015). The effect of simple nitrogen fertilizer recommendation strategies on product carbon footprint and gross margin of wheat and maize production in the North China Plain. *Journal of Environmental Management, 163*, 146–154.

Huang, X., Yang, X., Zhu, J., & Yu, J. (2020). Microbial interspecific interaction and nitrogen metabolism pathway for the treatment of municipal wastewater by iron carbon based constructed wetland. *Bioresource Technology, 315*, 123814.

Islam, S., Zhang, J., Zhao, Y., She, M., & Ma, W. (2021). Genetic regulation of the traits contributing to wheat nitrogen use efficiency. *Plant Science* 303, 110759.

Iwamoto, M., & Tagiri, A. (2016). Micro RNA-targeted transcription factor gene RDD 1 promotes nutrient ion uptake and accumulation in rice. *The Plant Journal, 85*(4), 466–477.

Johnston, A. M., & Bruulsema, T. (2014). 4R nutrient stewardship for improved nutrient use efficiency. *Procedia Engineering* 83, 365–370.

Jones, L., et al. (2014). A review and application of the evidence for nitrogen impacts on ecosystem services. *Ecosystem Services* 7, 76–88.

Ladha, J. K., et al. (2016). Global nitrogen budgets in cereals: A 50-year assessment for maize, rice, and wheat production systems. *Scientific Reports* 6, 1–9.

Le Gouis, J., Béghin, D., Heumez, E., & Pluchard, P. (2000). Genetic differences for nitrogen uptake and nitrogen utilisation efficiencies in winter wheat. *European Journal of Agronomy* 12, 163–173.

Lee, S. (2021). Recent advances on nitrogen use efficiency in rice. *Agronomy, 11*(4), 753.

Li, Y., Huang, L., Zhang, H., Wang, M., & Liang, Z. (2017). Assessment of ammonia volatilization losses and nitrogen utilization during the rice growing season in alkaline salt-affected soils. *Sustainability* 9, 132.

Lunde, C., Zygadlo, A., Simonsen, H. T., Nielsen, P. L., Blennow, A., & Haldrup, A. (2008). Sulfur starvation in rice: The effect on photosynthesis, carbohydrate metabolism, and oxidative stress protective pathways. *Physiologia Plantarum* 134, 508–521.

Mahjourimajd, S., Kuchel, H., Langridge, P., & Okamoto, M. (2016). Evaluation of Australian wheat genotypes for response to variable nitrogen application. *Plant Soil* 399, 247–255.

Mahmud, K., Panday, D., Mergoum, A., & Missaoui, A. (2021). Nitrogen losses and potential mitigation strategies for a sustainable agroecosystem. *Sustainability* 13, 2400.

Martin, C., & Li, J. (2017). Medicine is not health care, food is health care: Plant metabolic engineering, diet and human health. *New Phytologist* 216, 699–719.

Matres, J. M., Hilscher, J., Datta, A., Armario-Nájera, V., Baysal, C., He, W., ... & Slamet-Loedin, I. H. (2021). Genome editing in cereal crops: an overview. *Transgenic research* 30(4), 461–498.

Nazari-Sharabian, M., Ahmad, S., & Karakouzian, M. (2018). Climate change and eutrophication: a short review. *Engineering, Technology and Applied Science Research*, 8(6), 3668.

Njira, K. O., & Nabwami, J. (2015). A review of effects of nutrient elements on crop quality. *African Journal of Food, Agriculture, Nutrition and Development*, 15(1), 9777–9793.

Plett, D., Toubia, J., Garnett, T., Tester, M., Kaiser, B. N., & Baumann, U. (2010). Dichotomy in the NRT gene families of dicots and grass species. *PLoS One* 5, e15289.

Reid, K., Schneider, K., & McConkey, B. (2018). Components of phosphorus loss from agricultural landscapes, and how to incorporate them into risk assessment tools. *Frontiers in Earth Science* 6, 1–15.

Rurinda, J., et al. (2020). Science-based decision support for formulating crop fertilizer recommendations in sub-Saharan Africa. *Agricultural Systems* 180, 102790.

Sandhu, N., Sethi, M., Kumar, A., Dang, D., Singh, J., & Chhuneja, P. (2021). Biochemical and genetic approaches improving nitrogen use efficiency in cereal crops: A review. *Frontiers in Plant Science* 12, 657629.

Schlesinger, W. H., & Hartley, A. E. (1992). A global budget for atmospheric NH3. *Biogeochemistry* 15, 191–211.

Sharma, S., et al. (2019). Overlapping transcriptional expression response of wheat zinc-induced facilitator-like transporters emphasize important role during Fe and Zn stress. *BMC Molecular Biology* 20, 1–17.

Shewry, P. R., & Hey, S. J. (2015). The contribution of wheat to human diet and health. *Food and Energy Security* 4, 178–202.

Shrawat, A. K., Carroll, R. T., DePauw, M., Taylor, G. J., & Good, A. G. (2008). Genetic engineering of improved nitrogen use efficiency in rice by the tissue-specific expression of alanine aminotransferase. *Plant Biotechnol. Journal* 6, 722–732.

Stein, A. J. (2010). Global impacts of human mineral nutrition. *Plant and Soil* 335, 133–154.

Sun, C. H., et al. (2018). Chrysanthemum MADS-box transcription factor CmANR1 modulates lateral root development via homo-/heterodimerization to influence auxin accumulation in Arabidopsis. *Plant Science* 266, 27–36.

Thomsen, H. C., Eriksson, D., Møller, I. S., & Schjoerring, J. K. (2014). Cytosolic glutamine synthetase: A target for improvement of crop nitrogen use efficiency? *Trends Plant Science* 19, 656–663.

Tongwane, M., et al. (2016). Greenhouse gas emissions from different crop production and management practices in South Africa. *Environmental Development* 19, 23–35.

Valasai, G. D., Harijan, K., Uqaili, M. S., & Memon, H. R. (2005). Impact of greenhouse gases on agricultural productivity in Pakistan. In *Proceedings of the first international conference on environmentally sustainable development v.* 1–3.

Vitousek, P. M., Naylor, R., Crews, T., David, M. B., Drinkwater, L. E., Holland, E., Johnes, P. J., Katzenberger, J., Martinelli, L. A., Matson, P. A., Nziguheba, G., Ojima, D., Palm, C. A., Robertson, G. P., Sanchez, P. A., Townsend, A. R., & Zhang, F. S. (2009). Nutrient imbalances in agricultural development. *Science* 324, 1519–1520.

Wang, D., et al. (2020). Overexpression of OsMYB305 in rice enhances the nitrogen uptake under low- nitrogen condition. *Frontiers in Plant Science* 11, 1–19.

Wang, J., et al. (2013). A phosphate starvation response regulator Ta-PHR1 is involved in phosphate signalling and increases grain yield in wheat. *Annals of Botany* 111, 1139–1153.

Yamaji, N., Takemoto, Y., Miyaji, T., Mitani-Ueno, N., Yoshida, K. T., & Ma, J. F. (2017). Reducing phosphorus accumulation in rice grains with an impaired transporter in the node. *Nature* 541, 92–95.

Yan, X., Wei, Z., Hong, Q., Lu, Z., & Wu, J. (2017). Phosphorus fractions and sorption characteristics in a subtropical paddy soil as influenced by fertilizer sources. *Geoderma* 295, 80–85.

Zhang, F., Lee, J., Liang, S., & Shum, C. (2015). Cyanobacteria blooms and non-alcoholic liver disease: Evidence from a county level ecological study in the United States. *Environmental Health: A Global Access Science Source* 14, 1–11.

Zhang, Z., Gao, S., & Chu, C. (2020). Improvement of nutrient use efficiency in rice: Current toolbox and future perspectives. *Theoretical and Applied Genetics* 133, 1365–1384.

6 Genetic Resources of Cereal Crops
Collection, Characterization, and Conservation

P.E. Rajasekharan and G.S. Anil Kumar

CONTENTS

DOI: 10.1201/9781003250845-6

6.1 INTRODUCTION

India is endowed with diverse agroecologies suitable for different crops such as cereals, pulses, oil crops, fruits, vegetables, and flower crops. Cereals have been used throughout history and are unquestionably the most important sources of plant food for humans and livestock. All of the original ancestors of cereals have been lost over the millennia they have been cultivated. The development of all the major cereals occurred long before recorded history and the oldest civilizations were already familiar with several kinds of barley, wheat, and other grains. Additionally, the actual origin of cereals had been so long forgotten that they were given supernatural powers and played a part in the religious ceremonies of the various nations of antiquity. In ancient Rome, they held festivals at seeding time and harvest in honor of the goddess of Ceres, whom they worshipped as the giver of grain. They brought offerings of wheat and barley to these festivals, the *cerelia munera* or gifts of Ceres, from which the name "cereals" was derived. Similar religious ceremonies were observed in ancient Greece.

The major cereals grown in India include rice (*Oryza sativa* L.), wheat (*Triticum aestivum* L.), maize (*Zea mays* L.), barley (*Hordeum vulgarae* L.), sorghum (*Sorghum bicolor* L.), and finger millet (*Eleusine corocana* L.). All these cereals belong to the Poaceae family of grasses (formerly Graminae). It has been estimated that more than 282 million metric tons of cereals were produced in India by the end of 2021. These cereals include rice, wheat, barley, millet, and ragi. India is the second largest producer of rice and wheat in the world (www.statista.com/2021).

Cereals are the principal crops of India in terms of both area coverage and volume of production. The huge demand for cereals in the global market is creating an excellent environment for the export of Indian cereal products. India's exports of cereals stood at Rs. 74,490.83 Crore or 10,064.04 million USD during the year 2020–2021. Rice (including basmati and non-basmati) occupies the major share of India's total exports of cereals with 87.6% during the same period, whereas other cereals including wheat represent only a 12.37% share of total cereals exported from India during this period (http://apeda.gov.in/).

India's agroecosystem is one of the few agroecosystems where traditional varieties of rice, wheat, maize, and barley are cultivated and exhibit a wide range of genetic diversity. Although the diversity at species level is low for these crops, it is very high at the varietal level especially in rice and maize. This is attributed to the occurrence of a large number of agriculturally marginal environments in the hills, and modern varieties have not been a sufficiently attractive option for farmers to replace landraces in such environments. There are many reasons why cereals are such important crops. One or more of these grasses is available in each of the different world climates. The northern regions have barley and rye, the temperate regions have wheat, and the tropics and warmer temperate areas have rice and maize. Cereals also have a wide range of soil and moisture requirements. They can be cultivated with little effort and give a high yield. The grains are relatively easy to handle and store because of their low water content, and they are very high in food value. Cereals contain a higher percentage of carbohydrates than any other food plants as well as a considerable amount of proteins and some fats.

6.2 TAXONOMY

Taxonomically, all cereals belong to the Poaceae (grass) family, distributed across four of its five main subfamilies: sorghum (*S. bicolor*), maize (*Z. mays*), pearl millet (*Pennisetum glaucum*), and foxtail millet (*Setaria italica*) belong to the Panicoideae subfamily; finger millet (*E. indica*) belongs to the Chloridoideae subfamily; rice (*O. sativa*) belongs to the Ehrhartoideae subfamily; and wheat (*Triticum* spp.), barley (*H. vulgare*), oats (*Avena sativa*), and rye (*Secale cereale*) belong to the Pooideae subfamily. Analysis of the bibliographical records, available databases, and metadata information reveals that the use of different taxonomic nomenclature systems creates a major problem in the analysis of data for cereal crop collections. At the same time, the taxonomic classification itself is not sufficient to eliminate the duplication or misidentification of accessions (Dobrovolskaya et al. 2005). This problem mainly affects wheat, a major world cereal, and its economic significance, and is reflected in the number of existing collections as well as the total number of accessions held in gene banks. The genus *Triticum* includes the wild and domesticated species usually thought of as wheat. The number of taxonomic systems used to classify wheat resources – 14 different systems are used by the gene bank curators worldwide – also reflects this limitation. These nomenclature systems have been successively reviewed by Dorofeev et al. (1979), Gandilyan (1980), Kimber and Sears (1987), Kimber and Feldman (1987), Love (1984), Mac Key (1988), and Van Slageren (1994), and differ in the number of recognized species, subspecies, and botanical varieties. The inability to undisputedly identify botanical varieties within the wheat species

(*T. monoccocum* L., *T. dicoccum* Schrank, *T. durum* L., *T. turgidum* L., *T. aestivum* L., and *T. compactum* L.) creates additional difficulties in accurately classifying wheat landraces (Figure 6.1).

Carl Linnaeus recognized five wheat species, all domesticated:

- *T. aestivum* – Bearded spring wheat
- *T. hybernum* – Beardless winter wheat
- *T. turgidum* – Rivet wheat
- *T. spelta* – Spelt wheat
- *T. monococcum* – Einkorn wheat

Later classifications added to the number of species described, but continued to give species status to relatively minor variants, such as winter vs. spring forms. Wild wheats were not described until the mid-19th century because of the poor state of botanical exploration in the Near East, where they grow (Morrison and Laura 2001).

6.2.1 RICE

The genus *Oryza* L. is classified under the tribe Oryzeae, subfamily Oryzoideae, of the grass family Poaceae (Gramineae). This genus has two cultivated species (*O. sativa* L. and *O. glaberrima* Steud.) and more than 20 wild species distributed throughout the tropics and subtropics. The Asian cultivated rice (*O. sativa*) is an economically important crop and is the staple food for more than one-half of the world's population. All the wild relative species of the genus *Oryza*, together with weedy rice and different rice varieties, serve as an extremely valuable gene pool that can be used to broaden the genetic background of cultivated rice in breeding programs (Brar and Khush 1997, Bellon et al. 1998). There is tremendous diversity in *Oryza* and this is

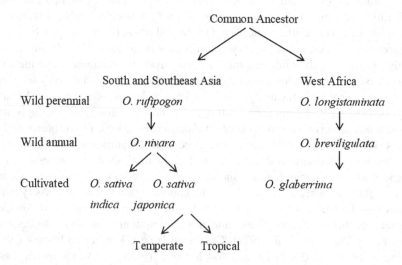

FIGURE 6.1 Evolutionary pathway of two cultivated species of rice.

reflected in the different genomes and genomic combinations of the genus, and in the significant morphological variation within and between species. On the other hand, the great morphological variation in this genus also causes certain taxonomic difficulties, leading to ambiguous delimitation between some *Oryza* taxa. In addition, different classification systems or taxonomic treatments have been proposed by authors who had access to herbarium specimens representing only certain geographic regions. This makes the taxonomy of *Oryza* species even more complicated. No single system has been generally accepted by scientists from different parts of the world to date. According to the US Department of Agriculture (USDA) Plants Database (2006), there are seven accepted species of rice:

Kingdom: Plantae – plants
Subkingdom: Tracheobionta – vascular plants
Superdivision: Spermatophyta – seed plants
Division: Magnoliophyta – flowering plants
Class: Liliopsida – monocotyledons
Subclass: Commelinidae
Order: Cyperales
Family: Poaceae – grass family
Genus: *Oryza* L. – rice
Species: *O. barthii* – Barth's rice
Species: *O. glaberrima* – African rice
Species: *O. latifolia* – broadleaf rice
Species: *O. longistaminata* – longstamen rice
Species: *O. punctata* – red rice
Species: *O. rufipogon* – brownbeard rice
Species: *O. sativa* – rice

6.2.2 Maize

Maize belongs to the tribe Maydeae of the grass family Poaceae. *Zea* derives from an old Greek name for a food grass. The genus *Zea* consists of four species of which *Z. mays* L. is economically important. The other *Zea* sp., referred to as teosinte, is largely wild grass native to Mexico and Central America. It is generally agreed that maize phylogeny was largely determined by the American genera *Zea* and *Tripsacum*; however, it is accepted that the genus *Coix* contributed to the phylogenetic development of the species *Z. mays*.

Kingdom: Plantae
Subkingdom: Tracheobionta
Superdivision: Spermatophyta
Division: Magnoliophyta
Class: Liliopsida
Subclass: Commelinidae
Order: Cyperales
Family: Poaceae – grass family

Genus: *Zea* – corn
Species: *Z. mays* – corn
Subspecies: *Z. mays* ssp. *mays*
Subspecies: *Z. mays* ssp. *parviglumis*

6.2.3 BARLEY

The taxonomy of barley is much simpler, with two major taxonomic systems recognized, i.e., the Nevski (1941) system with 28 species – *H. vulgare* L. and *H. spontaneum* C. Koch, as well as *H. agriocrithon* Aberg, are considered separate species – and Bothmer and Von Jacobsen (1985) with 40 taxa, where the three aforementioned species are considered subspecies of *H. vulgare* (NCBI 2012). The classification of varieties of barley is less developed than for the wheat taxonomy and is applied only within the same gene bank.

Traditionally, morphological differences have led to different forms of barley being classified into two-row and six-row barley. Two-row barley with shattering spikes (wild barley) is classified as *H. spontaneum* while two-row barley with non-shattering spikes is classified as *H. distichum*. Six-row barley with non-shattering spikes is classified as *H. vulgare* L. (or *H. hexastichum* L.) and six-row barley with shattering spikes is classified as *H. agriocrithon* Aberg. These differences were driven by single gene mutations, coupled with cytological and molecular evidence, and have led most recent classifications to treat these forms as a single species, *H. vulgare* (De Carvalho et al. 2013).

6.2.4 SORGHUM

The genus *Sorghum* is a member of the tribe Andropogonae of the grass family Poaceae. *Sorghum* comprises approximately 25 species and is divided into five subgenera: *Chaetosorghum*, *Heterosorghum*, *Parasorghum*, *Stiposorghum*, and *Eusorghum* (Garber 1950; De Wet et al. 1976). In addition to *S. bicolor*, the subgenus *Eusorghum* contains the agronomically important species *S. propinquum* (Kunth) Hitchc. and *S. halepense* L. (Pers.) (derived from past hybridizations between *S. bicolor* and *S. propinquum* [Paterson et al. 1995]). Some of these species are grown as cereals for human consumption and some in pastures for animals.

Kingdom: Plantae
Subkingdom: Tracheobionta
Superdivision: Spermatophyta
Division: Magnoliophyta
Class: Liliopsida
Subclass: Commelinidae
Order: Cyperales
Family: Poaceae (Grass)
Genus: *Sorghum*
Species: *S. bicolor*
Subspecies: *S. bicolor* ssp. *arundinaceum* – common wild sorghum

Subspecies: *S. bicolor* ssp. *bicolor* – grain sorghum
Subspecies: *S. bicolor* ssp. *Drummondii* – sudangrass

6.3 BOTANY OF MAJOR CEREALS

6.3.1 WHEAT

Wheat is an annual grass of the genus *Triticum* that comprises a large number of wild and cultivated species. The wild species are often weeds. Cultivated wheat, *T. aesticum*, reaches a height of 2–4 ft. The flower is a terminal spike or head consisting of 15–20 spikelets that are borne on a zigzag axis. The individual spikelets are sessile and solitary, consisting of one to five flowers each. The mature grain consists of the embryo (6%), a starchy endosperm (82%–86%), the nitrogenous aleurone layer (3%–4%), and the husk or bran (8%–9%). The husk is made up of the remains of the nucellus, the integuments of the seed coat, and the ovary walls or pericarp.

6.3.2 MAIZE

The largest of all cereals, maize is a tall annual grass that can reach heights of 3 to 15 ft. The jointed stem is solid and contains a considerable amount of sugar when not mature. The leaves are large and narrow with wavy margins. Maize has an extensive fibrous root system with aerial prop roots at the base of the stem. It has two kinds of flowers: the tassel at the top of the stem, which bears the staminate flowers, and the cob or ear with pistillate flowers. The ear is produced lower down the stalk and thus is protected by the leaves. Each ovary has a long silky style, the corn silk. The ovaries, which become the mature grains, are produced in rows on the cob. A husk composed of leafy bracts surrounds the cob. The grains have a hull (6%), a protein or aleurone layer (8%–14%), endosperm (70%), and embryo (11%). Two kinds of endosperm are usually present: a hard yellow endosperm and a soft white starchy endosperm.

6.3.3 RICE

Rice is a large annual grass that grows to a height of 2 to 4 ft. Instead of bearing an ear, rice produces a panicle, an inflorescence composed of a number of fine branches, each terminating in a single grain surrounded by a husk. The grains are easily detached together with the brown husk. In this condition, it is known as paddy. Innumerable varieties of rice have been developed, which differ in color, shape, size, flavor, and other traits of the grain. One of these types contains a sugary substance instead of starch, which forms a soft, sticky palatable mass after boiling. Other species of *Oryza* occur as wild plants in the tropics of both hemispheres.

6.3.4 SORGHUM

S. bicolor is a C_4 annual or short-lived perennial grass that typically has one generation per growing season. It can produce tillers (adventitious stems originating from the plant base) but not rhizomes. The culms (stalks) typically reach heights from 50 to over 240 cm and are 1–5 cm thick, sometimes branching above the base. The nodes are glabrous or have appressed pubescent hairs. The internodes are glabrous. The ligules are 1–4 mm long. The leaf blades are 5–100 cm long and 5–100 mm wide. The inflorescence (panicle) is 5–60 cm long and 3–30 cm wide, and may be open or contracted. The primary panicle branches are compound, terminating in racemes with two to seven spikelet pairs. The sessile spikelets are bisexual and 3–9 mm in length. The glumes are coriaceous to membranous and glabrous to densely hirsute or pubescent. The keels on the glumes (bracts) are usually winged. The upper lemmas vary from being awnless to having a geniculate, twisted, 5–30 mm awn. The anthers are 2–3 mm long. The pedicels are 1–3 mm long. The pedicellate spikelets are 3–6 mm long and are usually shorter than the sessile spikelets. The pedicellate spikelets may be staminate or sterile (Doggett 1988).

6.3.5 BARLEY

Barley is an annual grass that stands 60–120 cm tall. Barley has two types of root systems: seminal and adventitious. The depth of the roots depends on the condition, texture, and structure of the soil, as well as the temperature. The deepest roots are usually of seminal origin and the upper layers of the soil tend to be packed with the later-developing adventitious roots. If the grain is deeply planted, a "rhizomatous stem" is formed, which throws out leaves when it reaches the surface. The "rhizome" may be one or several internodes in length, and may carry adventitious roots (Briggs 1978).

The stems are erect and made up of hollow, cylindrical internodes, separated by nodes, which bear the leaves (Gomez-Macpherson 2001). A mature barley plant consists of a central stem and two to five branch stems, called tillers. The apex of the main stem and each fertile tiller carry a spike. At, or near, the soil surface, the part of the stem carrying the leaf bases swells to form the crown. It is from the crown that the adventitious roots and tillers develop (Briggs 1978).

Barley leaves are linear, 5–15 mm wide, and are produced on alternate sides of the stem. The leaf structure consists of the sheath, blade, auricles, and ligule. The sheath surrounds the stem completely. The ligule and auricles distinguish barley from other cereals as they are smooth, envelope the stem, and can be pigmented with anthocyanins (Gomez-Macpherson 2001).

6.4 ORIGIN, DOMESTICATION, DISTRIBUTION, AND SPREAD

6.4.1 WHEAT

Wheat is an important cereal crop of cool climates. It is widely grown across the world and stands first among cereals in area and production. Wheat is cultivated in the United States, the United Kingdom, Russia, Ukraine, China, Japan, Argentina, Canada, Mexico, India, and Pakistan, among many other countries.

In India, Uttar Pradesh, Madhya Pradesh, Punjab, Haryana, Rajasthan, Bihar, and Gujarat are the major wheat-growing states. Southwestern Asia (Iran, Iraq, and Afghanistan) is considered the center of origin and genetic diversity of wheat. Archaeological analysis of wild emmer indicates that it was first cultivated in the Southern Levant, with finds dating as far back as 9600 BCE. Genetic analysis of wild einkorn wheat suggests that it was first grown in the Karacadag Mountains in southeastern Turkey. Dated archaeological remains of einkorn wheat in settlement sites near this region, including those at Abu Hureyra in Syria, suggest the domestication of einkorn near the Karacadag mountain range.

6.4.2 BARLEY

The genus *Hordeum* has centers of diversity in central and southwestern Asia, western North America, southern South America, and in the Mediterranean (Von Bothmer 1992). *Hordeum* species occur in a wide range of habitats. The majority of the wild perennial species grow in moist environments whereas the annual species are mostly restricted to open habitats and disturbed areas. Many species have adapted to extreme environments and many are tolerant to cold and saline conditions (Von Bothmer 1992).

Barley was first domesticated about 10,000 years ago from its wild relative, *H. vulgare* ssp. *spontaneum*, in the area of the Middle East known as the Fertile Crescent. *H. vulgare* ssp. *spontaneum* still grows in the Middle East and adjacent regions of North Africa, in both natural and disturbed habitats, such as abandoned fields and roadsides. In the Fertile Crescent, central populations are often continuously and massively distributed. Peripheral populations have become increasingly sporadic and isolated and are largely restricted to disturbed habitats (Nevo 1992).

Until the late 19th century, all barleys existed as highly heterogeneous landraces adapted to different environments. Over the past 100 years, the landraces have mostly been displaced in agriculture by pure line varieties with reduced genetic diversity (Nevo 1992).

The progenitor of cultivated barley, *H. vulgare* ssp. *spontaneum*, has a brittle two-row spike and a hulled grain. Six-row barley appeared about 8,000 years ago (Komatsuda et al. 2007). The small, one seed arrow-like spikelets of *H. vulgare* ssp. *spontaneum* are adapted to reach the soil through stones and pebbles. However, the spontaneous six-row mutants, which produce larger three-seed spikelets, do not have this evolutionary advantage and do not reach the soil as easily; therefore, they are naturally eliminated from wild barley populations. Thus, six-row barley occurs primarily as cultivars or weeds in agricultural systems (Komatsuda et al. 2007).

6.4.3 RICE

O. sativa is believed to be associated with a wet, humid climate, though it is not a tropical plant. It is probably a descendent of wild grass that was most likely cultivated in the foothills of the Far Eastern Himalayas. Another school of thought believes that the rice plant may have originated in southern India, then spread to the north of the country and then onwards to China. It subsequently arrived in Korea and the Philippines (about 2000 BC) and then Japan and Indonesia (about 1000 BC). When Alexander the Great invaded India in 327 BC, it is believed that he took rice back to Greece. Arab travelers took it to Egypt, Morocco, and Spain, and that is how it travelled across Europe. Portugal and the Netherlands took rice to their colonies in West Africa and then it travelled to America through the "Columbian Exchange" of natural resources. However, as is traditionally known, rice is a slow starter and this is also true to the fact that it took close to two centuries after the voyages of Columbus for rice to take root in the Americas. Thereafter, the journey of rice continued with the Moors taking it to Spain in 700 AD and then the Spanish brought rice to South America at the beginning of the 17th century.

South India was the place where cultivated rice was originated. India and Burma should be regarded as the center of origin of cultivated rice.

The leading rice-producing countries are Japan, Brazil, China, India, Indonesia, Bangladesh, Vietnam, Thailand, Myanmar, and the Philippines. In India, rice is grown in almost all states, including Andhra Pradesh (AP), Bihar, Uttar Pradesh (UT), Madhya Pradesh (MP), and West Bengal. West Bengal and Uttar Pradesh have the highest rice production.

6.4.4 MAIZE

Most authorities consider the primary center of origin of maize to be Central America and Mexico, where many diverse types of maize are found. The discovery of fossil maize pollen with other archaeological evidence in Mexico indicates Mexico to be the origin of maize. The United States, China, Brazil, Mexico, India, Romania, the Philippines, and Indonesia are some of the main cultivators of maize crop. In India, Rajasthan, UP, MP, Bihar, Karnataka, Gujarat, AP, Jammu and Kashmir, Himachal Pradesh (HP), and Maharashtra are important states for the production of maize.

6.4.5 SORGHUM

Sorghum originated in the northeastern quadrant of Africa, where the greatest variability in wild and cultivated species is found to this day. It was probably domesticated in Ethiopia by selection from wild sorghum between 5,000 and 7,000 years ago. From this center of origin, it was distributed along trade and shipping routes throughout Africa, and through the Middle East to India at least 3,000 years ago. It reached China along the Silk Route. Sorghum was first taken to the Americas through the slave trade from West Africa. It was reintroduced in the late 19th century for commercial cultivation and has subsequently been introduced into South America and Australia. Sorghum is now widely found in the drier areas of Africa, Asia, the Americas, and Australia.

6.5 GERMPLASM CONSERVATION

6.5.1 WILD GENETIC RESOURCES OF CEREAL CROPS

Crop wild relatives (CWR) are wild plant species that are genetically related to cultivated crops. Untended by humans, they continue to evolve in the wild, developing traits such as drought tolerance and pest resistance that farmers and breeders can cross with domesticated crops to produce new varieties. They have been used to improve the yields and nutritional quality of crops since the beginnings of agriculture. Cultivated plants have been derived from their wild relatives through the process of selection followed by hybridization, resulting in desired improvements. Wild-related species have contributed significantly to the improvement of crop plants such as paddy, wheat, and maize (Pandey et al. 2005).

The Indian gene center holds about 166 species of native cultivated plants. Crops with primary, secondary, and regional centers of diversity represent a part of native and introduced species which account for over 480 species (Nayar et al. 2003). Diverse agroclimates and agricultural practices have led to a rich diversity of crop species in the form of landraces and cultivars. Additionally, the center has over 320 wild relatives (Arora and Nayar 1984; Arora 2000). The floristic diversity of the wild relatives of cultivated weedy types and related taxa constitutes a useful gene pool. The crop species are accessible for collection in fields, orchards, gardens, and markets and from farmers. On the contrary, the wild relatives are difficult to locate as they grow in their natural habitats with other wild plants. Except for some weedy species, most of the wild relatives are confined to specific habitats (Arora and Chandel 1972; Arora 1993). Their identity requires the skills of a systematic botanist. This is perhaps the reason why the wild relatives are meagerly represented in germplasm collections.

In general, the wild relatives of *crop* plants occur as a component of disturbed habitats within the major vegetation types. The information on their occurrence is available from different herbaria, floristic accounts/floras, etc. This genetic wealth is largely distributed in warm, humid, tropical and subtropical regions and in the Western Himalayas with low representation in the drier northwestern region (Arora and Nayar 1984; Arora 1993; 2000). Species of wild relatives are in abundance in Western Ghats, the northeastern region, and in the high altitudes of the Western Himalayas (Arora and Pandey 1996). However, disturbed grasslands, scrub vegetation, and open degraded forest areas on the one hand and farmers' fields as weedy components on the other, represent suitable habitats for rich diversity.

There has been scattered information on the distribution, collection, and conservation of the wild relatives of crop plants from India, presenting difficulties in identifying their potential and utilization. Some efforts have been made in the past to collect, analyze, and document the diversity of the wild relatives of crop plants of India. Negi et al. (1991), Singh and Pandey (1996), Pandey and Padhye (2000), Subudhi et al. (2000), and Patra et al. (2002) have studied distribution patterns and collected important information on the diversity of wild relatives of crop plants from the Himalayas, Rajasthan Desert, Gujarat, Orissa, and the West Bengal regions of India, respectively.

Crop plants and their wild progenitors may spontaneously hybridize with each other when grown side by side (Govindaswami et al. 1966). This may result in the infiltration of wild genes into the cultivated germplasm and vice versa. This process is called introgression or introgressive hybridization (Anderson 1949). It is more frequently observed in cross-pollinated crops than in self-pollinated crops. The achievement of inter-specific crossing between crops and their related types has opened up new dimensions in the utilization of wild relatives in various crop

TABLE 6.1

Wild Relatives of Some Cereal Crop Plants with Important Traits (Pandey et al. 2005)

Crops	Wild relatives	Important traits	References
Triticum aestivum and other cultivated wheat	*Aegilops comosa, A. umbeulata*, other wild relatives of *Triticum*	Resistance to yellow and leaf rust and other diseases	Sears (1956), Sharma and Knott (1966), Riley et al. (1968)
Oryza sativa	*Oryza nivara, O. officinalis, O. rhizomatis, O. granulate*	Resistance to insects, grassy stunt virus, tropical strain of bacterial blight, brown plant hopper, green leaf hopper	Govindaswami et al. (1966), Chang et al. (1975), Khush (1977), Khush et al. (1977), Jena and Khush (1990), Lore stand Jackson (1996)
Zea mays	*Zea diploperennis*	Immunity to major diseases of maize	Iltis (1979)
Hordeum vulgare	*Hordeum bulbosum*	Resistance to diseases and pests	Lange and Jochemsen (1976), Jie and Snape (1989)

improvement programs. The value of wild relatives was better recognized with the findings of *Z. diploperennis*, a new teosinte from Mexico (lltis 1979) and *O. nivara* from India (Govindaswami et al. 1966). The wild relatives of cereal crop plants with some important traits are shown in Table 6.1.

6.5.2 COLLECTIONS AND CONSERVATION STRATEGIES

There is a common understanding that ex situ germplasm collections have limited use and primarily serve only conservation purposes. However, despite the recent changes in the functions of gene banks highlighted in the case study of the US National Plant Germplasm System (NPGS), Smale and Day-Rubenstein (2002) reported that requests for cereal genetic resources depend on the country, crop, culture, and profile of the final users, while breeders are still the main users of germplasm. Globally, plant germplasm, including cereals, is requested for trait evaluation, breeding or pre-breeding, basic research, and the enhancement of other collections (Day-Rubenstein et al. 2006). Only the latter two are related to conservation gene bank functions. Often, there is an urgent need to preserve neglected genetic resources in order to prevent their disappearance (Hammer et al. 1999) as they are frequently underestimated sources of new crop traits (Smale and Day-Rubenstein 2002) and indispensable elements required to restore agricultural activities, for example, after disaster situations. The increasing concern about the loss of these cereal genetic resources over the last decades has led to a heightened concentration on methods for the conservation of genetic resources in gene banks (ex situ conservation) (Bommer 1991). Gene banks play an invaluable role in the conservation of cereal genetic resources, maintaining them as a crucial source of plant material for agriculture and food purposes, as well as a source of information about existing agrobiodiversity (Johnson 2008).

TABLE 6.2
Regional Distribution of Gene Banks, Their Cereals, and Their Cereal Landraces Collections (WIEWS 2012)

Geographical region	Number of gene banks maintaining cereals and their landrace (LR) collections								
	Total	Wheat	LR	Barley	LR	Oats	LR	Rye	LR
Africa	196	23	11	20	8	14	5	10	3
Asia	468	62	37	52	31	28	13	29	9
Europe	481	88	44	75	38	50	27	46	24
Latin America and Caribbean	425	32	9	24	6	15	2	7	2
Oceania	57	4	2	6	0	3	1	2	1
North America	88	11	7	11	4	6	3	5	3
Total	1715	220	110	188	87	116	51	99	42

The current situation of cereal collections, including cereal landraces, in worldwide gene bank systems is presented in Table 6.2. Data show that cereal and landrace genetic resources have a wide distribution with a large number of the collection holders in European gene banks.

6.5.3 Conservation Strategies

Initially, gene banks were established to conserve collected plant material to support breeding programs. Only after 1974, when heavy erosion of genetic resources was recognized as a serious problem, did the function of gene banks evolve to include the conservation of crop genetic resources and, in some cases, of their crop wild relatives (Damania 2008). This broadening of the gene bank mandate dictated the need for subsequent changes in sampling strategies and additional requirements for the thorough documentation of the collected accessions. Modern gene banks have diversified their research tasks to encompass plant genetic resources surveys, collection, conservation, characterization, documentation, and valorization, and the promotion of germplasm use. Usually, accessions conserved in gene bank collections result from direct sampling of farmers' fields, farm seed stores, and even local seed markets (Damania 2008), with one accession representing a specific crop population. Some authors (Zeven 1998; Negri 2003) consider an accession to be a crop landrace if it has been locally cultivated by farmers for several decades or generations in the absence of formal or conventional selection. Others consider accessions to be populations belonging to a particular landrace, if they share common key traits and have a history of traditional use (Camacho Villa et al. 2005; Dos Santos et al. 2009). The first approach allows the classification of an accession as a landrace in the absence of its characterization. The second approach, however, requires previous characterization of landraces and the determination of their genetic structure and sample sizes for proper conservation of their diversity. According to Dobrovolskaya et al. (2005), the use of the common or cultivar name is not sufficient to determine the presence of duplicates and to downsize germplasm collections, because of the existence of genetically different accessions under the same common designation. Demissie and Bjornstad (1996) proposed the use of the phenotypic variation of some traits as a methodology to sample and conserve barley landraces. Plant accessions of locally grown cereals have complex structures, are characterized by their plasticity, are adapted to local agroecological conditions, and their diversity can be increased by farmers' seed exchange (Zeven 1998; Ladizinsky 1998; Bellon 1996; Damania 2008; Newton et al. 2010). Due to the diversity and heterogeneity of landraces, germplasm sampling requires specific and extensive care during both the gathering (Allard 1970; Brown and Marshall 1995; Hawkes et al. 2000; Farias and Bettencourt 2006) and regeneration of landraces, in order to maintain rare alleles and genotypes and to avoid genetic drift.

6.6 STATUS OF CEREAL CROP GENETIC RESOURCES

6.6.1 STATUS OF WHEAT GENETIC RESOURCES

The gene pool for wheat consists of modern and obsolete cultivars and breeding lines, landraces, related species (both wild and domesticated) in the Triticeae tribe, and genetic and cytogenetic stocks. The primary pool consists of the biological species, including cultivated, wild, and weedy forms of the crop species which can be easily hybridized. In the secondary gene pool are species from which gene transfer is possible but with greater difficulty, typically species of *Triticum* and *Aegilops*. The tertiary gene pool is composed of other species of the tribe (primarily annual species) from which gene transfer is possible only with great difficulty. "Ease" of gene transfer is a technology-dependent concept and is subject to change as are the taxonomic delimitations within the tribe. The wild relatives of wheat have proven to be highly useful sources of resistance to biotic and abiotic stresses in wheat breeding over the last two decades and this trend is expected to accelerate in the future. Similarly, genetic stocks are finding increasing use as tools in the sophisticated application of modern biotechnologies in wheat improvement.

6.6.2 IN SITU CONSERVATION STATUS

One of the few global examples of a protected area created specifically for the conservation of the wild relatives of annual cereal crop is the Erebuni State Reserve in Armenia, an 89 ha region in the transition area between semi-desert and mountain-steppe zones. Three of the four known species of wild-growing wheat occur here (wild one-grain wheat, *T. boeticum*, wild two-grain Ararat wheat, *T. araraticum*, and wild urartu wheat, *T. urartu*) along with several species of *Aegilops*, in addition to a number of CWR of other cereal species (barley and rye). Succession with other indigenous species and invasive species (both plants and animals) is a threat to the integrity of the CWR species in this reserve as well as any other reserve where cereal CWR may be found. In general, any protected areas in countries with Mediterranean climates are likely to include some wheat CWR taxa. Whether the genetic integrity of such populations is being maintained in these reserves is the key question.

6.6.3 Ex Situ Conservation Status

In total, over 235,000 accessions are maintained in more than 200 ex situ collections. Landraces and modern and obsolete improved cultivars are generally well conserved in wheat germplasm collections, while the wild relatives of wheat are poorly represented. Because of the specialized needs and conditions for developing and reliably maintaining genetic and cytogenetic stocks, these are not well represented in germplasm collections (probably in fewer than 90 collections) and are most likely to be found in research institutions. Regeneration progress is lacking in many countries' collections and is probably the single greatest threat to the safety of wheat accessions held in globally important gene banks. Lack of funding is the principle limitation.

6.6.4 Gaps and Priorities

The major gaps in collections relate to landraces and cultivars. Key users of wheat genetic resources, however, indicated the need for more mapping populations, mutants, and genetic stocks, and a wider range of wild relatives. This divergence of perceptions of the major function of collections between gene bank managers and germplasm users complicates the evaluation of the status of diversity. CWR are relatively poorly represented in collections and more collecting is needed. The level of genetic diversity and the breadth of provenance of the wild-related species maintained in existing collections are small. One of the scenarios of climate change is increased regional temperatures. This could be beneficial for the wheat crop in some regions, but it could reduce productivity in regions where temperatures are optimal for wheat. New wheat cultivars will be needed to adapt the crop to changing environments and still meet the nutritional needs of people. The identification and deployment of heat-tolerant germplasm are a high priority.

6.7 STATUS OF RICE GENETIC RESOURCES

The primary gene pool of rice has been a source of useful genes for breeding and research. It consists of the other domesticated species, *O. glaberrima* and *O. rufipogan*, and several other wild species, all with a common genome (A), that can hybridize naturally with *O. sativa*. The secondary and tertiary gene pools, *Oryza* species with genome constitutions other than A, have potential as gene sources, but introgression of genes into rice is proving difficult. However, anther culture and embryo rescue techniques can be used effectively to overcome hybrid sterility. At the Centro International de Agricultural Tropical (CIAT), advanced breeding lines from crosses between *O. sativa* and *O. latifolia* (CCDD genomes) have been generated and distributed to the National Agricultural Research System (NARS) in Latin America.

6.7.1 In Situ Conservation Status

Potential genetic reserve locations in Asia and the Pacific have been identified for *O. longiglumis*, *O. minuta*, *O. rhizomatis*, and *O. schlechteri*, which are high priority CWR for in situ conservation. Efforts to conserve landraces and CWR outside protected areas aimed at preserving the globally significant agrobiodiversity of rice have been reported in Vietnam.

6.7.2 Ex Situ Conservation Status

Overall, about 775,000 accessions of rice are maintained in more than 175 ex situ collections; however, about 44% of these total holdings is conserved in five gene banks located in Asia. Landraces and obsolete and modern improved cultivars, as well as genetic and cytogenetic stocks, are generally well represented in rice germplasm collections. In general, CWR are poorly represented in the ex situ collections with the exception of those held at the International Rice Research Institute (IRRI) and at the National Institute of Agricultural Biotechnology in the Republic of Korea.

6.7.3 Gaps and Priorities

Further collection of better wild species representation in gene banks from all levels of gene pools, as well as the regeneration of existing wild accessions and networks for sharing the conservation responsibility for wild species among the several gene banks and research centers that maintain them are needed.

6.8 STATUS OF MAIZE GENETIC RESOURCES

The primary gene pool of maize includes *Z. mays* and teosinte, with which maize hybridizes readily with the production of fertile progeny. The secondary gene pool includes *Tripsacum* species (~16 species), some of which are endangered. The variability among maize landraces (some 300 have been identified) exceeds that for any other crop. Great variation exists in plant height, days to maturity, ears per plant, kernels per ear, yield per hectare, and latitudinal and elevational ranges of cultivation. Teosinte is represented by annual and perennial diploid species ($2n = 2x = 20$) and by a tetraploid species ($2n = 4x = 40$). They are found within the tropical and subtropical areas of Mexico, Guatemala, Honduras, and Nicaragua as isolated populations of variable sizes, occupying from less than 1 ha to several hundreds of square kilometers. The distribution of teosinte extends from the southern part of the cultural region known as Arid America, in the Western Sierra Madre of Chihuahua and the Guadiana Valley in Durango in Mexico, to the western part of Nicaragua, including practically the entire western part of Mesoamerica.

6.8.1 In Situ Conservation Status

It is extremely important to act now to complete ecogeographic sampling for New World maize, since economic and demographic changes are eroding the genetic diversity of maize in many areas that were once untouched by modern agricultural, horticultural, forestry, and industrial practices.

6.8.2 Ex Situ Conservation Status

While there are relatively few areas where no comprehensive collection has already been made, maize from portions of the Amazon Basin and parts of Central America and waxy maize in Southeast Asia have never been adequately collected. Public or private tropical inbred lines are not well represented in collections, nor are important hybrids (or their bulk increases). Wild *Zea* and *Tripsacum* species are potentially important sources of genetic variation for maize, but they are not well represented in collections and existing accessions are in small quantities. The Maize Genetic Cooperation Stock Center at the University of Illinois is the primary gene bank holding maize mutants, genetic stocks, and chromosomal stocks. Teosinte representation is uneven and incomplete in major gene banks. The major teosinte collections are those of Mexico's National Institute for Forestry, Agriculture and Livestock Research (INIFAP), the University of Guadalajara, and the International Maize and Wheat Improvement Center (CIMMYT) in Mexico, and in the USDA-ARS collections in the United States.

6.8.3 Gaps and Priorities

The major gaps identified in existing ex situ maize collections include hybrids and tropical inbred lines, in addition to gaps resulting from the loss of accessions from collections; for example, the entire collection of Dominica has been lost because much of the material was collected by the International Board for Plant Genetic Resources (IBPGR) in the 1970s. The Global Crop Diversity Test (GCDT) maize strategy specifically emphasized that hybrids and private inbred lines (not those now with plant variety protection [PVP] or with recently expired PVP) are missing from gene banks. There is a need to identify core subsets of the maize races; however, this depends on expertise not only in statistical procedures, but also more critically, in racial and accession classification and the availability of the type of data needed to develop reasonable classification decisions. While coverage of New World maize is good in gene banks, about 10% of those New World holdings are in need of regeneration. The distribution of teosinte extends from the southern part of the cultural region known as Arid America, in the Western Sierra Madre of Chihuahua and the Guadiana Valley in Durango in Mexico, to the western part of Nicaragua, including practically the entire western part of Mesoamerica.

National and international reserves need to be established to protect the remaining fragments of the Balsas, Guatemala, Huehuetenango, and Nicaraguan races of teosinte. The CIMMYT's current ex situ *Tripsacum* garden at Tlaltizapan, Morelos, should continue to be maintained, with a duplicate garden established in Veracruz (or some equivalent lowland, tropical environment). Another *Tripsacum* garden could be established near the International Institute of Tropical Agriculture (IITA) headquarters in Africa. In situ monitoring of *Tripsacum* populations should be conducted in Mexico and Guatemala, the centers of diversity for the genus, and in other countries in Central and South America, where both widespread and endemic species are found. Ex situ *Tripsacum* gardens at CIMMYT and USDA in Florida should be enriched with the diversity found from the wild, and more collaboration should occur between these two unique sites.

6.9 STATUS OF SORGHUM GENETIC RESOURCES

Over the period 1996–2018, the yield of sorghum (*S. bicolor*) did not change significantly. In 2018, sorghum was cultivated over a harvested area of 45 million hectares with global production of 66 million tonnes. Sorghum is mainly used for human consumption in Africa and India and for animal feed in China and the United States. The five highest producers of sorghum are the United States (18% of global production), Nigeria (14%), India (12%), Mexico (10%), and the Sudan (6%). The primary gene pool consists of *S. bicolor* and its many races and several other species, the number of which depends on the taxonomic treatments.

6.9.1 Ex Situ Conservation Status

The major sorghum collections are at the International Crops Research Institute for the Semi-Arid Tropics (ICRISAT) and at the USDA Plant Genetic Resources Conservation Unit, Southern Regional Plant Introduction Station, followed by those at the Institute of Crop Germplasm Resources (ICGR) in China and at the National Bureau of Plant Genetic Resources (NBPGR) in India. In addition, there are about 30 other institutions holding ex situ sorghum collections (primarily national collections). In total, over 235,000 accessions are maintained, of which 4,700 accessions are wild materials. A high number of duplicate accessions among collections is suspected, except for the Chinese collection which consists primarily of Chinese landraces.

6.9.2 Gaps and Priorities

A massive number (28,000) of accessions urgently need regeneration. Bottlenecks include quarantines and day length issues, labor costs, and capacities. Eco sampling of the wild progenitors and landraces of *S. bicolor* in each of its primary, secondary, and tertiary centers of diversity is needed. Further collection and conservation of wild close relatives are needed. Gaps in geographic coverage were noted for West Africa, Central America, Central Asia and the Caucasus, and Sudan in Darfur and the South.

6.10 GERMPLASM USE

6.10.1 Major Constraints on Cereal Crop Production

a. **Relatively low productivity of cereals:** The national average yield of cereals in India is relatively low, amounting to 2.1 t ha^{-1} for barley, 2.7 t ha^{-1} for wheat, 3.8 t ha^{-1} for maize, 2.5 t ha^{-1} for sorghum, and 2.8 t ha^{-1} for rice. This is due to the widespread use of low-yielding varieties coupled with unimproved traditional practices that ultimately contribute to the low national average yield of major cereals in India.

b. **Diseases:** Several diseases are constraining cereal production and productivity in different parts of the country. The impact of these biotic factors on the general performance, yield, and grain quality varies depending on the genetic, environmental, and management conditions and the interactions of these factors. The magnitude of the yield loss associated with various diseases varies with varieties, locations, seasons, and planting dates (Table 6.3).

c. **Insect pests:** Large numbers of insect pests that attack cereals under field and storage conditions have been identified (Table 6.4). Depending on the incidence and damage, some insect pests are known to be economically important.

TABLE 6.3
Important Diseases of Major Cereals in India

Crop	Major diseases
Barley	Scald (*Rhynchosporium secalis*), net blotch (*Helminthosporium* spp.), stripe rusts (*Puccinia* spp.), powdery mildew (*Erysiphae graminis*), head blight (*Fusarium hetrosporium*), covered smut (*Ustilago hordei*), barley yellow dwarf virus (BYDB)
Maize	Turcicum leaf blight (*Exserohilum turcicum*), gray leaf spot (*Carpospora zea-maydis*), common leaf rust (*Puccinia sorghi*), maize streak virus (MSV), maize lethal necrosis disease (MLND)
Rice	Rice blast (*Pyricularia oryzae*), brown spot (*Cochliobolus miyabenus*), sheath rot (*Sarocladium oryzea*), and sheath blight (*Thanatephorus cucumeris*)
Sorghum	Anthracnose (*Colletotrichum sublineolum*), grain mold (*Fusarium* spp., *Alternaria* spp., *Helminthosporium* spp., *Curvularia* spp.), gray leaf spot (*Cercospora sorghi*), rust (*Puccinia purpurea*), smut (*Sphacelotheca spp*), ergot (*Claviceps sorghi*), downy mildew (*Peronosclerospora sorghi*), and leaf blight (*Helminthosporium turccium*)
Wheat	Yellow/stripe rust (*Puccinia striiformis* Westrnd.), stem/black rust (*P. graminis fsp. tritici*), leaf/brown rust (*P. ricondite fsp. tritici*), Septoria tritici (*Microsphearelia graminicola*)

TABLE 6.4
Important Insect Pests of Major Cereals in India

Crop	Major insect pests
Barley	Barley shoot fly (*Delia arambourgi* Seguy, *D. flavibasis* Stein.), Russian aphid (*Diuraphis noxia* Mordvilko), chafer grub (*Melolontha* spp.)
Maize	Stalk borer (*Busseola fusca*), Spotted stalk borer (*Chile partellus*), termites (*Macrotermes* and *Microtermes* spp.), maize weevels (*Sitotroga zeamias*), large grain borer (*Postephanus turncatus*)
Rice	Termites, stem borer (Pyraliae), stalked-eyed flies (*Diopsis thoracica*)
Sorghum	Stark borer (*Chilo partellus*), shoot fly (*Atherigona soccata*), midge (*Contarinia sorghicola*), weevil (*Curculionidae*)
Wheat	Shoot fly (*D. steiniella* Emden), Russian aphids (*Diuraphis noxia* Mordvilko)

d. **Biotic constraints:** The most important abiotic stresses in the production of cereals are drought, cold, water logging, low soil fertility, soil acidity and salinity, hail, and frost. In the coastal lowland and rift valley areas, moisture stress significantly limits maize production and productivity. Late onset and early cessation of rainfall are predominant phenomena in this area, and occasionally across the country. Also, rainfall distribution is often erratic resulting in a shortage of precipitation during the critical growth stage of the plants. Sorghum is critically affected by recurrent drought due to its inherent adaptation to warmer lowlands, and the erratic nature of rainfall in such agro-ecologies of the country is frequent particularly in arid and semi-arid lowland areas where over half of the country's sorghum is produced under severe to moderate drought stress conditions. The most important abiotic stresses in barley farming systems include low soil fertility, low soil pH, poor soil drainage, frost, and drought. Likewise, the major abiotic stresses in teff and rice husbandry include drought, soil acidity, waterlogging, cold, and frost. Wheat is severely affected by abiotic factors such as low and poor distribution of rainfall in lowland areas and water logging in half of the highlands, and soil erosions are perceived as the cause of significant wheat yield losses in the country. Recent estimates indicate that low rainfall, poor distribution problems, and plant lodging can cause 10%–30% of yield losses (Dessi, A. 2018).

6.10.2 TRAITS DESIRED

a. **Wheat:** Grain yield in wheat is influenced by several agronomic traits (Chen et al. 2012; Liu et al. 2015) which have been widely explored in wheat improvement programs to accelerate cultivar development. Due to their high heritability and correlation with grain yield, agronomic traits can be used as indirect selection criteria during breeding and cultivar development (Chen et al. 2012; Abdolshahi et al. 2015; Liu et al. 2015; Gao et al. 2017). Moreover, it has been suggested that genetic progress in yield can be achieved if several traits conferring a better agronomic and physiological performance with biotic and abiotic stress tolerance are simultaneously selected and introgressed in a single variety (Lopes et al. 2012). Some important agronomic traits that have been exploited in wheat improvement programs to aid cultivar development and increase grain yield potential and genetic gains are discussed below.
  - **Early flowering and maturity:** Understanding the genetic factors controlling flowering time is essential to manipulate phenological development processes to improve yield potential in wheat. Most modern wheat genotypes incorporate vernalization and photo-period insensitive genes to promote early flowering and maturity.
  - **Plant height:** Many wheat improvement programs have developed wheat genotypes incorporating the dwarfing/height-reducing genes namely: *Rht1* (*Rht-B1b*), *Rht2* (*Rht-D1b*), *Rht-D1c*, and *Rht8*. These genes reduce coleoptile and internode length and plant height resulting in increased grain yield by increasing assimilate partitioning to the ear. This has resulted in a higher harvest index and lodging resistance.

- **Biomass production:** Increased biomass has resulted in grain yield improvement in wheat. The increase in biomass has been largely attributed to a higher photosynthetic rate, stomatal conductance, leaf chlorophyll content, and improved radiation-use efficiency. It has been suggested that further improvements in grain yield can be achieved by increasing the photosynthetic capacity by optimizing biomass production while maintaining lodging resistance.
- **Number of productive tillers:** The number of productive tillers defined as the number of tillers that produce spikes and seeds is a key agronomic trait that affects biomass production and grain yield potential in wheat. Wheat genotypes with reduced tillering capacity are more productive than free-tillering genotypes under drought-stressed conditions due to reduced sterile spikelets.
- **Other traits:** The number of grains per spike, spike fertility, leaf morphology and its component traits, root and root-related traits, and tolerance to biotic and abiotic stress are some of the other traits to be considered in the improvement of wheat.

b. Rice

In addition to the high-yielding varieties, under global climate variability scenarios, it is a challenging task to meet the rice food demand of the growing population. Identifying green traits (tolerance to biotic and abiotic stresses, nutrient-use efficiency, weed competitive ability traits, and nutritional grain quality) and stacking them in high-yielding elite genetic backgrounds is one promising approach to increase rice productivity. Developing multi-stress-tolerance varieties such as biotic and abiotic stress are highly desirable in rice improvement work.

c. Maize

The preferred characteristics of maize for profitable cultivation and higher quality grains are high yield and prolificacy, drought- and heat-resistant types, disease and pest resistance, early maturity, white grain color, and drying and shelling qualities.

d. Barley

The initial breeding and selection objectives in barley were focused on agronomic traits such as grain morphology, yield, and disease resistance. Crop management plays an important role in barley spreading and yield improvement. Breeders need to develop cultivars for markets that demand clean bright grain with low moisture, better resistance to biotic and abiotic stresses, and high-quality parameters. When breeding, significant consideration must be given to the ability of the barley cultivar to achieve quality in different environmental conditions.

e. Sorghum

Sorghum can withstand severe droughts, making it suitable to grow in regions where other major crops cannot be grown. With this added advantage, some of the other traits desired are increased grain yield, increased height, earliness in development, tolerance to cold, and resistance to ergot, shoot fly, and nematodes for better quality sorghum production.

6.10.3 EVALUATION OF GENETIC DIVERSITY

The genetic diversity of crop plants is the foundation for the sustainable development of new varieties. Thus, there is a need to characterize diverse genetic resources using different statistical tools and utilize them in breeding programs. Different sets of morphological traits are taken into consideration for different groups of crop plants. For cereals, stem pigmentation, panicle length, grain color, grain shape, and tolerance to biotic and abiotic stress are preferably considered traits. Many studies have focused on understanding the genetic diversity through the morphological, molecular, cytological, and biochemical characterization of the germplasm. Malik et al. (2014) evaluated 258 wheat varieties developed in the last 50 years (1961–2010) for six different agroclimatic wheat-sowing regions of India that were characterized for distinctness, uniformity, and stability (DUS) using 20 agro-morphological characteristics. DUS descriptors such as waxiness on plant parts, plant height, and growth habit were identified as prominent morphological determinants for genetic diversity in Indian wheat. Likewise, Prasad et al. (2020) evaluated a collection of 208 aromatic rice germplasm at ICAR–IIRR, Hyderabad, and documented the genetic variations in 46 morphological, agronomical, and grain quality traits as revealed by SSR markers. A parent can be selected from these 46 core set genotypes to improve aromatic rice depending on the breeding objective. Similarly, the morphological and molecular characterization of different types of sorghum, such as sweet types, grain types, and forage types, was evaluated under temperate conditions by Kanbar et al. (2020) and they identified a wide range of variability in terms of sugar yield and related traits, which provide valuable resources for sorghum improvement by breeding programs in temperate zones. Sweet sorghum genotypes (especially ICSSH30-11-ADP) generated the most fresh biomass and had the highest sugar yield, compared to grain and forage sorghum genotypes.

6.10.4 SOURCES OF DESIRABLE TRAITS

The characterization and evaluation of the genetic resources of different cereals at various locations over the past few decades have led to the identification of superior genotypes/species, which can be utilized in crop improvement programs. Some of the genetic resources/wild relatives with desirable traits are listed in Table 6.5.

6.10.5 BREEDING OPTIONS

The ultimate goal of crop breeding is to develop varieties with high yield potential and desirable agronomic characteristics. In cereal crops breeding, the most important qualities sought by breeders have been high yield potential; resistance to major diseases and insects; and improved grain and eating quality. However, there seems to be some conflict between these aims. An emphasis on high grain quality tends to result in unstable yields. Conversely, too much emphasis on disease and insect resistance and stable yields leads to poor grain quality. Hence, breeding efforts should concentrate on varieties with the potential to minimize yield losses under unfavorable conditions and to maximize yields when conditions are favorable (Khan et al. 2015). The conventional breeding methods used to improve cereal crops are introduction, selection, pedigree method, and backcrossing, but the development of hybrids and population

TABLE 6.5
Genetic Resources of Wheat, Maize, Rice, and Sorghum for Desirable Traits

Crop	Source	Desirable trait	References
Wheat	*Aegilops comosa, A. umbeulata*, other wild relatives of *Triticum*	Resistance to yellow and leaf rust and other diseases	Sears (1956), Sharma and Knott (1966), Riley et al. (1968)
Maize	*Z. mays* subsp. *parviglumis, Z. diploperennis*	Tolerance to fall armyworm (*Spodoptera frugiperda*)	Mammadov et al. (2018)
		Gray leaf spot resistance	
	Z. mays ssp. *Mexicana, Z. mays* spp. *parviglumis*	Tolerance to maize spotted stalk bore	
	Eastern gamagrass (*Tripsacum dactyloides*)	Tolerance to western corn rootworm (*Diabrotica v. virgifera*)	
		Drought and salinity tolerance	
		Acid soil and aluminum tolerance	
	Z. diploperennis	Maize chlorotic dwarf virus resistance	
		Maize chlorotic mottle virus resistance	
		Maize streak virus resistance	
		Maize bushy stunt mycoplasma resistance	
		Maize stripe virus resistance	
		Northern corn leaf blight resistance	
		Tolerance to *Striga hermonthica* weed	
	Z. nicaraguensis *Z. luxurians*	Waterlogging tolerance	
Rice	*O. nivara*	Brown plant hopper resistance	Mammadov et al. (2018)
	O. punctate		
	O. longistaminata		
	O. barthii		
	O. rufipogon		
	O. officinalis		
	O. austaliensis		
	O. minuta		
	O. latifolia		
	O. glaberimma		
	O. minuta	Blast resistance	
	O. autraliensis		
	O. rufipogon		
	O. rhizomatis		
	O. longistaminata	Bacterial blight resistance	
	O. rufipogon		
	O. minuta		
	O. officinalis		
	O. nivara		
	O. brachyantha		
	O. glaberrima	Drought and heat tolerance	
	O. barthii		
	O. meridionalis		
	O. australiensis		
	O. longistaminata		
	O. rufipogon	Cold tolerance	

(Continued)

TABLE 6.5 (CONTINUED)

Genetic Resources of Wheat, Maize, Rice, and Sorghum for Desirable Traits

Crop	Source	Desirable trait	References
Sorghum	S. halepense	Resistance to green bug, chinch bug, and sorghum shoot fly	Ananda et al. (2020)
	S. angustum	Resistance to egg laying by sorghum	
	S. amplum	midge	
	S. bulbosum		
	S. macrospermum	Insect and disease resistance, higher growth rate, and an insignificant aboveground dhurrin content under drought conditions	
	S. exstans	Resistance to shoot fly	
	S. stipoideum		
	S. matarankense		
	S. leiocladum	Cold tolerance	

improvement are added to the breeder's portfolio. Breeders have been taking advantage of biotechnology tools to enhance their breeding capacity; however, many programs are still struggling with how to integrate them into breeding programs and how to balance the allocation of resources between conventional and modern tools. Various breeding options for increasing the yield potential in cereal crops include

1. Conventional hybridization
2. Ideotype breeding
3. Heterosis breeding
4. Male sterility
5. Wide hybridization
6. Genetic engineering

6.10.6 GENOMICS-ASSISTED BREEDING

Plant genomics technologies have made many breakthroughs in today's agriculture. They have helped breeders in achieving targeted objectives to improve the quality and productivity of crops. DNA-based molecular markers have made the selection process easier by enabling early generation selection for key traits, thereby overcoming the drawbacks of conventional methods. Using this technology, both the genotype and phenotype of new varieties can be analyzed. The goal of these technologies is to integrate molecular tools into classical breeding and attain sustainability (Salgotra et al. 2017). Some of the genomics-based methods used in cereal crop improvement are as follows:

1. Marker-assisted selection (MAS)
2. Marker-assisted gene pyramiding (MAGP)

3. Marker-assisted backcross selection (MABC)
4. Advanced backcross QTL (AB-QTL) analysis
5. Genomic selection

6.10.7 Present Status of Use or Incorporation of Desired Traits

The germplasm of cereals collected have been evaluated for various economically important traits and a few superior types have been identified. However, few attempts have been made to transfer these specific traits from donor plants through agrobacterium-mediated transfer. But, plant genetic engineering methods were developed over 30 years ago, and since then, genetically modified (GM) crops or transgenic crops have become commercially available and widely adopted in many countries. In these plants, one or more genes coding for desirable traits have been inserted. The genes may come from the same or another plant species, or from totally unrelated organisms. The traits targeted through genetic engineering are often the same as those pursued by conventional breeding. However, because genetic engineering allows for direct gene transfer across species boundaries, some traits that were previously difficult or impossible to breed can now be developed with relative ease.

First-generation GM crops have improved traits such as herbicide resistance (soybeans and maize) and pest resistance (cotton and corn). Second-generation GM crops have enhanced quality traits, such as a higher nutrient content. "Golden Rice", one of the very first GM crops, is biofortified to address vitamin A deficiency. Crops can also be modified to ward off plant viruses or fungi. Even though the seed is more expensive, these GM crops lower the costs of production by reducing the inputs of machinery, fuel, and chemical pesticides. Important environmental benefits, such as controlling farm runoff that otherwise pollutes water systems, are associated with reduced spraying of chemical insecticides and highly toxic herbicides (Usha Kiran Betha 2019).

Apart from the above-discussed technologies, genome editing has been used to achieve important agronomic and quality traits in cereals. These include adaptive traits to mitigate the effects of climate change, tolerance to biotic stresses, higher yields, more optimal plant architecture, improved grain quality and nutritional content, and safer products. Genome editing is a disruptive technology with profound applications in many sectors including agriculture for crop improvement (Bortesi et al. 2016; Zhu et al. 2017; Zhang et al. 2018; Armario Najera et al. 2019). Genome editing can improve many crops through precise targeted mutagenesis and gene targeting (GT) (Sedeek et al. 2019). The application of genome editing techniques that complement other modern breeding methods can lead to yield gain in a sustainable way. The advancement of a relatively simple editing approach by the clustered regularly interspaced short palindromic repeats (CRISPR)/Cas system combined with the availability of open-source data on genes and single nucleotide polymorphisms (SNPs) involved in important traits in cereals has resulted in a surge of publications in genome editing for crop improvement.

The products of genome editing are often classified as site-directed nuclease SDN-1, SDN-2, and SDN-3 (Grohmann et al. 2019). All three mechanisms utilize double-strand break (DSB) repair mechanisms. SDN1 relies on the error-prone non-homologous end-joining (NHEJ) pathway to introduce point mutations at the specific target site resulting in insertion and deletion of a few bases (Figure 6.2). SDN-2 relies on an alternative repair mechanism called homology-directed repair (HDR) and utilizes a template sequence that differs by only a few nucleotides from the existing sequence. SDN-3 uses the same mechanism as SDN-2; however, longer DNA sequences are included in the template (Grohmann et al. 2019). Different genome editing techniques have been applied in cereals: meganucleases, zinc finger nucleases (ZFNs), transcription activator-like effector nucleases (TALENs), and CRISPR/Cas9 (reviewed in Zhu et al. 2017). In addition to techniques that utilize DSB repair mechanisms, base editors are also used for precise cereal editing (Matres et al. 2021).

6.10.8 RESEARCH NEEDS

With the continued rapid increase in the population, there is a great urgency to increase the crop productivity of staple food grain crops; however, their productivity is greatly affected by various environmental and genetic factors. Even though many elite genotypes/varieties have been identified in cereals, there is a need to develop cereal crop varieties/hybrids with all the desirable traits (tolerance to biotic and abiotic stresses, higher yields, more optimal plant architecture [resistance to lodging], and improved grain quality and nutritional content) for commercial cultivation. With the improved technology, attempts must be made for the successful transformation of desired traits from donor plants to receptors.

Although significant progress has been made on the physiological and biochemical basis of resistance to abiotic stress factors, very little progress has been made in increasing productivity under sustainable agriculture. Therefore, there is a great necessity for inter-disciplinary research to address this issue and to evolve efficient technology and its transfer to the farmers' fields (Maiti and Satya 2014).

6.11 FUTURE PERSPECTIVE

Achieving food security has been the overriding goal of agricultural policy in India. The introduction and rapid spread of high-yielding rice and wheat varieties in the late 1960s and early 1970s resulted in the steady output growth of food grains. Public investment in irrigation and other rural infrastructure and research and extension, together with improved crop production practices, has significantly helped to expand the production and stocks of food grains. Food grain production, which was 72 million metric tons in 1965/1966, rose to 308.65 million tons in 2020–2021.

In the years to come, higher economic growth as well as sizable population growth will increase the demand for food. The structure of the demand is also changing, as diets are diversifying from the basic cereal staples to fruits, vegetables, and other higher-valued foods. These evolving scenarios will change the supply and demand prospects for food in the next century. What then are the prospects for India's trade in grains? Will India become a big importer or exporter?

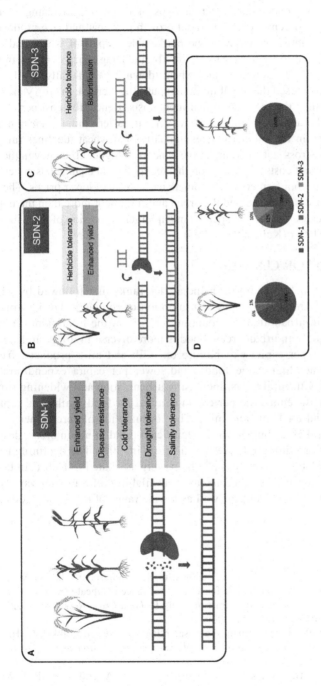

FIGURE 6.2 (A) Site-directed nuclease (SDN)-1 editing with non-homologous end-joining (NHEJ) DSB repair mechanism and traits developed in rice, maize, and wheat. (B) SDN-2 editing, mainly through homology-directed repair and traits developed in rice, maize, and wheat. (C) SDN-3 editing, insertion in targeted locus, mainly through homology-directed repair mechanism and traits developed in rice, maize, and wheat. (D) The current percentage of products developed through SDN-1, 2, and 3 in rice, maize, and wheat (Matres et al. 2021).

6.12 RECENT TRENDS IN SUPPLY AND DEMAND FOR CEREALS

Indian agriculture has undergone different rates of technological change across regions and among different crops. The rapid growth in wheat and rice production has resulted in substantial increases in the marketable surplus of wheat and rice. These have contributed to food security mainly by inducing sharp declines in real prices of rice (down 2.2% annually) and wheat (down 3.3% annually).While new technologies have increased the use of modern inputs, the increase in crop yields has been much higher than the increase in real input costs. Hence, the cost per unit of output has declined dramatically for rice (down 1.1% in eastern India, 2.1% in northern India, and 3.9% in the southern states annually) and wheat (declines ranging from 2.0% to 2.8% across states). Many of the benefits of higher efficiency in the use of inputs and lower unit costs of production that technological change has generated have been passed on from farmers to consumers in the form of lower prices. The fall in rice and wheat prices has benefited the urban and rural poor more than the upper-income groups, because the former spend a much larger proportion of their income on cereals than the latter (Kumar et al. 1995).

6.13 DEMAND FOR CEREALS

Cereals dominate food expenditure in India; rice ranks first followed by wheat and coarse grains. Annual per capita consumption of cereals is 176 kg in rural areas and 136 kg in urban areas, accounting for 73% of the total calorie intake in rural areas and 62% in urban areas. Food is more diverse in urban areas, with significantly higher per capita expenditure on milk and milk products, fruits, vegetables, and other high-value foods, and lower per capita expenditure on cereals, compared with rural areas. Increasing urbanization and widening rural–urban disparities will reduce the per capita demand for food grains and rapidly increase the demand for fruits and milk. The consumption of cereals per capita has declined in rural areas but shows no significant trend in urban areas, despite rising incomes and declining relative cereal prices in cities. The declining trend in cereal consumption in rural areas can be largely attributed to a shift in tastes and preferences resulting from the increasing availability of a greater variety of food items other than food grains as well as a wide range of non-food goods and services.

REFERENCES

Abdolshahi, R., Nazari, M., Safarian, A., Sadathossini, T. S., Salarpour, M., and Amiri, H. (2015). Integrated selection criteria for drought tolerance in wheat (*Triticum aestivum* L.) breeding programs using discriminant analysis. *Field Crops Research* 174, 20–29. https://doi.org/10.1016/j.fcr.2015.01.009.

Allard, R. W. (1970). Population structure and sampling methods. In Frankel, O. H., and Bennett, E. (eds), *Genetic resources in plants-their exploration and conservation.* Blackwell, Oxford, pp. 97–107.

Ananda, G. K., Myrans, H., Norton, S. L., Gleadow, R., Furtado, A., and Henry, R. J. (2020). Wild sorghum as a promising resource for crop improvement. *Frontiers in Plant Science* 11, 1108.

Anderson, E. R. (1949). *Introgressive hybridization.* John Wiley & Sons, New York.

Armario Najera, V., Twyman, R. M., Christou, P., et al. (2019). Applications of multiplex genome editing in higher plants. *Current Opinion in Biotechnology* 59, 93–102. https://doi.org/10.1016/j.copbio.2019.02.015.

Arora, R. K. (1993). Biodiversity in crop plants and their wild relatives. In *Proceedings of the First National Agricultural Science Congress*, pp. 160–180. National Academy of Agricultural Sciences, New Delhi.

Arora, R. K. (2000). Wild relatives of cultivated plants. In *Flora of India. Introductory Volume, pt II*, pp. 218–234. Botanical Survey of India, Calcutta.

Arora, R. K., and Chandel, K. P. S. (1972). Botanical source areas of wild herbage legumes in India. *Tropical Grasslands* 6(3), 213–221.

Arora, R. K., and Nayar, E. R. (1984). *Wild relatives of crop plants in India. Sci. Monogr. 7*. National Bureau of Plant Genetic Resources, New Delhi.

Arora, R. K., and Pandey, A. (1996). *Wild edible plants of India: Diversity, conservation and use*. National Bureau of Plant Genetic Resources, New Delhi.

Bellon, M. R. (1996). The dynamic of crop infraspecific diversity: A conceptual frame work at the farmer level. *Economic Botany* 50, 26–37.

Bellon, M. R., Brar, D. S., Lu, B. R., and Pham, J. L. (1998). Rice genetic resources. In Dowling, N. G., Greenfield, S. M., and Fischer, K. S. (eds), *Sustainability of rice in the global food system*. Pacific Basin Study Centerand Manila (Philippines): International Rice Research Institute, Davis, CA, pp. 251–283.

Betha, U. K. (2019). *Transgenic crops-biosafety concerns and regulations in India*. ICAR-Indian Institute of Oilseeds Research, Hyderabad. https://vikaspedia.in/agriculture/crop-production/advanced-technologies/transgenic-crops-biosafety-concerns-and-reg-ulations-in-india.

Bommer, D. F. R. (1991). The historical development of international collaboration in plant genetic resources. In van Hintum, T. J. L., Frese, L., and Perret, P. M. (eds), *Searching for new concepts for collaborative genetic resources management*. Papers of the EUCARPIA/IBPGR Symposium, Wageningen, The Netherlands, 3–6 December 1990. International Crop Networks Series no. 4. International Board for Plant Genetic Resources, Rome, pp. 3–12.

Bortesi, L., Zhu, C., Zischewski, J., et al. (2016). Patterns of CRISPR/Cas9 activity in plants, animals and microbes. *Plant Biotechnology Journal* 14, 2203–2216. https://doi.org/10.1111/pbi.12634.

Bothmer, R., and von Jacobsen, N. (1985). Origin, taxonomy, and related species. In Rasmusson, D. C. (ed), *Barley*. American Society of Agronomists, Madison, WI, pp. 19–56.

Brar, S. D., and Khush, G. S. (1997). Alien introgression in rice. *Plant Molecular Biology* 35, 35–47.

Briggs, D. E. (1978). *Barley*. Chapman and Hall Ltd, London.

Brown, A. H. D., and Marshall, D. R. (1995). A basic sampling strategy: Theory and practice. In Guarino, L., Ramanatha Rao, V., and Reid, R. (eds), *Collecting plant genetic diversity technical guidelines*. CABI, Wallingford, pp. 75–91.

Camacho Villa, T. C., Maxted, N., Scholten, M. A., and Ford-Lloyd, B. V. (2005). Defining and identifying crop landraces. *Plant Genetic Resources: Characterisation and Utilisation* 3, 373–384. https://doi.org/10.1079/PGR200591.

Chang, T. T., Ou, S. H., Pathak, M. D., Ling, K. C., and Kauffmann, H. E. (1975). The search for disease and insect resistance in rice germplasm. In O. H. Frankel and J. G. Hawkes (eds), *Crop genetic resources for today and tomorrow*. Cambridge University Press, Cambridge, pp. 183–200.

Chen, X., Min, D., Yasir, T. A., and Hu, Y. G. (2012). Evaluation of 14 morphological, yield-related and physiological traits as indicators of drought tolerance in Chinese winter bread wheat revealed by analysis of the membership function value of drought toler-ance (MFVD). *Field Crops Research* 137, 195–201. https://doi.org/10.1016/j.fcr.2012.09.008.

Damania, A. B. (2008). History, achievements, and current status of genetic resources conservation. *Agronomy Journal* 100, 9–21.

Day Rubenstein, K., Smale, M., and Widrlechne, M. P. (2006). Demand for genetic resources and the U.S. National Plant Germplasm System. *Crop Science* 46, 1021–1031.

De Carvalho, M. A. P., Bebeli, P. J., Bettencourt, E., Costa, G., Dias, S., Dos Santos, T. M., and Slaski, J. J. (2013). Cereal landraces genetic resources in worldwide Gene Banks. A review. *Agronomy for Sustainable Development* 33(1), 177–203.

De Wet, J., Gupta, S., Harlan, J., and Grassl, C. (1976). Cytogenetics of introgression from *Saccharum* into *Sorghum*. *Crop Science* 16(4), 568–572.

Demissie, A., and Bjornstad, A. (1996). Phenotypic diversity of Ethiopian barleys in relation to geographical regions, altitudinal range, and agro-ecological zones: As an aid to germplasm collection and conservation strategy. *Hereditas* 124, 17–29.

Dessi, A. (2018). Cereal crops research achievements and challenges in Ethiopia. *International Journal of Research Studies of Agricultural Science* 4(6), 23–29.

Dobrovolskaya, O., Saleh, U., and Malysheva-Otto, L. (2005). Rationalising germplasm collections: A case study for wheat. *Theoretical and Applied Genetics* 111, 1322–1329.

Doggett, H. (1988). *Sorghum*. Longman Scientific and Technical, Essex, England, UK.

Dorofeev, V. F., Filatenko, A. A., Migushova, E. F., Udaczin, R. A., and Jakubziner, M. M. (1979). Wheat. In Dorofeev, V. F., and Korovina, O. N. (eds), *Flora of cultivated plants*, vol. 1. Kolos, Leningrad.

Dos Santos, T. M. M., Gananca, F., Slaski, J. J., and Pinheiro de Carvalho, M. A. A. (2009). Morphological characterization of wheat genetic resources from the Island of Madeira, Portugal. *Genetic Resources and Crop Evolution* 56, 363–375. https://doi.org/10.1007/s10722-008-9371-5.

Farias, R. M., and Bettencourt, E. (2006). *Estratégia para missões sistemáticasde colheita de espécies vegetais para conservação ex situ*. Editores: Instituto Nacional de Investigação Agrária e das Pescas (INIAP); Direcção Geral de Agricultura de Entre Douro e Minho (DRAEDM), p. 39.

Gandilyan, P. A. (1980). *Key to wheat, Aegilops, rye and barley*. Academy of Science, Armenian SSR, Erevan.

Gao, F., Ma, D., Yin, G., Rasheed, A., Dong, Y., Xiao, Y., et al. (2017). Genetic progress in grain yield and physiological traits in Chinese wheat cultivars of southern Yellow and Huai Valley since 1950. *Crop Science* 57, 760–773. https://doi.org/10.2135/cropsci2016.05.0362.

Garber, E. D. (1950). Cytotaxonomic studies in the genus Sorghum. *University of California Publications in Botany* 23, 283–362.

Gomez-Macpherson, H. (2001). *Hordeum vulgare*. http://ecoport.org/ep?Plant=1232&entityType=PL & entity Display Category=full.

Govindaswami, S., Krishnamurty, A., and Sastry, N. S. (1966). The role of introgression in the varietal variability in rice in the Jeypore tract of Orissa. *Oryza* 3, 74–85.

Grohmann, L., Keilwagen, J., Duensing, N., et al. (2019). Detection and identification of genome editing in plants: Challenges and opportunities. *Frontiers in Plant Science* 10, 236. https://doi.org/10.3389/fpls.2019.00236.

Hammer, K., Diederichsen, A., and Spahillar, M. (1999). Basic studies toward strategies for conservation of plant genetic resources. In Serwinski, J., and Faberova, I. (eds), *Proceedings of technical meeting on the methodology of the FAO World Information and EarlyWarning System on Plant Genetic Resources*. FAO, Rome, pp. 29–33. https://apps3.fao.org/wiews/Prague/Paper1.htm.

Hawkes, J. G., Maxted, N., and Ford-Lloyd, B. V. (2000). *The ex situ conservation of plant genetic resources*. Kluwer, Dordrecht.

Iltis, H. H. (1979). *Zea diploperennis* (Gramineae): A new teosinte from Mexico. *Science* 203(4376), 186–188.

Jena, K. K., and Khush, G. S. (1990). Introgression of genes from *Oryza ojjicinalis* Wail. ex Watt to cultivated rice, *O. sativa* L. *Theoretical and Applied Genetics* 80, 737–745.

Jie, X., and Snape, J. W. (1989). The resistance of *Hordeum bulbosum* and its hybrid with *H. vulgare* to common fungal pathogens. *Euphytica* 41, 273–276.

Johnson, R. C. (2008). Gene banks pay big dividends to agriculture, the environment and human welfare. *PLoS Biology* 6, e148.

Kanbar, A., Shakeri, E., Alhajturki, D., Horn, T., Emam, Y., Tabatabaei, S. A., and Nick, P. (2020). Morphological and molecular characterization of sweet, grain and forage sorghum (*Sorghum bicolor* L.) genotypes grown under temperate climatic conditions. *Plant Biosystems-An International Journal Dealing with all Aspects of Plant Biology* 154(1), 49–58.

Khan, M. H., Dar, Z. A., and Dar, S. A. (2015). Breeding strategies for improving rice yield— A review. *Agricultural Sciences* 6(05), 467.

Khush, G. S. (1977). Disease and insect resistance in rice. *Advances in Agronomy* 29, 265–341.

Khush, G. S., Ling, K. C., Aquino, R. C., and Aquino, V. M. (1977). Breeding for resistance to grassy stunt in rice. In *Proc. 3rd Intl. Congr. SABRAO*. Vol 1.4(b), pp. 3–9.

Kimber, G., and Feldman, M. (1987). *Wild wheat, an introduction*. Special Report 353, College of Agriculture, University of Missouri, Columbia.

Kimber, G., and Sears, E. R. (1987). Evolution in the genus Triticum and the origin of cultivated wheat. In Heyne, E. G. (ed), *Wheat and wheat improvement*, 2nd ed. American Society of Agronomy, Madison, pp. 154–164.

Komatsuda, T., Pourkheirandish, M., He, C., Azhaguvel, P., Kanamori, H., Perovic, D., Stein, N., Graner, A., Wicker, T., Tagiri, A., Lundqvist, U., Fujimura, T., Matsuoka, M., Matsumoto, T., and Yano, M. (2007). Six-rowed barley originated from a mutation in ahomeodomain-leucinezipperI-classhomeoboxgene. *Proceedings of the National Academy of Sciences* 104, 1424–1429.

Kumar, P., Rosegrant, M. W., and Hazell, P. B. (1995). *Cereals prospects in India to 2020: Implications for policy* (No. 567-2016-39015).

Ladizinsky, G. (1998). *Plant evolution under domestication*. Kluwer, Dordrecht.

Lange, W., and Iochemsen, G. (1976). Karyotypes, nucleoli and amphiplasty in hybrids between *Hordeum vulgare* L. and *H. bulbosum* L. *Genetica* 46, 217–233.

Liu, H., Searle, I. R., Mather, D. E., Able, A. J., and Able, J. A. (2015). Morphological, physiological and yield responses of durum wheat to pre-anthesis water-deficit stress are genotype-dependent. *Crop and Pasture Science* 66, 1024–1038. https://doi.org/10.1071/CP15013.

Lopes, M. S., Reynolds, M. P., Manes, Y., Singh, R. P., Crossa, J., and Braun, H. J. (2012). Genetic yield gains and changes in associated traits of CIMMYT spring bread wheat in a "historic" set representing 30 years of breeding. *Crop Science* 52, 1123–1131. https://doi.org/10.2135/cropsci2011.09.0467.

Lorest, G., and Jackson, M. (1996). South Asia partnerships forged to conserve rice genetic resources. *Diversity* 12(3), 60–61.

Love, A. (1984). Conspectus of the Triticeae. *Feddes Repert* 95, 425–521.

Mac Key, J. (1988). A plant breeder's perspective on taxonomy of cultivated plants. *Biologisches Zentralblatt* 107, 369–379.

Maiti, R. K., and Satya, P. (2014). Research advances in major cereal crops for adaptation to abiotic stresses. *GM Crops & Food* 5(4), 259–279.

Malik, R., Sharma, H., Sharma, I., Kundu, S., Verma, A., Sheoran, S., Kumar, R., and Chatrath, R. (2014). Genetic diversity of agro-morphological characters in Indian wheat varieties using GT biplot. *Australian Journal of Crop Science* 8(9), 1266–1271.

Mammadov, J., Buyyarapu, R., Guttikonda, S. K., Parliament, K., Abdurakhmonov, I. Y., and Kumpatla, S. P. (2018). Wild relatives of maize, rice, cotton, and soybean: Treasure troves for tolerance to biotic and abiotic stresses. *Frontiers in Plant Science* 9, 886.

Matres, J. M., Hilscher, J., Datta, A., Armario-Nájera, V., Baysal, C., He, W., Huang, X., Zhu, C., Valizadeh-Kamran, R., Trijatmiko, K. R., and Capell, T. (2021). Genome editing in cereal crops: An overview. *Transgenic Research* 30(4), 461–498.

Morrison, L. A. (2001). The Percival Herbarium and wheat taxonomy: yesterday, today and tomorrow. *Wheat taxonomy: the legacy of John Percival. Linnean special issue*, 3rd ed. Linnean Society, London, pp. 65–80.

Nayar, E. R., Pandey, A., Venkateswaran, K., Gupta, R., and Dhillon, B. S. (2003). *Crop plants of India: A check-list of scientific names*. National Bureau of Plant Genetic Resources, New Delhi.

NCBI. (2012). Available from http://www.ncbi.nlm.nih.gov/Taxonomy/Browser/wwwtax.cgi ?id04513. Accessed February 2012.

Negi, K. S., Pant, K. C., Koppar, M. N., and Thomas, T. A. (1991). Wild relatives of genus *Allium* L. in Himalaya. *Indian Journal of Plant Genetic Resources* 4(1), 73–77.

Negri, V. (2003). Landraces in central Italy: Where and why they are conserved and perspectives for their on-farm conservation. *Genetic Resources and Crop Evolution* 50, 871–885.

Nevo, E. (1992). Origin, evolution, population genetics and resources for breeding of wild-barley, *Hordeum spontaneum*, in the fertile crescent. Chapter 2. In P. R. Shewry (ed), *Barley: Genetics, biochemistry, molecular biology and biotechnology*. C.A.B International, Wallingford, Oxon, pp. 19–43.

Nevski, A. (1941). Beiträge zur Kenntniss der wildwachsenden Gersten in Zusammenhang mit der Frage über den Ursprung von Hordeumvulgare L. und Hordeum distichon L. (Versucheiner Monographie der Gattung Hordeum). *Trudy Eot Imr Akad Nauk SSSR* 1, 64–255.

Newton, A. C., Akar, T., Baresel, J. P., Bebeli, P. J., Bettencourt, E., Bladenopoulos, K. V., Czembor, J. H., Fasoula, D. A., Katsiotis, A., Koutis, K., Koutsika-Sotiriou, M., Kovacs, G., Larsson, H., Pinheiro de Carvalho, M. A. A., Rubiales, D., Russell, J., dos Santos, T. M. M., and VazPatto, M. C. (2010). Cereal landraces for sustainable agriculture. A review. *Agronomy for Sustainable Development* 30, 237–269. https://doi.org/10.1051/agro/2009032.

Pandey, A., Bhandari, D. C., Bhatt, K. C., Pareek, S. K., Tomer, A. K., and Dhillon, B. S. (2005). *Wild relatives of crop plants in India: Collection and Conservation*. National Bureau of Plant Genetic Resources, New Delhi, India.

Pandey, R. P., and Padhye, P. M. (2000). Wild relatives and related species of cultivated crop plants and their diversity in Gujarat, India. *Journal of Economic and Taxonomic Botany* 24(2), 339–348.

Paterson, A. H., Schertz, K. F., Lin, Y.-R., Liu, S.-C., and Chang, Y.-L. (1995). The weediness of wild plants: Molecular analysis of genes influencing dispersal and persistence of johnsongrass, *Sorghum halepense* (L.) Pers. *Proceedings of the National Academy of Sciences* 92(13), 6127–6131.

Patra, B. c., Saha, R. K., Patnaik, S. S. C., Marandi, B. C., Nayak, P. K., and Dhua, S. R. (2002). Wild rice and related species in Orissa and West Bengal. In S. Sahu et al. (eds), *Proceedings of. plant resource utilization for backward areas development*. Regional Research Laboratory, Bhubaneshwar, 28–29 December, 2002. Allied Publ. Ltd., New Delhi, p. 87.

Prasad, G. S. V., Padmavathi, G., Suneetha, K., Madhav, M. S., and Muralidharan, K. (2020). Assessment of diversity of Indian aromatic rice Germplasm collections for morphological, agronomical, quality traits and molecular characters to identify a core set for crop improvement. *CABI Agriculture and Bioscience* 1(1), 1–24.

Riley, R., Chapman, V., and Johnson, R. (1968). Introduction of yellow rust resistance of *Aegilops comosa* into wheat by genetically induced homeologous recombination. *Nature* 217, 383–384.

Salgotra, R. K., Gupta, B. B., and Raina, M. (2017). Strategies for breeding cereal crops to attain sustainability with major emphasis on rice. In *Plant omics and crop breeding*. Apple Academic Press, England, pp. 443–459.

Sears, E. R. (1956). The transfer of leaf rust resistance from *Aegilops umbellulata* to wheat. *Brookhaven Symposium in Biology* 9, 1–22.

Sedeek, K. E. M., Mahas, A., and Mahfouz, M. (2019). Plant genome engineering for targeted improvement of crop traits. *Frontiers in Plant Science* 10, 114. https://doi.org/10.3389/fpls.2019.00114.

Sharma, D., and Knott, D. R. (1966). The transfer of leaf rust resistance from *Agropyron* to *Triticum* by irradiation. *Canadian Journal of Genetics and Cytology* 8, 137–143.

Singh, V., and Pandey, R. P. (1996). An assessment of wild relatives of cultivated plants in Indian desert and their conservation. In S. P. Singh (ed), *Scientific horticulture*. Scientific Publishers, Jodhpur, pp. 155–162.

Smale, M., and Day-Rubenstein, K. (2002). The demand for crop genetic resources: International use of the US national plant germplasm system. *World Development* 30, 1639–1655.

Subudhi, H. N., Saha, D., and Choudhury, B. P. (2000). Collection of *Desmodium* Desf. and wild relatives from Orissa. *Journal of Economic and Taxonomic Botany* 24(3), 695–699.

Van Slageren, M. W. (1994). *Wild wheats: A monograph of Aegilops L. and Amblyopyrum (Jaub. & Spach) Eig (Poaceae)*. Wageningen Agriculture University Papers, 7. Wageningen Agricultural University, Wageningen, 513 pp.

Von Bothmer, R. (1992). The wild species of *Hordeum*: Relationships and potential use for improvement of cultivated barley. Chapter 1. In P. R. Shewry (ed), *Barley: Genetics, biochemistry, molecular biology and biotechnology*. C.A.B International, Wallingford, Oxon, pp. 3–18.

WIEWS. (2012). *Regional distribution of gene banks, their cereals, and their cereal landraces collections*.

Zeven, A. C. (1998). Landraces: A review of definitions and classifications. *Euphytica* 104, 127–139.

Zhang, Y., Li, D., Zhang, D., et al. (2018). Analysis of the functions of TaGW2 homoeologs in wheat grain weight and protein content traits. *Plant Journal for Cell and Molecular Biology* 94(5), 857–866. https://doi.org/10.1111/tpj.13903.

Zhu, C., Bortesi, L., Baysal, C., et al. (2017). Characteristics of genome editing mutations in cereal crops. *Trends in Plant Science* 22(1), 38–52. https://doi.org/10.1016/j.tplants.2016.08.009.

7 Resistance Identification and Implementation
Genomics-Assisted Use of Genetic Resources for Breeding against Abiotic Stress

A.M. Shackira, Nair G. Sarath,
Mathew Veena, Riya Johnson, S. Jeyaraj,
K.P. Aswathy Raj, and Jos T. Puthur

CONTENTS

7.1 INTRODUCTION

Climate changes raise biotic and abiotic pressures on living organisms and may adversely affect the stability of the ecosystem. Plant exposure to abiotic and biotic stress, especially crop plants, greatly threatens food security that is unable to meet the demands of the world's rising population (Zhou et al., 2017). Plants are continuously subjected to excessive light intensity, heat, ultraviolet (UV) radiation, drought, cold,

DOI: 10.1201/9781003250845-7

salt, nutritional inadequacy, and various kinds of pollutants (Gharechahi et al., 2019). Crop plants have an inherent mechanism of tolerance against these adverse climatic constraints but may not be able to sustain it in the future. Trait alterations in the characteristics of plants, including at the morphological, physiochemical, and molecular levels, assist plants in coping with the changing climatic conditions (Song et al., 2019). The development of stress-tolerant crop cultivars either through identifying a stress-tolerant genotype or through manipulating stress-sensitive varieties is required to feed the world's growing population in the face of climate change. The availability of natural genetic variations in a given crop species plays a big role in developing such tolerant cultivars through conventional breeding methods. However, the genetic variability of stress-tolerant genotypes is limited in most crops and must be increased in order to improve crop tolerance to climate change (Hartung and Joachim, 2014).

Critical steps for enhancing plant performance and productivity under severe environmental conditions include screening for stress-tolerant plants, identifying and characterizing their genomes, and successfully manipulating the target species. A variety of procedures and protocols, ranging from simple ancient plant breeding to advanced biotechnology, can be used to manipulate a stress-sensitive plant. Nowadays, increasing emphasis is placed on genetic engineering tools, and screening for stress tolerance can be accomplished using a variety of strategies, including structural, functional, and comparative genomic approaches. Tools to access stress-resistance genes in plants include sequence-based approaches, hybridization-based approaches, tilling and molecular markers, genome sequencing, and comparative genomics. The main approaches to crop enhancement in modern agriculture are cross-breeding, mutation breeding, and transgenic breeding. Both cross-breeding and genetic recombination take a long time to introduce beneficial alleles and increase the resistance of a particular variety. A large segment of the genomes of important crops has been fixed, and genetic variability has been considerably decreased, restricting the possibility to improve many features, thanks to thousands of years of directed evolution through breeding. Using chemical mutagens or physical irradiation, mutation breeding has increased genetic diversity to a greater extent by introducing random mutations. However, because of their stochastic nature and the appearance of undesirable characteristics, these approaches are limited in their ability to generate and test large numbers of mutants. Even if marker-assisted breeding procedures are used to improve selection efficiency, such time-consuming, tedious, untargeted breeding processes will not be able to keep up with the need for greater agricultural yield. The bottleneck of reproductive isolation can be broken by transgenic breeding, which develops desirable features by transferring exogenous genes into elite background varieties. However, long and expensive regulatory evaluation processes, as well as public concerns, impede the marketing of genetically modified crops (Chen et al., 2019).

7.2 ABIOTIC STRESS AND PLANT METABOLISM

Abiotic stress is generally defined as environmental conditions that reduce the growth and yield of plants below optimum levels. Plant responses to abiotic stresses are dynamic and complex, and may be reversible or irreversible. Abiotic factors such as high light, salinity, drought, heavy metal, cold, and temperature (sub and super optimal) may cause harm to the plants and animals in the area affected (Ali et al., 2017).

7.2.1 DROUGHT

Osmotic stress is one of the most severe environmental stresses and has a major impact on plant development and productivity, causing serious agricultural yield losses. One of the major environmental factors limiting the worldwide production and distribution of crops is osmotic stress resulting from drought. Polyethylene glycol (PEG) compounds have been used to simulate water stress effects in plants. Several works have studied the effects of sodium chloride (NaCl) and PEG on growth and ion absorption in rice (Demir and Mavi, 2008), bean and cowpea (Berli et al., 2010), bean (Costa-Franca et al., 2000), and tomato (Alian et al., 2000) to better understand and unravel the mechanism of plant resistance to osmotic stress. Osmotic adjustment plays an important role in sustaining growth under water-deficit conditions. The response of plants differs significantly at various organizational levels, depending on the intensity and duration of the stress as well as the plant species and its stage of development. The genotypic response to osmotic stress has been identified for a range of morphological and physiological characteristics, including root development, stomatal activity, osmotic adjustment, and abscisic acid and proline levels. Osmotic stress results in the increased generation of reactive oxygen species (ROS) due to energy accumulation in stressed plants which absorb more light energy than is consumed through photosynthetic carbon fixation (Munne-Bosch, 2005). Depending on the timing and the magnitude of the water deficit, plants induce stomatal regulation of water loss, reducing the leaf area, hastening or delaying the reproductive cycle, or by developing a deep root system (Hall, 2004). Osmotic stress inhibits or slows down photosynthetic carbon fixation mainly through limiting the entry of CO_2 into the leaf or directly inhibiting metabolism (Apel and Hirt, 2004). To counteract the toxicity of ROS, plant cells have developed an antioxidative system, consisting of low-molecular-weight antioxidants such as ascorbate, tocopherol, glutathione, and carotenoids, as well as protective enzymes such as peroxidase, catalase, superoxide, and dismutase (Vinocur and Altman, 2005).

7.2.2 SALINITY

The salinity problem is common in arid and semi-arid regions where rainfall is insufficient to leach salts out of the root zone. These areas often have high evaporation rates, which can cause an increase in salt concentration at the soil surface (Demir et al., 2004). Under saline conditions, it is important to maintain and/or improve soil water availability to crops (Kaya et al., 2006). This can be accomplished through several strategies such as leaching salts from the soil profile, maintaining high soil water content in the root zone, selecting more salt-tolerant plants, and improving cropping systems (Blum, 2018). Salt and water stresses are responsible for both inhibition or delayed seed germination and seedling establishment (Siddiqui et al., 2006). Under these stress conditions there is a decrease in water uptake, during both imbibition and seedling establishment, and in the case of salt stress, this can be followed by excessive uptake of ions (Murillo-Amador, 2000).

Increased salinity induces osmotic as well as toxic ionic effects, thereby affecting growth and other physiological and biochemical processes including photosynthesis and ion homeostasis. Altered photosynthesis is associated with perturbed carbon and nitrogen metabolism. In legumes, salinity alters nitrogen fixation and hence growth as well as yield. Understanding how plants respond to salt and co-occurring stresses can play a major role in stabilizing crop performance under saline conditions and in protecting natural vegetation. Adequate management techniques and plant genetic breeding are tools to improve resource use efficiency (including water) in plants using new technologies (Klein et al., 2007). The effect of salinity on seedling establishment is more conspicuous than on seed germination (Kaya et al., 2008).

7.2.3 TEMPERATURE

Plants cope with adverse temperature stress by altering their molecular mechanisms involving proteins, antioxidants, metabolites, regulatory factors, other protectants, and membrane lipids. Temperature-stressed plants show low germination rates, growth retardation, and reduced photosynthesis, and often die in extreme cases. The elucidation of mechanisms by which temperature stress causes disorders is important to reveal responses by which plants cope with adverse temperature conditions. However, plants respond to temperature stress by regulating their membrane lipid composition, stress-related transcription factors, metabolite synthesis, and detoxification pathways (Kai and Iba, 2014).

7.2.4 UV LIGHT

Ultraviolet B (UV-B) radiation (280–315 nm) is a minor component of the solar spectrum, but it is the most energetic one that reaches the earth's surface and is strongly absorbed by stratospheric ozone (O_3). Exposure of living organisms to UV-B causes harmful effects and due to the necessary requirement for sunlight, the exposure of plants to UV-B is potentially detrimental (León-Chan et al., 2017). In plants, UV-B irradiation can increase the level of ROS (Frohnmeyer and Staiger, 2003). They subsequently oxidize the biomolecules in the cells and negatively influence the functionality and integrity of proteins and cell membranes. UV-B-induced oxidative stress in plants results in alterations in plant growth and developmental processes (Dotto and Casati, 2017), impaired photosynthetic pigments and processes (Kataria, 2017), changes in secondary metabolites (Escobar-Bravo et al., 2017), alteration in anatomical characteristics, changes in cuticular wax deposition (Nascimento et al., 2016; Willick et al., 2018), and damage to the photosystem (PSI and PSII) proteins and electron transport activities (Zhang et al., 2016; Rai et al., 2018).

7.2.5 FLOOD

As an abiotic stress, flooding strongly affects almost 16% of agricultural production areas worldwide (Ahsan et al., 2007). The primary causes of waterlogging are heavy rainfall events and poor soil drainage (Sundgren et al., 2018). Unfortunately, areas subjected to waterlogging are anticipated to increase along with climatic changes (Bailey-Serres et al., 2012). Flooding directly affects the diffusion of oxygen in plant tissues and gas diffusion between cells limits oxygen exchange and mitochondrial respiration, thus seriously affecting normal plant physiological and biochemical activities (Voesenek and Bailey-Serres, 2013). A low oxygen level significantly affects most plant developmental stages and shifts the energy-related metabolic pathways from aerobic respiration to anaerobic fermentation (Xu et al., 2014). Limited energy production causes an excessive accumulation of in-tissue toxins, such as alcohols and aldehydes (Tamang et al., 2014). Most plants are sensitive to flooding (Bailey-Serres and Colmer, 2014), thus they have developed different strategies to survive this stressful condition (Fukao et al., 2006; Xu et al., 2006; Hattori et al., 2009).

7.2.6 Heavy Metals

A number of essential and non-essential heavy metals are present in soil, having reached the soil through various natural and anthropogenic activities such as smelting, the discharge of industrial as well as domestic wastes, and the application of fertilizers and pesticides (Yadav, 2010). From the soil these metal ions, including non-essential ones, gain entry into the plant tissues through the root system and may become toxic if absorbed above the optimal level required by the plant. In response to metal toxicity, plant cells generate reactive oxygen species leading to membrane peroxidation, and to counteract this the plant activates the antioxidant machinery as a first line of defense strategy. Upon severe stress, these metal ions are sequestered in the vacuole through small low-molecular-weight thiols such as phytochelatins or metallothioneins, which are highly specific to various heavy metals. There is high affinity between metal ions and these metal-binding molecules, so they are effectively detoxified inside the plant cell without much deleterious effect on the plant (Andresen et al., 2018).

7.3 ENGINEERING ABIOTIC STRESS-TOLERANT PLANTS

Abiotic stresses such as drought, waterlogging, salinity, and temperature are major environmental constraints that negatively influence plant growth and development, ultimately causing a reduction in crop quality and quantity. Engineering plants for abiotic stress tolerance can be achieved through conventional breeding approaches or advanced genome editing techniques such as clustered regularly interspaced short palindromic repeat (CRISPR). Such techniques serve as a promising tool for the sustainable production of climate-resilient crops. In the present scenario, advancements in genetic engineering and genome editing techniques draw more attention from conventional plant breeding methods, thus enabling the engineering of plants with targeted traits. The relative importance of developing abiotic stress-tolerant varieties mainly depends on the geographical region and, to a greater extent, the adaptability of a particular crop to that area.

7.3.1 CONVENTIONAL TECHNIQUES

Conventional breeding approaches are widely employed to improve cultivars for better quality and tolerance to various biotic and abiotic stresses by means of traditional tools. Of important concern to plant breeders are genes that control traits of interest, since the transfer of traits that are controlled by a single gene is easier than that controlled by multiple genes (Acquaah, 2015). Many resources are available for improving abiotic stress tolerance in crop plants, and such desired traits can be found in landraces and wild relatives. The selection of candidate genes conferring higher tolerance to various abiotic stresses was critical until the recent advancement of molecular tools. In earlier times, plant breeders screened plants chiefly based on morphological parameters. The enhancement of plant tolerance to various abiotic stresses can be attained through various breeding approaches, such as mass selection, pure line selection, bulk method, backcrossing, recurrent selection, and mutation breeding. The selection of suitable breeding approaches is an important task for developing tolerant cultivars to a defined environment.

Among the major abiotic stresses, drought is considered the most challenging environmental constraint on crop productivity. The three important factors to be considered in breeding for drought tolerance are timing, duration, and intensity of stress. So far, a number of drought-tolerant varieties of crops such as peanut, safflower, common bean, maize, wheat, and barley have been developed through conventional breeding techniques (Meena et al., 2017). Similarly, salinity stress is a major threat to agricultural productivity since higher levels of salt are harmful to plant growth and development. Different breeding strategies for engineering salt tolerance in cross-pollinating species through recurrent selection were already described by Dewey (1962). A few varieties have been developed for salinity tolerance with conventional breeding approaches such as selection and introduction, pedigree, modified bulk method, and anther culture. Pure line selection was found to be successful in establishing three salt-tolerant rice varieties (CSR-1, CSR-2, and CSR-3) from traditional cultivars (Meena, 2017). Similarly, some rice varieties such as Jaladhi-1, 2, Jalaprabha, Neeraja, Dinesh, and Hangseswari were developed through selection for tolerance to deep water (Indrani et al., 2013). Tammam et al. (2004) suggested pedigree selection as an effective method to produce drought-tolerant lines. Some of the achievements of conventional breeding approaches are the development of seven salt-resistant rice varieties (CSR-43, CSR-36, CSR-30, CSR-27, CSR-23, CSR-13, and CSR-10), three salt-tolerant varieties of mustard (CS-52, CS-54, and CS-56), and one chickpea (Karnal Chana-1). However, the development of crop plants with improved tolerance to various abiotic stresses through traditional breeding approaches has limited scope in certain crops due to their incompatibility (Figure 7.1).

7.3.2 TRANSGENIC APPROACHES

Conventional breeding methods have limited application in attaining abiotic stress-tolerant crops, mainly due to unwanted linkages that require repeated backcrosses to eliminate such undesirable traits. In such cases, the transgenic approach can be used to transfer desirable traits in engineering abiotic stress tolerance in plants. A

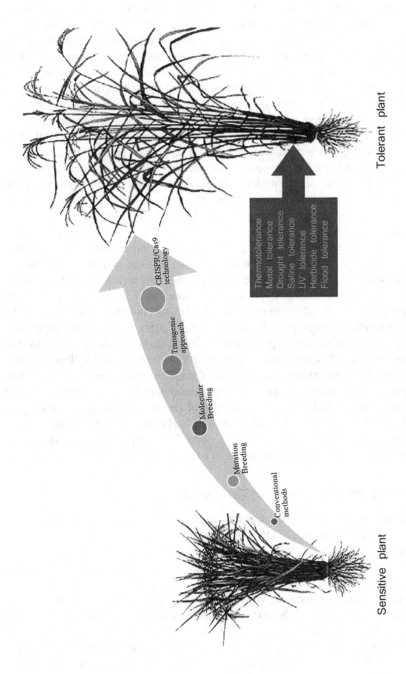

FIGURE 7.1 Various methods for imparting abiotic stress tolerance in plants.

functional screening strategy based on bacteria or yeast experimental systems is used to uncover genes involved in plant stress tolerance. Yeast is favored over bacteria because it is a eukaryote with posttranslational modification that is virtually identical to that of higher plants. This method can also be used to pick the most significant genes of stress tolerance from a vast group of genes, while also having the potential to find unknown genes, resulting in the discovery of new stress-tolerant genes. With the availability of a huge quantity of genomic, transcriptomic, and proteomic data, it is also possible to predict the potential stress-tolerant gene (Kappachery et al., 2013).

Potatoes are very susceptible to drought stress. The putative drought-tolerant gene in potato was discovered using a yeast functional screening technique. A cDNA library was created from hyperosmotic-challenged potato plants, and a total of 69 drought-tolerant genes were discovered. Based on relative tolerant data, 12 genes were chosen which may be important for drought tolerance (Kappachery et al., 2013). Yeast functional screening for salinity-tolerant genes in *Salicornia europaea* suggested the participation of three genes (Nakahara et al., 2015). In a yeast system, 32 genes were found to be involved in *Jatropha curcas* under salinity stress (Eswaran et al., 2010).

Transgenic plants having genes conferring abiotic stress tolerance are mainly developed through *Agrobacterium*-mediated gene transfer and biolistic methods. Such transgenics have been developed in many crop plants such as rice, wheat, maize, sugarcane, tobacco, *Arabidopsis*, potato, and tomato (Wani et al., 2016). Researchers introduced a gene (HVAII) from barley to wheat that showed higher tolerance under low water levels for longer periods (Khurana et al., 2011). Similarly, the superoxide dismutase (SOD) gene from a pea plant to tobacco improved drought tolerance in transgenic tobacco (Sengupta and Majumder, 2009). Transgenic rice (Pusa Basmati 1) carrying HVAI exhibited tolerance against abiotic stresses (Rohilla et al., 2000). Sometimes, abiotic tolerance can be achieved through engineering plants for higher levels of cellular osmolytes. Overexpression of the aldehyde dehydrogenase gene in *Arabidopsis* imparted higher tolerance to drought and salt stress (Sunkar et al., 2003). Even though genetic transformation methods are helpful in developing abiotic stress tolerance in plants, their low frequency limits their scope in some crops.

7.3.3 CRISPR/Cas 9-Mediated Genome Editing

Genome editing is a tool to manipulate the genome to make desirable changes in predetermined specific sites of DNA. CRISPR/Cas is a more recent, simpler, easier to design, robust, flexible, cost-effective, multiplexed, and efficient genome editing tool. CRISPR DNA sequences are separated by non-repeating DNA sequences called spacers (Adli, 2018). They are part of the prokaryotic immune system to obtain protection from invading viruses. This system helps bacteria to identify the specific DNA sequences of the conquering phages and target them for destruction through RNA-guided DNA cleavage with the help of specialized enzymes called Cas (CRISPR-associated proteins). To do so, the system requires two RNA: CRISPR RNA (CrRNA, transcribed from spacers) and trans-activating CrRNA (tracrRNA). This CrRNA, which can form a complex with the Cas protein (having helicase and nuclease activity), acts as a guiding sequence and can pair with the tracrRNA. Thus, the formed Cas protein–RNA complex slices DNA forming DSBs (double-stranded breaks) at target sites. It also requires a short two to six nucleotide long sequence immediately after the target sequence, called a protospacer adjacent motif (PAM) which is indispensable for this activity. One of the best studied is the Type II CRISPR system consisting of the nuclease Cas 9 (Gaj et al., 2016).

In the case of CRISPR/Cas 9, which relies on DNA–RNA interaction, all that is necessary is the creation of an 18–20 bp oligonucleotide. This makes adapting the CRISPR/Cas 9 technology for genome editing applications so easy. Cas 9 and gRNA must attach to a certain protospacer adjacent motif sequence, which is a short nucleotide sequence positioned at the 3′ end of the target sequence, in order to operate as genome editing tools. CRISPR/Cas 9 has been used effectively in a variety of plant species, including crops such as rice, tobacco, sorghum, wheat, maize, soybean, tomato, potato, poplar, apple, and banana, as well as model plants like *Arabidopsis* (Table 7.1). The ability to edit many target genes at the same time is a significant benefit of the CRISPR/Cas 9 system (Wada et al., 2020).

The CRISPR/Cas system can be employed in developing abiotic stress-tolerant plants. For the development of cold-tolerance crops, tomato C-repeat binding factor 1 (CBF1) mutant (cbf1) generated by the CRISPR/Cas technique was tolerant to chilling stress by reducing the electrolyte leakage, and the accumulation of indole acetic acid (IAA) and hydrogen peroxide (Li et al., 2018). The editing of an annexin (OsANN3), which codes a calcium-dependent phospholipid-binding protein in a *Japonica* cultivar, showed an increase in relative electrical conductivity and tolerance to cold (Shen et al., 2017). The editing of SAPK2 (ABA-activating protein kinase 2) and Osmyb 30 mutant in rice provided tolerance to cold stress (Lou et al., 2017; Zeng et al., 2019).

The CRISPR/Cas system was particularly effective in unraveling the molecular mechanism behind salt stress tolerance in plants. Drought resistance was reported in rice knockouts of osmotic stress/ABA-activated protein kinases (SAPK-1, SAPK-2) and the SnRK2 gene (Lou et al., 2017). The genes OsBBS1 and OsMIR528 were discovered to be salt stress-sensitive and positive regulators, while the genes OsNACO41 and OsRR22 boosted salt tolerance (Bhat et al., 2021). The editing tool also revealed the induction of the OsRAV2 gene, which confers salinity tolerance via the GT-1 element's regulatory function (Duan et al., 2016).

7.4 FUTURE PROSPECTS

Various kinds of abiotic stresses, such as drought, salinity, high and low light, UV, mechanical injuries, and flooding, negatively affect the quality and yield of numerous crops all over the world. In response to abiotic stress, plants trigger various defense strategies which may be the development of morphological, physiological, biochemical, anatomical, or molecular adaptations. Recently, much attention has been given to the molecular approaches for imparting abiotic stress tolerance in sensitive varieties of crops with a special focus on the transgenic lines. Transgenic endeavor may include the transfer of a single or a set of genes from the tolerant plants

TABLE 7.1

List of Plants Characterized for Tolerance to Various Abiotic Stresses through CRISPR-Based Technology

Name of plant	Name of target gene	Abiotic stress	References
Oryza sativa	OsSIT1	Salinity	Li et al. (2014)
Oryza sativa	OsCOLD1	Low temperature	Ma et al. (2015)
Oryza sativa	OsTT1	High temperature	Li et al. (2015)
Arabidopsis thaliana	OST2	Drought	Osakabe et al. (2016)
Arabidopsis thaliana	miR169a	Drought	Zhao et al. (2016)
Oryza sativa	OsMODD	Drought	Tang et al. (2016)
Zea mays	ARGOS8	Drought	Shi et al. (2017)
Arabidopsis thaliana	AVP1	Drought	Park et al. (2017)
Oryza sativa	OsCTB4a	Low temperature	Zhang et al. (2017)
Oryza sativa	OsPT4	Heavy metal (As)	Ye et al. (2017)
Oryza sativa	OsANN3	Low temperature	Shen et al. (2017)
Oryza sativa	OsNRAM5	Heavy metal (Cd)	Tang et al. (2017)
Oryza sativa	OsARM1	Heavy metal (As)	Wang et al. (2017a)
Oryza sativa	OsHAK1	Radioactive metal (Cs)	Nieves-Cordones et al. (2017)
Solanum lycopersicum	SlMAPK3	Drought	Wang et al. (2017b)
Oryza sativa	SAPK2	Drought and salinity	Wang et al. (2017b)
Zea mays	ARGOS8	Drought	Shi et al. (2017)
Oryza sativa	BBS1	Drought	Zeng et al. (2018)
Oryza sativa	OsNCED	Drought and salinity	Huang et al. (2018)
Oryza sativa	OsNAC14	Drought	Shim et al. (2018)
Arabidopsis thaliana	AREB1	Drought	Roca Paixão et al. (2019)
Oryza sativa	OsRAV2	Salinity	Duan et al. (2016)
Oryza sativa	OsRR22	Salinity	Zhang et al. (2019)
Glycine max	GmWRKY54	Drought	Wei et al. (2019)
Oryza sativa	OsALS	Herbicide (Bispyribac sodium)	Butt et al. (2020)
Oryza sativa	HIS1	Herbicide (beta-triketone)	Sandeep et al. (2021)
Glycine max	GmAITR	Salinity	Tianya et al. (2021)
Arabidopsis thaliana	AITR	Drought and salinity	Chen et al. (2021)
Glycine max	GmNAC06	Salinity	Li et al. (2021)

to the sensitive plants which confer resistance to abiotic stress. Many successful attempts have been carried out in the past few decades through the pyramiding of tolerant genes, and CRISPR/CAS 9 technology has paved a new way of producing abiotic stress-tolerant crops through targeted gene editing. However, the exposure of plants to a multitude of abiotic stress factors under natural environmental conditions may pose severe challenge to the transgenic variety as it may be designed to impart tolerance to a specific stress factor. In this perspective, more field-level assessment of transgenic lines is essential as this will help in establishing a successful stress-resistant candidate against a number of stress factors.

REFERENCES

Acquaah, G. (2015). *Conventional plant breeding principles and techniques. Advances in plant breeding strategies: Breeding, biotechnology and molecular tools* (pp. 115–158). Cham: Springer.

Adli, M. (2018). The CRISPR tool kit for genome editing and beyond. *Nature Communications* 9(1): 1–13.

Alian, A., Altman, A., Heuer, B. (2000). Genotypic difference in salinity and water stress tolerance of fresh market tomato cultivars. *Plant Science* 152(1): 59–65.

Ali, F., Bano, A., & Faza, A. (2017). Recent methods of drought stress tolerance in plants. *Plant Growth Regulation, 82*(3), 363–375.

Andresen, E., Peiter, E., Küpper, H. (2018). Trace metal metabolism in plants. *Journal of Experimental Botany* 69(5): 909–954. doi:10.1093/jxb/erx465.

Apel, K., Hirt, H. (2004). Reactive oxygen species: Metabolism, oxidative stress, and signal transduction. *Annual Review of Plant Biology* 55: 373–399.

Ahsan, N., Lee, D. G., Lee, S. H., Kang, K. Y., Bahk, J. D., Choi, M. S., ... Lee, B. H. (2007). A comparative proteomic analysis of tomato leaves in response to waterlogging stress. *Physiologia Plantarum, 131*(4), 555–570. https://doi.org/10.1111/j.1399-3054.2007.00980.x

Bailey-Serres, J., Fukao, T., Gibbs, D. J., Holdsworth, M. J., Lee, S. C., Licausi, F., ... van Dongen, J. T. (2012). Making sense of low oxygen sensing. *Trends in Plant Science, 17*(3), 129–138.

Bailey-Serres, J., & Colmer, T. D. (2014). Plant tolerance of flooding stress—Recent advances. *Plant, Cell and Environment, 37*(10), 2211–2215.

Berli, F. J., Moreno, D., Piccoli, P., Hespanhol Viana, L., Silva, M. F., Bressan Smith, R., Bottini, R. (2010). Abscisic acid is involved in the response of grape (*Vitis vinifera* L.) cv. Malbec leaf tissues to ultraviolet-B radiation by enhancing ultraviolet absorbing compounds, antioxidant enzymes and membrane sterols. *Plant, Cell and Environment* 33(1): 1–10.

Bhat, M. A., Mir, R. A., Kumar, V., Shah, A. A., Zargar, S. M., Rahman, S., Jan, A. T. (2021). Mechanistic insights of CRISPR/Cas-mediated genome editing towards enhancing abiotic stress tolerance in plants. *Physiologia Plantarum, 172*(2), 1255–1268.

Blum, A. (2018). *Plant Breeding, for Stress Environments*. CRC Press, Boca Raton, 17.

Butt, H., Rao, G. S., Sedeek, K., Aman, R., Kamel, R., Mahfouz, M. (2020). Engineering herbicide resistance via prime editing in rice. *Plant Biotechnology Journal* 18(12): 2370–2372. doi:10.1111/pbi.13399.

Chen, K., Wang, Y., Zhang, R., Zhang, H., Gao, C. (2019). CRISPR/Cas genome editing and precision plant breeding in agriculture. *Annual Review of Plant Biology* 70: 667–697.

Chen, S., Zhang, N., Zhou, G., Hussain, S., Ahmed, S., Tian, H., Wang, S. (2021). Knockout of the entire family of AITR genes in *Arabidopsis* leads to enhanced drought and salinity tolerance without fitness costs. *BMC Plant Biology* 21(1): 137. doi:10.1186/s12870-021-02907-9.

Costa-Franca, M. G., Thi, A. T. P., Pimentel, C., Rossiello, R. O. P., Zuily-Fodil, Y., & Laffray, D. (2000). Differences in growth and water relations among Phaseolus vulgaris cultivars in response to induced drought stress. *Environmental and Experimental Botany, 43*(3), 227–237.

Demir, N., Acar, J., & Bahceci, K. S. (2004). Effects of storage on quality of carrot juices produced with lactofermentation and acidification. *European Food Research and Technology, 218*(5), 465–468.

Demir, I., Mavi, K. (2008). Effect of salt and osmotic stresses on the germination of pepper seeds of different maturation stages. *Brazilian Archives of Biology and Technology* 51(5): 897–902.

Dewey, P. R. (1962). Breeding crested wheatgrass for salt tolerance. *Crop Science, 2*(5), 403–407.

Dotto, M., Casati, P. (2017). Developmental reprogramming by UV-B radiation in plants. *Plant Science* 264: 96–101.

Duan, Y. B., Li, J., Qin, R. Y., Xu, R. F., Li, H., Yang, Y. C., Ma, H., Li, L., Wei, P. C., Yang, J. B. (2016). Identification of a regulatory element responsible for salt induction of rice OsRAV2 through ex situ and in situ promoter analysis. *Plant Molecular Biology* 90(1–2): 49–62. doi:10.1007/s11103-015-0393-z.

Eswaran, N., Parameswaran, S., Sathram, B., Anantharaman, B., Tangirala, S. J. (2010). Yeast functional screen to identify genetic determinants capable of conferring abiotic stress tolerance in Jatropha curcas. *BMC Biotechnology* 10(1): 1–15.

Frohnmeyer, H., Staiger, D. (2003). Ultraviolet-B radiation-mediated responses in plants. Balancing damage and protection. *Plant Physiology* 133(4): 1420–1428.

Fukao, T., Xu, K., Ronald, P. C., & Bailey-Serres, J. (2006). A variable cluster of ethylene response factor-like genes regulates metabolic and developmental acclimation responses to submergence in rice. *Plant Cell, 18*(8), 2021–2034.

Gaj, T., Sirk, S. J., Shui, S. L., Liu, J. (2016). Genome-editing technologies: Principles and applications. *Cold Spring Harbor Perspectives in Biology* 8(12): a023754.

Gharechahi, J., Sharifi, G., Mirzaei, M., Zeinalabedini, M., Hosseini Salekdeh, G. (2019). Abiotic stress responsive microRNome and proteome: How correlated are they? *Environmental and Experimental Botany* 165: 150–160.

Hall, A. E. (2004). Breeding for adaptation to drought and heat in cowpea. *European Journal of Agronomy* 21(4): 447–454.

Hartung, F., Joachim, S. (2014). Precise plant breeding using new genome editing techniques: Opportunities, safety and regulation in the EU. *The Plant Journal* 78(5): 742–752.

Hattori, Y., Nagai, K., Furukawa, S., Song, X. J., Kawano, R., Sakakibara, H., ... Ashikari, M. (2009). The ethylene response factors SNORKEL1 and SNORKEL2 allow rice to adapt to deep water. *Nature, 460*(7258), 1026–1030.

Huang, Y., Guo, Y., Liu, Y., Zhang, F., Wang, Z., Wang, H., Wang, F., Li, D., Mao, D., Luan, S., Liang, M., Chen, L. (2018). 9-cis-epoxycarotenoid dioxygenase 3 regulates plant growth and enhances multi-abiotic stress tolerance in rice. *Frontiers in Plant Science* 9: 162. doi:10.3389/fpls.2018.00162.

Indrani, C., Singh, P., Bhattacharya, A., PriyaSingh, S. J., & Singhamahapatra, A. (2013). In vitro callus induction, regeneration and micropropagation of Solanum Lycopersicum. *International Journal of Current Microbiology and Applied Sciences, 2*(12), 192–197.

Kai, H., & Iba, K. (2014). Temperature stress in plants. In *ELS*. https://doi.org/10.1002/9780470015902.a0001320.pub2

Kappachery, S., Yu, J. W., Baniekal-Hiremath, G., Park, S. W. (2013). Rapid identification of potential drought tolerance genes from Solanumtuberosum by using a yeast functional screening method. *Comptesrendusbiologies* 336(11–12): 530–545.

Kataria, S. (2017). *Role of reactive oxygen species in Magnetoprimed induced acceleration of germination and early growth characteristics of seeds.* V. P. Singh, S. Singh, D. K. Tripathi, S. M. Prasad & D. K. Chauhan (Eds.). https://doi.org/10.1002/9781119324928.ch4

Kaya, M. D., Okçu, G., Atak, M., Çıkılı, Y., Kolsarıcı, Ö. (2006). Seed treatments to overcome salt and drought stress during germination in sunflower (*Helianthus annuus* L.). *European Journal of Agronomy* 24(4): 291–295.

Kaya, I., Yigit, N., & Benli, M. (2008). Antimicrobial activity of various extracts of Ocimum basilicum L. and observation of the inhibition effect on bacterial cells by use of scanning electron microscopy. *African Journal of Traditional, Complementary and Alternative Medicines*, 5(4), 363–369.

Khurana, P., Chauhan, H., & Khurana, N. (2011). Characterization and expression of high temperature stress responsive genes in bread wheat (Triticum aestivum L.). *Czech Journal of Genetics and Plant Breeding*, 47(Special Issue), S94–S97.

Klein, A. M., Vaissie`re, B. E., Cane, J. H., Steffan-Dewenter, I., Cunningham, S. A., Kremen, C., & Tscharntke, T. (2007). Importance of pollinators in changing landscapes for world crops. *Proceedings of the Royal Society of London, Series B*, 274(1608), 303–313.

Lee, H. Y., Byeon, Y., & Back, K. (2014). Melatonin as a signal molecule triggering defense responses against pathogen attack in *Arabidopsis* and tobacco. *Journal of Pineal Research*, 57(3), 262–268. https://doi.org/10.1111/jpi.12165

León-Chan, R. G., López-Meyer, M., Osuna-Enciso, T., Sañudo-Barajas, J. A., Heredia, J. B., León-Félix, J. (2017). Low temperature and ultraviolet-B radiation affect chlorophyll content and induce the accumulation of UV-B-absorbing and antioxidant compounds in bell pepper (Capsicum annuum) plants. *Environmental and Experimental Botany* 139: 143–151.

Li, M., Chen, R., Jiang, Q., Sun, X., Zhang, H., Hu, Z. (2021). GmNAC06, a NAC domain transcription factor enhances salt stress tolerance in soybean. *Plant Molecular Biology* 105(3): 333–345. doi:10.1007/s11103-020-01091-y.

Li, X. M., Chao, D. Y., Wu, Y., Huang, X., Chen, K., Cui, L. G., Su, L., Ye, W. W., Chen, H., Chen, H. C., Dong, N. Q., Guo, T., Shi, M., Feng, W., Zhang, P., Han, B., Shan, J. X., Gao, J. P., Lin, H. X. (2015). Natural alleles of a proteasome $\alpha2$ subunit gene contribute to thermotolerance and adaptation of African rice. *Nature Genetics* 47(7): 827–833. doi:10.1038/ng.3305.

Li, Y., Zhang, W., Ma, L., Wu, L., Shen, J., Davies, W. J., ... Dou, Z. (2014). An analysis of China's grain production: Looking back and looking forward. *Food and Energy Security*, 3(1), 19–32.

Lou, D., Wang, H., Liang, G., & Yu, D. (2017). OsSAPK2 confers abscisic acid sensitivity and tolerance to drought stress in rice. *Frontiers in Plant Science*, 8, 993. https://doi.org/10.3389/fpls.2017.00993

Ma, Y., Dai, X., Xu, Y., Luo, W., Zheng, X., Zeng, D., Pan, Y., Lin, X., Liu, H., Zhang, D., et al. (2015). COLD1 confers chilling tolerance in rice. *Cell* 160(6): 1209–1221. doi:10.1016/j.cell.2015.01.046.

Meena, K. K., Sorty, A. M., Bitla, U. M., Choudhary, K., Gupta, P., Pareek, A., ... Minhas, P. S. (2017). Abiotic stress responses and microbe-mediated mitigation in plants: The omics strategies. *Frontiers in Plant Science*, 8, 172.

Munne-Bosch, S. (2005). The role of α-tocopherol in plant stress tolerance. *Journal of Plant Physiology* 162(7): 743–748.

Murillo-Amador, B., Troyo-Die´guez, E., Jones, H. G., Ayala-Chairez, F., Tinoco-Ojanguren, C. L., & Lo´pez-Corte´s, A. (2000). Screening and classification of cowpea genotypes for salt tolerance during germination. *Phyton-International Journal of Experimental Botany*, 67, 71–84.

Nakahara, Y., Sawabe, S., Kainuma, K., Katsuhara, M., Shibasaka, M., Suzuki, M., Sakamoto, H. (2015). Yeast functional screen to identify genes conferring salt stress tolerance in Salicornia europaea. *Frontiers in Plant Science* 6: 920.

Nascimento, F. X., Brígido, C., Glick, B. R., & Rossi, M. J. (2016). The role of rhizobial ACC deaminase in the nodulation process of leguminous plants. *International Journal of Agronomy*, 2016, 1–9.

Nieves-Cordones, M., Mohamed, S., Tanoi, K., Kobayashi, N. I., Takagi, K., Vernet, A., Guiderdoni, E., Périn, C., Sentenac, H., Véry, A.-A. (2017). Production of low-Cs+ rice plants by inactivation of the K+ transporter OsHAK1 with the CRISPR-Cas system. *Plant Journal* 92(1): 43–56. doi:10.1111/tpj.13632.

Osakabe, Y., Watanabe, T., Sugano, S. S., Ueta, R., Ishihara, R., Shinozaki, K., Osakabe, K. (2016). Optimization of CRISPR/Cas9 genome editing to modify abiotic stress responses in plants. *Scientific Reports* 6: 26685.

Park, J. J., Dempewolf, E., Zhang, W., Wang, Z. Y. (2017). RNA-guided transcriptional activation via CRISPR/dCas9 mimics overexpression phenotypes in Arabidopsis. *PLOS ONE* 12(6): e0179410.

Rai, R. U. C. H. I., Singh, S. H. I. L. P. I., Yadav, S. H. I. V. A. M., Chatterjee, A. N. T. R. A., Rai, S. H. W. E. T. A., Shankar, A. L. K. A., Rai, L. C. (2018). *Impact of UV-B Radiation on Photosynthesis and Productivity of Crop. Environment and Photosynthesis: A Future Prospect*. Studium Press, New Delhi, 336–346.

Roca Paixão, J. F., Gillet, F. X., Ribeiro, T. P., Bournaud, C., Lourenço-Tessutti, I. T., Noriega, D. D., Melo, B. P., de Almeida-Engler, J., Grossi-De-Sa, M. F. (2019). Improved drought stress tolerance in Arabidopsis by CRISPR/dCas9 fusion with a histone acetyl transferase. *Scientific Reports* 9(1): 8080.

Rocio, E.-B., Klinkhamer Peter, G. L., & Leiss Kirsten, A. (2017). Interactive effects of UV-B light with abiotic factors on plant growth and chemistry, and their consequences for defense against arthropod herbivores. *Frontiers in Plant Science, 8.*

Rohilla, R., Singh, V. P., Singh, U. S., Singh, R. K., & Khush, G. S. (2000). Crop husbandry and environmental factors affecting aroma and other quality traits. *Aromatic Rices*, 201–216.

Sandeep, K., Ann, R. L., Hiroshi, E., Vladimir, N. (2021). CRISPR-Cas in agriculture: Opportunities and challenges. *Frontiers in Plant Science* 12, 672329 doi:10.3389/fpls.2021.672329.

Sengupta, S., & Majumder, A. L. (2009). Insight into the salt tolerance factors of a wild halophytic rice, Porteresia coarctata: A physiological and proteomic approach. *Planta, 229*(4), 911–929.

Shen, C., Que, Z., Xia, Y., Tang, N., Li, D., He, R., Cao, M. (2017). Knock out of the annexin gene OsAnn3 via CRISPR/Cas9-mediated genome editing decreased cold tolerance in rice. *Journal of Plant Biology* 60(6): 539–547. doi:10.1007/s12374-016-0400-1.

Shi, J., Gao, H., Wang, H., Lafitte, H. R., Archibald, R. L., Yang, M., Hakimi, S. M., Mo, H., Habben, J. E. (2017). ARGOS8 variants generated by CRISPR-Cas9 improve maize grain yield under field drought stress conditions. *Plant Biotechnology Journal* 15(2): 207–216. doi:10.1111/pbi.12603.

Shim, J. S., Oh, N., Chung, P. J., Kim, Y. S., Choi, Y. D., Kim, J.- K. (2018). Overexpression of osnac14 improves drought tolerance in rice. *Frontiers in Plant Science* 9: 310. doi:10.3389/fpls.2018.00310.

Siddiqui, Z. S., Shaukat, S. S., ZAMAN, A. U. (2006). Alleviation of salinity-induced dormancy by growth regulators in wheat seeds. *Turkish Journal of Botany* 30(5): 321–330.

Song, Y., Lv, J., Ma, Z., Dong, W. (2019). The mechanism of alfalfa (*Medicago sativa* L.) response to abiotic stress. *Plant Growth Regulation* 89(3): 239–249.

Sundgren, T. K., Uhlen, A. K., Lilemmo, M., Briese, C., & Wojciechowski, T. (2018). Rapid seedling establishment and a narrow root stele promotes waterlogging tolerance in spring wheat. *Journal of Plant Physiology.* http://doi.org/10.1016/j.jplph.2018.04.010

Sunkar, P., Bartels, D., & Kirch, H.-H. (2003). Overexpression of a stress-inducible aldehyde dehydrogenase gene from Arabidopsis thaliana in transgenic plants improves stress tolerance. *Plant Journal, 35*(4), 452–464.

Tamang, S., Paudel, K. P., & Shrestha, K. K. (2014). Feminization of agriculture and its implications for food security in rural Nepal. *Journal of Forest and Livelihood, 12*(1), 20–32.

Tammam, A. M., El-Ashmoony, M. S. F., El-Sherbeny, A. A., & Amin, I. A. (2004b). Breeding for drought tolerance and the association of grain yield and other traits of bread wheat. *Egyptian Journal of Agricultural Research*, 82(3), 1227–1241.

Tang, L., Mao, B., Li, Y., Lv, Q., Zhang, L., Chen, C., He, H., Wang, W., Zeng, X., Shao, Y., et al. (2017). Knockout of OsNramp5 using the CRISPR/Cas9 system produces low Cd-accumulating indica rice without compromising yield. *Scientific Reports* 7(1): 14438. doi:10.1038/ s41598-017-14832-9.

Tang, N., Ma, S., Zong, W., Yang, N., Lv, Y., Yan, C., Guo, Z., Li, J., Li, X., Xiang, Y., et al. (2016). MODD mediates deactivation and degradation of OsbZIP46 to negatively regulate ABA signaling and drought resistance in rice. *The Plant Cell* 28(9): 2161–2177. doi:10.1105/tpc.16.00171.

Tianya, W., Hongwei, X., Wei, W., Xiaoyang, D., Hainan, T., Saddam, H., Qianli, D., Yingying, L., Yuxin, C., Chen, W. et al. (2021). Mutation of GmAITR genes by CRISPR/Cas9 genome editing results in enhanced salinity stress tolerance in soybean. *Frontiers in Plant Science 12*, 2752 doi:10.3389/fpls.2021.779598.

Vinocur, B., Altman, A. (2005). Recent advances in engineering plant tolerance to abiotic stress: Achievements and limitations. *Current Opinion in Biotechnology* 16(2): 123–132.

Voesenek, L. A. C. J., & Bailey-Serres, J. (2013). Flooding tolerance: O_2 sensing and survival strategies. *Current Opinion in Plant Biology, 16*(5), 1–7.

Wada, N., Ueta, R., Osakabe, Y., Osakabe, K. (2020). Precision genome editing in plants: State-of-the-art in CRISPR/Cas9-based genome engineering. *BMC Plant Biology* 20(1): 1–12.

Wang, F.-Z., Chen, M.-X., Yu, L.-J., Xie, L.-J., Yuan, L.-B., Qi, H., Xiao, M., Guo, W., Chen, Z., Yi, K., et al. (2017a). Osarm1, an R2R3 MYB transcription factor, is involved in regulation of the response to arsenic stress in rice. *Frontiers in Plant Science* 8: 1868. doi:10.3389/fpls.2017.01868.

Wang, L., Chen, L., Li, R., Zhao, R., Yang, M., Sheng, J., Shen, L. (2017b). Reduced drought tolerance by CRISPR/ Cas9-mediated SlMAPK3 mutagenesis in tomato plants. *Journal of Agricultural and Food Chemistry* 65(39): 8674–8682. doi:10.1021/acs.jafc.7b02745.

Wani, S. H., Sah, S. K., Hussain, M. A., Kumar, V., & Balachandra, S. M. (2016). Transgenic approaches for abiotic stress tolerance in crop plants. In J. M. Al-Khayri, S. M. Jain & D. V. Johnson (Eds.), *Advances in plant breeding strategies, Vol. 2 agronomic, abiotic and biotic stress traits* (pp. 345–396). Gewerbestrasse: Springer international.

Wei, W., Liang, D. W., Bian, X. H., Shen, M., Xiao, J. H., Zhang, W. K., Ma, B., Lin, Q., Lv, J., Chen, X., Chen, S. Y., Zhang, J. S. (2019). GmWRKY54 improves drought tolerance through activating genes in abscisic acid and Ca^{2+} signaling pathways in transgenic soybean. *Plant Journal* 100(2): 384–398. doi:10.1111/tpj.14449.

Willick, I. R., Takahashi, D., Fowler, D. B., Uemura, M., & Tanino, K. K. (2018, February 20). Tissue-specific changes in apoplastic proteins and cell wall structure during cold acclimation of winter wheat crowns. *Journal of Experimental Botany, 69*(5), 1221–1234.

Xu, C. G., Tang, T. X., Chen, R., Liang, C. H., Liu, X. Y., Wu, C. L., ... Wu, H. (2014). A comparative study of bioactive secondary metabolite production in diploid and tetraploid *Echinacea purpurea* (L.) Moench. *Plant Cell, Tissue and Organ Culture, 116*(3), 323–332.

Xu, K., Xu, X., Fukao, T., Canlas, P., Maghirang-Rodriguez, R., Heuer, S., ... Mackill, D. J. (2006). Sub1A is an ethylene-response-factor-like gene that confers submergence tolerance to rice. *Nature, 442*(7103), 705–708.

Yadav, S. K. (2010). Heavy metals toxicity in plants: An overview on the role of glutathione and phytochelatins in heavy metal stress tolerance of plants. *South African Journal of Botany* 76(2): 167–179.

Ye, Y., Li, P., Xu, T., Zeng, L., Cheng, D., Yang, M., Luo, J., Lian, X. (2017). Ospt4 contributes to arsenate uptake and transport in rice. *Frontiers in Plant Science* 8: 2197. doi:10.3389/fpls.2017.02197.

Zeng, D.-D., Yang, C.-C., Qin, R., Alamin, M., Yue, E.-K., Jin, X.-L., Shi, C.-H. (2018). A guanine insert in OsBBS1 leads to early leaf senescence and salt stress sensitivity in rice (Oryza sativa L.). *Plant Cell Reports* 37(6): 933–946. doi:10.1007/s00299-018-2280-y.

Zeng, D., Li, X., Huang, J., Li, Y., Cai, S., Weizhi, Y., ... Tan, J. (2019). Engineered Cas9 variant tools expand targeting scope of genome and base editing in rice. *Plant Biotechnology Journal*, 1–3.

Zhang, A., Liu, Y., Wang, F., Li, T., Chen, Z., Kong, D., Bi, J., Zhang, F., Luo, X., Wang, J., Tang, J., Yu, X., Liu, G., Luo, L. (2019). Enhanced rice salinity tolerance via CRISPR/Cas9-targeted mutagenesis of the OsRR22 gene. *Molecular Breeding* 39: 47.

Zhang, Z., Li, J., Pan, Y., Li, J., Zhou, L., Shi, H., Zeng, Y., Guo, H., Yang, S., Zheng, W., et al. (2017). Natural variation in CTB4a enhances rice adaptation to cold habitats. *Nature Communications* 8: 14788. doi:10.1038/ ncomms14788.

Zhang, H., Mittal, N., Leamy, L. J., Barazani, O., & Song, B. H. (2016). Back into the wild-apply untapped genetic diversity of wild relatives for crop improvement. *Evolutionary Applications*, 10(1), 5–24.

Zhao, Y., Zhang, C., Liu, W., Gao, W., Liu, C., Song, G., Li, W. X., Mao, L., Chen, B., Xu, Y., Li, X., Xie, C. (2016). An alternative strategy for targeted gene replacement in plants using a dual-sgRNA/Cas9 design. *Scientific Reports* 6: 23890.

Zhou, R., Xiaqing, Yu, Ottosen, C.-O., Rosenqvist, E., Zhao, L., Wang, Y., Wengui, Y., Tongmin, Z., Wu, Z. (2017). Drought stress had a predominant effect over heat stress on three tomato cultivars subjected to combined stress. *BMC Plant Biology* 17: 1–13.

8 Genomics-Assisted Use of Genetic Resources for Environmentally Adaptive Plant Breeding
Salinity Tolerance

Murat Aycan, Marouane Baslam,
Toshiaki Mitsui, and Mustafa Yildiz

CONTENTS

DOI: 10.1201/9781003250845-8

8.1   INTRODUCTION

From the time that humans started making tools, farming had been the main livelihood of families, societies, and countries until the Industrial Revolution. The cereals we consume today have undergone a selection process over hundreds of years, from their wild-type forms to their current edible forms. Moreover, since this selection process is not carried out in one center, but in different centers in different climates and geographies, today's varieties have gained different characteristics. The invention of steam machines, the use of technology in agricultural areas, and the development of transportation are some of the most essential factors that have accelerated agricultural production (Jordan, 2016). In the past decades, the increase in the world's population and the decrease in agricultural areas due to industrialization have pushed breeders to develop more productive varieties. However, in the last 20 years, the increase in environmental problems, together with global climate change, has led to the formation of a center for breeding programs to develop high-yield new genotypes that are tolerant to abiotic stress factors (Gilliham et al., 2017).

Two of the most critical problems of the twenty-first century are climate change and food security. The world's population is expected to reach 9 billion by the end of 2050, and the demand for food is projected to increase by 85% (FAO, 2017). When these expectations are evaluated together with the decrease in agricultural areas, serious food problems are expected in the future. In addition, stresses such as increased drought, excessive precipitation, high or low temperatures, and salinity due to climate change threaten crop production to a great extent, and constitute an essential obstacle to meeting the food demands that will arise in the future (Dhankher & Foyer, 2018). One of the most permanent environmental stresses is salinity. Salinity, which is a permanent stress factor in the soil, poses a severe problem in approximately 33% of coastal lands, especially with the rise in sea level over the past 25 years (Rahman et al., 2018). In addition, salinity appears more prominently in semi-arid and arid regions. It has been reported that increasing temperatures and decreasing precipitation due to climate change have been positively correlated with salinity in arid lands in the past 30 years. By 2050, it is expected that approximately 50% of total arable agricultural lands will experience high salinity problems (Bannari & Al-Ali, 2020).

Salinity is a significant problem that needs to be addressed and prioritized, and new agricultural strategies should be developed in saline areas. An annual economic loss of US$27.3 billion is due to the decrease in plant yield and the increase in salinity (Qadir et al., 2014). It is crucial to determine the salinity tolerance capacities of crops and the salinity tolerance mechanisms of plants and develop new varieties with high salt stress tolerance capacity to ensure food security and global food equality. To develop new high salt–tolerant genotypes, knowledge of the current traits of the crops cultivated is essential so that parents can be chosen in breeding programs. Currently, new technologies that allow the selection of individuals to be used as parents and the rapid selection of offspring at breeding stages are being used in breeding programs that are aimed at developing new salt-tolerant varieties. The development of sequencing technology appears as markers based on identifying associated morphological features that reference nucleic acid sequence differences in the genome. The development and widespread use of this technology are some of the essential steps that will affect the success and speed of breeding studies in the future. In this chapter, we evaluate the genome-assisted genetic resources used in salt stress tolerance breeding studies.

8.1.1 Soil Salinity: Serious Threats to Cereal Crops Production and Global Food Security

The productivity of crops is determined by their ability to set high-quality seeds successfully. Environmental stress factors affect the ability of cereals to reach their yield potential and seed quality. The sensitivity of cereals to increased stress factors is measured by a reduction in their yield potential (Mueller et al., 2012). Salinity is a significant abiotic stress that severely affects crop production worldwide (Kumar et al., 2007). Out of the total cultivated and irrigated agricultural land, 50% is affected by high salinity at a global level (Martinez Beltran & Licona Manzur, 2005). A total of >831 Mha of agricultural land is affected by high salinity worldwide. Salinity affects 397 Mha, while sodicity affects up to 434 Mha of lands (FAO, 2008). Saltwater irrigation with low precipitation and high evaporation are the main factors that cause 10% of the annual salinization in agricultural lands. If these practices continue, >50% of usable land will be saline by 2050 (Jamil et al., 2011). At the continental level, the largest area of salt-affected soils with approximately 7.14 Mkm^2 is Asia (including the Middle East), followed by Africa, Australia and Oceania, South America, North America, and Europe (Figure 8.1A). In terms of the area of salt-affected lands, the top-ranking countries are China (211.7 Mha), Australia (131.4 Mha), Kazakhstan (93.3 Mha), Iran (88.3 Mha), and Saudi Arabia (68.2 Mha) (Figure 8.1B) (Hassani et al., 2020).

Saline soil is characterized by the accumulation of soluble salts in the root zone of plants. On the other hand, the main feature of sodic soils is the accumulation of high levels of sodium salts compared to other exchangeable cations. The reason behind soil salinization is classified as "primary" naturally sourced, such as rainfall, wind, and parent rock weathering, and "secondary" human-induced soil salinization (Hassani et al., 2020). Soil salinity is classified as non-saline (<2 ds/m), slight (2–4 ds/m), moderate (4–8 ds/m), high (8–16 ds/m), and extreme (>16 ds/m) based on electrical conductivity extract (ECe) values (the ability of water-saturated earth paste extract to conduct electric current) (Ivushkin et al., 2019). Based on the last data published by Hassani et al. (2020), the highest rate of annual increase in soil salinization was detected in Brazil, Peru, Sudan, Colombia, and Namibia (Figure 8.1C). Moreover, the highest rate of annual increase in soil sodicity was detected in Iran, Saudi Arabia, Argentina, Afghanistan, and the United States (Figure 8.1D).

The cereal species provide 90% of food demands; most of them show serious yield losses starting from medium salinity; this EC >4 dS/m; thereby ca. 40 mM NaCl and osmotic pressure of ca. 0.2 MPa, reducing crops yields (Hanin et al., 2016). Cotton, barley, and sugar beet are highly salt-tolerant species, while sweet potato, wheat, rice, and corn are highly sensitive to salt stress, with yields decreasing if soil salinity exceeds species-specific thresholds (Zörb et al., 2019). In the presence of 100 mM NaCl (10 dS/m), a decrease in crop yield is observed in wheat, while

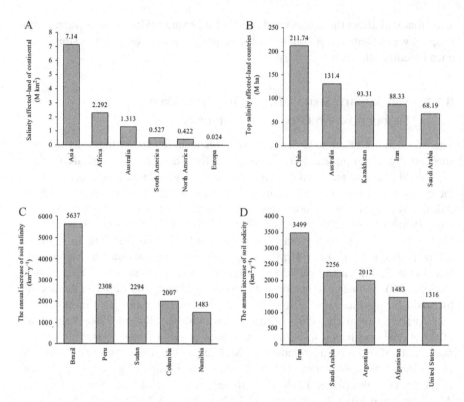

FIGURE 8.1 Soil salinity and sodicity conditions across the world. (A) Salinity-affected land by continent. (B) Top salinity-affected countries. (C) Annual increase in soil salinity levels. (D) Annual increase in soil sodicity levels.

rice (*Oryza sativa*) cannot survive when salt stress reaches above 75 mM NaCl in field conditions. Barley (*Hordeum vulgare*), the highest salt-tolerant crop plant, can tolerate around 250 mM NaCl. Other cereal crops such as durum wheat (*Triticum turgidum* ssp.), sorghum (*Sorghum bicolor*), and maize (*Zea mays*) are less tolerant to salinity stress (Aycan et al., 2021; Kim et al., 2021; Rana et al., 2019).

Yield-related traits change inversely with abiotic stress conditions such as salt or drought during plant growth. If salt stress reduces growth early in plant development, yields are severely reduced, and crops are compromised in terms of quality and quantity. Even though the effects of salinity are not observed on the plant, salinity can cause crop yield losses. Whether a particular crop is resistant or sensitive to salt stress depends on its ability to absorb water and nutrients from saline soils and prevent the accumulation of excess salt ions in tissues (Ahmad et al., 2017; Kaleem et al., 2018; Zörb et al., 2019).

8.1.2 Assessing Salinity Tolerance in the Current Cereal Crops

The susceptibility of plants to salt stress changes at various stages of plant growth. In general, previous studies have shown that plants at the germination and seedling stages are at their highest sensitive periods to salt stress than plants at the maturity stage. Tolerance is determined by the survival percentage at germination and sprouting, while at later developmental stages, tolerance is usually determined by relative growth (Läuchli & Grattan, 2007). Salt accumulation in the soil reduces water availability and increases water stress, reflected by the leaf water potential. This is the most crucial factor leading to osmotic stress, ion toxicity, nutrient imbalances, and water deficiency. The accumulation of excessive salt ions damages photosynthetically active leaves and can cause chlorosis and premature leaf senescence (Hanin et al., 2016). Due to the osmotic effect of salt ions, it is assumed that the most vigorous response occurs immediately after plants are exposed to salinity, with a decrease in stomatal opening. Salinity affects the stomatal structure and causes a decrease in stomata conductivity, size, and density. These lead to reductions in transpiration and photosynthesis rates. Previous studies linked a significant reduction in photosynthesis with a decrease in chlorophyll content in many crops exposed to salt stress (Zhang et al., 2014). Plants also use reactive oxygen species (ROS) molecules at different growth stages, such as development, and stress response signaling molecules (Aycan et al., 2021). All these physiological changes reduce plant height, root length, and biomass, resulting in plant yield reduction.

To assess the salinity tolerance of cereals, breeders utilize these physio-morphological characteristics to select suitable parents for breeding programs. Over two decades, the number of screening studies in several cereal plants has increased worldwide. Belkhodja et al. (1994) suggest that chlorophyll fluorescence can be used as a marker for salinity tolerance in barley genotypes. Also, Chen et al. (2005) screened barley genotypes with the K^+ flux method and found that 80 mM NaCl was the threshold for salinity tolerance in barley. Using osmoprotectants and antioxidative enzyme activities as screening tools for salinity tolerance in radish, Sanoubar et al. (2020) found that malondialdehyde (MDA), hydrogen peroxidase (H_2O_2), and ascorbate peroxidase (APX) content was not affected by salinity, while proline was significantly affected according to the results of the experiment. The accumulation of superoxide dismutase (SOD) and glutathione reductase (GR) activity is associated with salinity tolerance. Arteaga et al. (2020) used proline and water content to screen drought and salinity-tolerant common bean (*Phaseolus vulgaris* L.) genotypes. Arifuddin et al. (2021) screened five rice cultivars using a deep flow technique (DPF) for salinity tolerance at 0, 60, and 120 mM NaCl stress, with the 60 mM NaCl level recommended for the screening and selection of salt-tolerant rice genotypes. Tao et al. (2021) evaluated and screened the salinity tolerance of wheat cultivars at 150 mM NaCl through agrophysiological parameters at the seedling stages. They found that tolerant cultivars had higher water content and shoot dry weight, and lower stomatal density and leaf sap osmolality. The results suggested that the leaf K^+/Na^+ ratio and the stomatal density can be used as reliable screening indices for salt tolerance in wheat at the seedling stage. A similar study was checked by Yildiz et al. (2018) with 19 wheat cultivars under 150 mM NaCl stress conditions and they determined salt-tolerant and sensitive genotypes based on morpho-physiological analysis.

Although the selection process of salinity-tolerant genotypes may vary, researchers mainly use agronomic, physiological, and osmoprotectant content and antioxidant stress enzyme activities (Arifuddin et al., 2021; Tao et al., 2021). Despite extensive efforts, little success has been achieved to date in breeding genotypes tolerant of saline conditions, probably due to the salinity related lack of detailed indices on morphological, physiological, and agronomic characteristics and to the low genetic diversity of the plant germplasm (Ismail & Horie, 2017; Miransari & Smith, 2019).

8.1.3 ADVANCES AND CHALLENGES IN DEVELOPING SALT-TOLERANT CEREAL CROPS

The basic mechanisms of stress tolerance in plants are not fully understood, and improving crop performance under stressful conditions is still mainly at the research level. A prerequisite for the development of salt-tolerant crops is the identification of critical genetic determinants and the elucidation of the mechanism involved in stress tolerance (Yamaguchi & Blumwald, 2005). Additionally, improving stress tolerance with a less negative impact on yield and quality has been a significant challenge for growers. Genetics research into underlying salt tolerance and the use of biotechnology to produce salt-tolerant plants would accelerate salt-tolerant breeding varieties.

Salinity stress induces metabolite changes, and various physiological mechanisms contribute to the ability of plants to cope with salt excess. Studies revealed that Na^+/K^+ ratio, proline, hydrogen peroxide, peroxidase activity, sugars, etc., under salt stress are affected (Qin et al., 2020). The fact that the degree of susceptibility or tolerance to salt stress varies in different developmental stages indicates that salt tolerance has very complex genetic and physiological characteristics (Nam et al., 2015). Throughout the life cycle, salt tolerance in plants is a total reflection of the salt tolerance at each developmental stage of producing seeds and in the seedling stage. Determining salt-tolerant genotypes at the field level is more complex than laboratory and greenhouse studies.

For this reason, scanning under laboratory conditions is considered advantageous compared to field scanning. Since salt types in saline–alkaline fields are heavy salts, the salt tolerance determined in the laboratory may not always reflect the salinity in the natural environment. Therefore, the most reliable way to assess a plant's salt tolerance is by applying salt stress to laboratory and field conditions and comparing changes in the morphological and physiological parameters at various developmental stages under normal conditions. However, biochemical methods lack specific evaluation standards and require relevant instruments, knowledge, and kits to measure these indices, which are relatively laborious to work with (Qin et al., 2020).

Currently, two genetic approaches are used to improve salt stress tolerance: the mapping of quantitative trait loci (QTL) together with the use of natural genetic variations by natural selection under stress conditions followed by marker-assisted selection (MAS) and the identification of new genes, and the use of clustered, regularly interspaced short palindromic repeats (CRISPR/Cas 9) technology to provide salt tolerance through genome editing.

8.2 GENETIC RESOURCES FOR CEREALS IMPROVEMENTS: ROAD TOWARD DESIGNING OF SOPHISTICATED AND ENVIRONMENTALLY ADAPTIVE PLANT BREEDING

The identification of genetic resources is the basis of breeding programs. The variability of sufficient genetic sources is the main factor in the success of any crop breeding studies, but these genetic sources must be in usable form and preserved in suitable conditions in gene banks for present and future studies (Sharma et al., 2013). In the past century, breeders have been making plans for improving yield and quality by conventional breeding; and over the last two decades, one of the hottest research topics has been the development of stress-tolerant plants. Genetic identification and selection are needed for developing stress-tolerant plants, and are essential for parental selection in breeding programs.

The determination of currently cultivated or seed collections to salinity tolerance levels using agronomic characteristics is the classical way for agronomists. However, this method needs a long time for genotypes response identification, and it often depends on the geography and climate. Following the developments in molecular biology and genomics, classical methods were more accurate, usable, and independent of environmental factors. New methods were used for parental selection and helped breeders in offspring selection; in this way, they developed new and more accurate stress-tolerant genotypes. Next, we describe conventional and molecular breeding similarities and differences from the molecular perspective with example studies.

8.2.1 CONVENTIONAL BREEDING

Although significant progress has been made in the past century to improve cereals yield and quality by conventional or classical breeding methods, few studies have focused on improving crops' resilience to abiotic stresses, particularly salinity stress. Over the past century, plant breeders have developed several breeding programs in which they exploit the genetic variation of crops at the intraspecific, interspecies, and intergeneric levels to produce salt-tolerant varieties (Ashraf & Akram, 2009). As a result, several salt-resilient lines/varieties of crops by conventional breeding were tested in natural field conditions (Ansari et al., 2019).

One of the main problems facing conventional plant breeders is the low variation of genetically based variations in the gene pools of most crop species. Therefore, significant improvements in the salt tolerance of crops cannot be expected. Under such conditions, it is advisable to use the salt-tolerant wild relatives of crop plants as a gene source for crop improvement to increase salt tolerance (Ashraf & Akram, 2009). However, the transfer of salt-tolerant genes from wild relatives to domesticated crops is not easy because of reproductive barriers. The approach is time-consuming and labor-intensive; extra genes are often transferred in combination with desirable ones; and reproductive barriers limit the transfer of suitable alleles from interspecies sources. For these reasons, molecular breeding as an alternative strategy to conventional breeding is employed worldwide.

8.2.2 MOLECULAR BREEDING

The selection of highly salt-resistant genotypes in field conditions could be inhibited by the significant effects of environmental factors such as temperature, pH, and light. Additionally, the genetic factors controlling physiological responses could also have salinity responses (Johnson & Lenhard, 2011). With the development of new molecular methods, an acceleration in the development of DNA markers that can be used to identify QTLs has been observed. QTLs are known to increase selection efficiency, especially for traits controlled by many genes and affected by environmental factors (Flowers, 2004). Molecular breeding offers rapid and targeted selection in enhancing cultivar development for salt stress resistance.

Researchers initially used DNA markers to determine the salinity tolerance in plant genomes. However, developing technologies have provided great opportunities to screen the whole genome and map QTLs in main crop genomes, including rice, maize, wheat, and barley. Screening research is most abundant in the literature for identifying the localization and characterization of markers in the genome and linked traits. Breeders started to develop new tolerant and/or high-yield genotypes using those QTLs for tolerance. Breeding methods were also affected by new technologies and techniques.

8.2.3 BASIC GENOME SCREENING TECHNOLOGIES

In contrast to direct phenotypic screening, marker-assisted selection provides huge advantages. Polymerase chain reaction (PCR)-based methods are used to determine markers, shortening the time required to screen individuals and reducing the impact of environmental influences on traits (Yamaguchi & Blumwald, 2005). The development of high-density DNA maps containing microsatellite markers, simple-sequence repeats (SSR), restriction fragment length polymorphisms (RFLP), and amplified fragment length polymorphisms (AFLP) advances in MAS techniques make easy, relevant pyramid features to achieve significant improvement in plant salt tolerance (Figure 8.2).

Several studies were carried out to determine the QTLs and marker regions on the genome associated with salinity tolerance in many crop species. Identified marker–trait association (MTA) candidate genes underlying germination for salinity stress in barley (*H. vulgare* L.) using approximately 24,000 genetic markers and traits related to seed germination under stressful and control conditions in 350 different barley cultivars. Hussain et al. (2017) genotyped 154 wheat lines by QTL mapping and detected 988 single nucleotide polymorphisms (SNPs) related to salinity tolerance and mineral concentration. A total of 1293 distinctive SNPs localized in genes related to salt stress tolerance, various ion channels, signaling pathways, transcription factors (TFs), and metabolic pathways were identified, in which 258 were differentially expressed under salinity. Furthermore, marker-based screening studies are frequently carried out to describe the current condition of cultivars and the usability of markers. Eighteen wheat genotypes were screened and classified by phenotypic salinity response and then screened by several SSR markers, and 23 SSR markers were found highly useful in detecting tolerant and sensitive genotypes (Al-Ashkar et al., 2020). Tariq et al. (2019) screened 63 rice genotypes under salt stress conditions and evaluated their salinity tolerance at the molecular level by using 21 SSR markers, and they detected salt-tolerant and sensitive genotypes among 63 rice genotypes.

Using AFLP molecular markers for evaluating the association between markers and the investigated traits, Yazdizadeh et al. (2020) developed salt-tolerant genotypes of *Panicum miliaceum*. Using several SSR markers, Singh et al. (2018) enhanced the

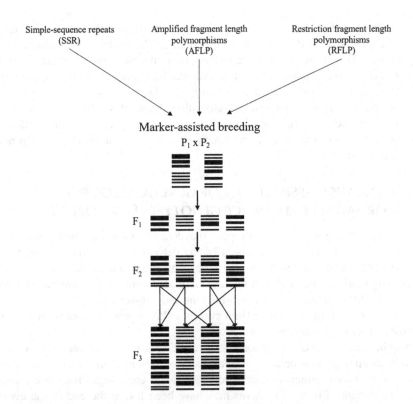

FIGURE 8.2   Marker-assisted breeding methods and schematic polymorphism view.

salt tolerance at the seedling stage in Pusa Basmati 1 (PB1) rice variety developed by marker-assisted transfer of a significant QTL, located on chromosome 1 in the rice genome Saltol, which is responsible for salinity tolerance at the seedling stage. Similarly, Krishnamurthy et al. (2020) used Saltol QTL to improve the salinity tolerance in rice cultivars using the marker-assisted backcross breeding (MABB) method. To obtain high-yield and salt-tolerant rice, Wu et al. (2020) produce crossed lines with Sea Rice 89 (SR89) and Dianjingyou 1 and checked the F_2 population with salt tolerance–related QTLs. They confirmed that the qST1.1 region was associated with salinity tolerance and high yield in rice.

8.2.4   ADVANCED GENOME SCREENING TECHNOLOGIES

A developing technology in science is next-generation sequencing (NGS), which has made DNA sequencing very cost- and time-effective (Barabaschi et al., 2015). NGS technology has opened up many opportunities to discover the connection between genetic and phenotypic variances. Reference genome sequences have been published on several plant species, including rice, maize, sorghum, barley, and wheat (Michael & Jackson, 2013). These reference sequences allow us to compare re-sequencing data to detect high-density SNPs in whole-genome scans and correlate them with quantitative trait variations.

The dissemination of low-cost sequencing technologies offers new opportunities to analyze genetic diversity in plant breeding. The analysis of whole-genome sequencing (WGS) instead of fragments has helped to understand the exact localization of SNPs on the genome. Thus, it has allowed us to understand the functions of genes in salinity tolerance mechanisms. For the marker-trait association in considerable genetic resources, genome-wide association studies (GWAS) have become an attractive approach for mapping QTLs. Moreover, genome sequences have allowed modification on specific genes using genome editing technologies to develop new salt-tolerant varieties.

8.3 GENOMICS-ASSISTED USE OF GENETIC RESOURCES FOR SALINITY-RESILIENCE GENOTYPES DEVELOPING

To develop salinity-resilient genotypes, the determination of genetic resources is vital for parental selection. Classically, the determination of the salinity tolerance response of genetic resources was done using agronomic characteristics. After developing markers and NGS technology, these agronomic traits have been related to genome differentiation among salt-tolerant and salt-sensitive genotypes. QTLs determined for salinity tolerance from current cultivar genomes can be applied in breeding programs to save time and costs.

Another advantage of the genome structure determination of genetic resources is understanding genetic heredity. The ancestry of various crops plants is currently living in wild environments under severely stressful conditions. Tolerance-related QTLs or mutations of wild relatives may have been lost at the end of the evolution progress. However, thanks to genomic-assisted technologies, there are ways of using or activating unused or lost genomic parts for salinity tolerance which could be employed in salinity-tolerance breeding programs. From these perspectives, several studies have been successfully completed and have developed salinity-resilient genotypes.

8.3.1 NEXT-GENERATION BREEDING OF SALT-RESILIENT CEREAL GERMPLASMS

The development of NGS technology opened up a new and precise way to understand genome structures, QTLs, and SNPs for salinity tolerance. Also, developing new methods such as GWAS based on NGS formed the foundation for next-generation breeding (Figure 8.3A). As the technology has recently been developed, studies are primarily carried out on the identification and mapping of salt tolerance regions across the genome in cereals (Nayyeripasand et al., 2021). Ren et al. (2005); after QTL mapping studies, SKC1 was revealed to maintain K^+ homeostasis against the salinity in the rice genome. Luo et al. (2021) found that QTLs for 15 agronomic traits were mapped in a newly constructed population using an SNP array under several levels of salt stress in a saline field. Moreover, eight QTLs were validated in a screened natural population related to salt stress in wheat. Chen et al. (2020) detected 23 QTLs for different salt tolerance, including the previously reported Saltol.

New genotypes resistant to salt stress have been developed using high-yielding, salinity-tolerant QTLs and SNPs (Atieno et al., 2021; Rana et al., 2019). Many

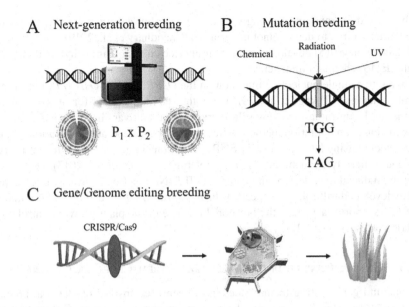

A Next-generation breeding

P_1 x P_2

B Mutation breeding

Chemical Radiation UV

TGG
↓
TAG

C Gene/Genome editing breeding

CRISPR/Cas9

FIGURE 8.3 Genomics-assisted use of genetic resources for developing salinity-resilience genotypes. (A) Next-generation breeding. (B) Mutation breeding. (C) Genome/gene-editing breeding approaches to develop salt-tolerant genotypes.

genomic markers can be used to identify plant genes by spatial cloning. Large-scale single nucleotide polymorphism has been used to screen thousands of cereal germplasms (Razzaq et al., 2021). Recently, Rana et al. (2019) transferred a single nucleotide mutation (*hst1*), which supports salinity tolerance (Takagi et al., 2015), into a high-yielding rice cultivar using the speed breeding method. NGS identified the transferred mutation on the new genome to detect SNPs. Aycan et al. (2021) crossed two salt-tolerant bread wheat cultivars and developed two new highly salt-tolerant bread wheat genotypes. After the F_2 generation genotypes have been screened under high salt stress conditions and a promising QTL marker, localized chromosome 2D was identified sequenced in both parents and offspring genomes (Aycan et al., 2021).

8.3.2 MUTATIONAL APPROACHES TO DEVELOP SALT-RESILIENT GENOTYPES

Mutational approaches are commonly used to develop salt-tolerant crops. Desiring some functional mutations on the genome for salinity tolerance using some mutagen agents (biological, chemical, and physical), more than 3200 mutant cultivars were produced worldwide (MDV, 2021). Mutagens provide better opportunities to obtain the desired physiological variation and can also be used to study genomic variation related to phenotypes as well as the expression of gene function (Chaudhary et al., 2019). Generating a vast number of mutants to identify subsequent phenotypic changes is relatively "easier" than identifying transient variation induced by mutagenesis (Figure 8.3B). Linkage analysis using molecular markers and separating the population developed from the mutant line and genotype with diverse genetic backgrounds has been used for random mapping variation. Even after genetic mapping, extensive efforts are required to detect random variation (Chaudhary et al., 2019).

Takagi et al. (2015) developed a salt-tolerant rice cultivar using ethyl methanesulfonate (EMS) chemical mutagens.

Owing to a single nucleotide mutation on the *OsRR22* gene (*OsHST1*), providing salinity tolerance (Takagi et al., 2015). Another technique used for making mutations is ion beam radiation. New salt-tolerant wheat cultivars have been developed using the ion beam mutation method. After obtaining M_6 generation, mutation points were detected using NGS to identify SNPs. Mutations in genes related to Na^+ transport could directly contribute to salinity tolerance (Xiong et al., 2017). Using the targeting induced local lesions in genomes (TILLING) method, Navarro-León et al. (2020) developed salt-tolerant rapeseed (*Brassica rapa*) influenced by an SNP in the CAX1a transporter affecting the hormonal balance of the plant growth control and ion homeostasis under salinity.

8.3.3 GENOME/GENE EDITING TO ACCELERATE CEREAL CROPS' SALT TOLERANCE

Genome and gene-editing technologies have been used instead of classical breeding techniques. The development of transgenic technology was a perfect tool to overcome salinity problems in plants (Yin & J. Zhang, 2010). The most significant advantage of transgenic technology was that only the target region could be modified. Although many salt stress–related genes have been successfully cloned (Huang et al., 2009), the inability of these studies to go beyond laboratory and/or greenhouse studies due to legal regulations is another difficulty in the use of transgenic technology (Zimny et al., 2019).

Recently, CRISPR-Cas 9 technology was developed as a faster, cheaper, more precise, and highly efficient editing genome technology at the multiplex level. Clustered, regularly interspaced short palindromic repeats has been effectively applied in several cereals over the last decades (Figure 8.3C). The CRISPR-related endonuclease Cas 9 targets the location of genes and site-specific nucleases (SSNs) used for genome editing in the subsequent repair of a DNA double-stranded break (DSB) for a cell's specific inherent repair mechanism (W. A. Ansari et al., 2020). This pathway-mediated homologous sequence exchange makes it a valuable tool for gene knock-in or gene replacement. Recently, CRISPR/Cas 9 has been considered the most effective and most accessible genome editing tool.

The CRISPR/Cas 9-mediated genome editing technology has been successfully applied in various cereals, including wheat, rice, maize, and cotton, for improving abiotic stress tolerance and yield (X. Zhang et al., 2012), drought tolerance (Shi et al., 2017), herbicide tolerance (Sun et al., 2017), disease resistance (Y. Zhang et al., 2017), and salinity tolerance (Yue et al., 2020). Plants response mechanisms to abiotic stress factors are highly complex and hundreds and/or thousands of genes are regulated up or down under stress conditions (Joshi et al., 2016). However, recently the AGROS8 gene was edited using the CRISPR/Cas 9 system to test the drought tolerance of maize and edited plants at the field level (Shi et al., 2017). This success was very motivating for breeders working for those who want to develop abiotic stress tolerance.

8.4 CONCLUDING REMARKS: WAY FORWARD AND CHALLENGES AHEAD

The explosion of the human population and the adverse environment due to climate changes are posing severe challenges to food security. Globally, soil salinity is one of the severely yield-reducing factors. To overcome the salinity problem, breeders have attempted to develop new salt-tolerant crops. Developing technologies have had a serious impact on breeding studies. The cheaper sequencing technique and the widespread use of next-generation sequencing technology in breeding studies have paved the way for more successful results in the development of new salt-tolerant genotypes. The determination of stress tolerance in current cultivars and the acceleration of intergenerational selection owing to reference genomes and QTL maps have shortened breeding times and resulted in salt-tolerant genotypes.

Next-generation CRISPR/Cas 9 systems offer growers unprecedented ease in editing plant genomes with greater precision. Genome editing under rapid breeding conditions has provided a controlled system to accelerate crop growing cycles with minimal cost and has been a technological breakthrough in obtaining non-transgenic plants without tissue culture. With multiscale knowledge, next-generation approaches to using artificial intelligence and data fusion to establish big data handling approaches are needed. Consequently, a robust and complex strategy to deal with these large data sets will also be required. Multidisciplinary integration, knowledge sharing, and the development of mega projects under a single umbrella are necessary to speedily resolve all current challenges. Well-integrated multidisciplinary approaches from physiology to genomics must be carried out to overcome current challenges and barriers in plant science. Using integrated next-generation approaches will undoubtedly help to achieve substantial success in ensuring future food security and meet the food demands of the rapidly growing population in the coming decades.

REFERENCES

Ahmad, R., Jamil, S., Shahzad, M., Zörb, C., Irshad, U., Khan, N., Younas, M., & Khan, S. A. (2017). Metabolic profiling to elucidate genetic elements due to salt stress. In *Clean - Soil, Air, Water* (Vol. 45, Issue 12). https://doi.org/10.1002/clen.201600574

Al-Ashkar, I., Alderfasi, A., Romdhane, W. ben, Seleiman, M. F., El-Said, R. A., & Al-Doss, A. (2020). Morphological and genetic diversity within salt tolerance detection in eighteen wheat genotypes. *Plants, 9*(3). https://doi.org/10.3390/plants9030287

Ansari, M., Shekari, F., Mohammadi, M. H., Juhos, K., Végvári, G., & Biró, B. (2019). Salt-tolerant plant growth-promoting bacteria enhanced salinity tolerance of salt-tolerant alfalfa (*Medicago sativa* L.) cultivars at high salinity. *Acta Physiologiae Plantarum, 41*(12). https://doi.org/10.1007/s11738-019-2988-5

Ansari, W. A., Chandanshive, S. U., Bhatt, V., Nadaf, A. B., Vats, S., Katara, J. L., Sonah, H., & Deshmukh, R. (2020). Genome editing in cereals: Approaches, applications and challenges. In *International Journal of Molecular Sciences* (Vol. 21, Issue 11, pp. 1–32). MDPI AG. https://doi.org/10.3390/ijms21114040

Arifuddin, M., Musa, Y., Farid, M., Anshori, M. F., Nasaruddin, N., Nur, A., & Sakinah, A. I. (2021). Rice screening with hydroponic deep-flow technique under salinity stress. *Sabrao Journal of Breeding and Genetics, 53*(3), 435–446.

Arteaga, S., Yabor, L., Díez, M. J., Prohens, J., Boscaiu, M., & Vicente, O. (2020). The use of proline in screening for tolerance to drought and salinity in common bean (*Phaseolus vulgaris* L.) genotypes. *Agronomy, 10*(6). https://doi.org/10.3390/agronomy10060817

Ashraf, M., & Akram, N. A. (2009). Improving salinity tolerance of plants through conventional breeding and genetic engineering: An analytical comparison. In *Biotechnology Advances* (Vol. 27, Issue 6, pp. 744–752). https://doi.org/10.1016/j.biotechadv.2009.05.026

Atieno, J., Colmer, T. D., Taylor, J., Li, Y., Quealy, J., Kotula, L., Nicol, D., Nguyen, D. T., Brien, C., Langridge, P., Croser, J., Hayes, J. E., & Sutton, T. (2021). Novel salinity tolerance loci in chickpea identified in glasshouse and field environments. *Frontiers in Plant Science, 12*. https://doi.org/10.3389/fpls.2021.667910

Aycan, M., Baslam, M., Asiloglu, R., Mitsui, T., & Yildiz, M. (2021). Development of new high-salt tolerant bread wheat (*Triticum aestivum* L.) genotypes and insight into the tolerance mechanisms. *Plant Physiology and Biochemistry, 166*, 314–327. https://doi.org/https://doi.org/10.1016/j.plaphy.2021.05.041

Bannari, A., & Al-Ali, Z. M. (2020). Assessing climate change impact on soil salinity dynamics between 1987-2017 in arid landscape using Landsat TM, ETM+ and OLI data. *Remote Sensing, 12*(17). https://doi.org/10.3390/RS12172794

Barabaschi, D., Tondelli, A., Desiderio, F., Volante, A., Vaccino, P., Valè, G., & Cattivelli, L. (2015). Next generation breeding. *Plant Science, 242*, 3–13. https://doi.org/10.1016/j.plantsci.2015.07.010

Belkhodja, R., Morales, F., Abadía, A., Gómez-Aparisi, J., & Abadía, J. (1994). Chlorophyll fluorescence as a possible tool for salinity tolerance screening in barley (*Hordeum vulgare* L.). *Plant Physiology, 104*(2), 667–673. https://doi.org/10.1104/pp.104.2.667

Chaudhary, J., Alisha, A., Bhatt, V., Chandanshive, S., Kumar, N., Sonah, H., Deshmukh, R., Mir, Z., Kumar, A., Yadav, S., & Shivaraj, S. M. S. (2019). Mutation breeding in tomato: Advances, applicability and challenges. *Plants, 8*(5). https://doi.org/10.3390/plants8050128

Chen, T., Zhu, Y., Chen, K., Shen, C., Zhao, X., Shabala, S., Shabala, L., Meinke, H., Venkataraman, G., Chen, Z. H., Xu, J., & Zhou, M. (2020). Identification of new QTL for salt tolerance from rice variety Pokkali. *Journal of Agronomy and Crop Science, 206*(2), 202–213. https://doi.org/10.1111/jac.12387

Chen, Z., Newman, I., Zhou, M., Mendham, N., Zhang, G., & Shabala, S. (2005). Screening plants for salt tolerance by measuring K+ flux: A case study for barley. *Plant, Cell and Environment, 28*(10), 1230–1246. https://doi.org/10.1111/j.1365-3040.2005.01364.x

Dhankher, O. P., & Foyer, C. H. (2018). Climate resilient crops for improving global food security and safety. In *Plant Cell and Environment* (Vol. 41, Issue 5). https://doi.org/10.1111/pce.13207

FAO. (2008). *FAO Land and Plant Nutrition Management Service*. http://www.fao.org/Ag/Agl/Agll/Spush.

FAO. (2017). *FAO, The Future of Food and Agriculture: Trends and Challenges*. In Food and Agriculture Organization of the United Nations. http://www.fao.org/3/a-i6583e.pdf (accessed on January 8, 2018).

Flowers, T. J. (2004). Improving crop salt tolerance. *Journal of Experimental Botany, 55*(396), 307–319.

Gilliham, M., Able, J. A., & Roy, S. J. (2017). Translating knowledge about abiotic stress tolerance to breeding programmes. *Plant Journal, 90*(5). https://doi.org/10.1111/tpj.13456

Hanin, M., Ebel, C., Ngom, M., Laplaze, L., & Masmoudi, K. (2016). New insights on plant salt tolerance mechanisms and their potential use for breeding. In *Frontiers in Plant Science* (Vol. 7, Issue November 2016). https://doi.org/10.3389/fpls.2016.01787

Hassani, A., Azapagic, A., & Shokri, N. (2020). Predicting long-term dynamics of soil salinity and sodicity on a global scale. *Proceedings of the National Academy of Sciences of the United States of America, 117*(52). https://doi.org/10.1073/PNAS.2013771117

Huang, X. Y., Chao, D. Y., Gao, J. P., Zhu, M. Z., Shi, M., & Lin, H. X. (2009). A previously unknown zinc finger protein, DST, regulates drought and salt tolerance in rice via stomatal aperture control. *Genes and Development, 23*(15). https://doi.org/10.1101/gad.1812409

Hussain, B., Lucas, S. J., Ozturk, L., & Budak, H. (2017). Mapping QTLs conferring salt tolerance and micronutrient concentrations at seedling stagein wheat. *Scientific Reports, 7*(1). https://doi.org/10.1038/s41598-017-15726-6

Ismail, A. M., & Horie, T. (2017). Genomics, physiology, and molecular breeding approaches for improving salt tolerance. *Annual Review of Plant Biology, 68*(1), 405–434. https://doi.org/10.1146/annurev-arplant-042916-040936

Ivushkin, K., Bartholomeus, H., Bregt, A. K., Pulatov, A., Kempen, B., & de Sousa, L. (2019). Global mapping of soil salinity change. *Remote Sensing of Environment, 231*. https://doi.org/10.1016/j.rse.2019.111260

Jamil, A., Riaz, S., Ashraf, M., & Foolad, M. R. (2011). Gene expression profiling of plants under salt stress. *Critical Reviews in Plant Sciences, 30*(5), 435–458. https://doi.org/10.1080/07352689.2011.605739

Johnson, K., & Lenhard, M. (2011). Genetic control of plant organ growth. In *New Phytologist* (Vol. 191, Issue 2). https://doi.org/10.1111/j.1469-8137.2011.03737.x

Jordan, C. F. (2016). The farm as a thermodynamic system: Implications of the maximum power principle. *BioPhysical Economics and Resource Quality, 1*(2). https://doi.org/10.1007/s41247-016-0010-z

Joshi, R., Wani, S. H., Singh, B., Bohra, A., Dar, Z. A., Lone, A. A., Pareek, A., & Singla-Pareek, S. L. (2016). Transcription factors and plants response to drought stress: Current understanding and future directions. *Frontiers in Plant Science, 7*(2016JULY). https://doi.org/10.3389/fpls.2016.01029

Kaleem, F., Shabir, G., Aslam, K., Rasul, S., Manzoor, H., Shah, S. M., & Khan, A. R. (2018). An overview of the genetics of plant response to salt stress: Present status and the way forward. In *Applied Biochemistry and Biotechnology* (Vol. 186, Issue 2). https://doi.org/10.1007/s12010-018-2738-y

Kim, S. H., Kim, D. Y., Yacoubi, I., & Seo, Y. W. (2021). Development of single-nucleotide polymorphism markers of salinity tolerance for *Tunisian durum* wheat using RNA sequencing. *Acta Agriculturae Scandinavica Section B: Soil and Plant Science, 71*(1). https://doi.org/10.1080/09064710.2020.1843701

Krishnamurthy, S. L., Pundir, P., Warraich, A. S., Rathor, S., Lokeshkumar, B. M., Singh, N. K., & Sharma, P. C. (2020). Introgressed Saltol QTL lines improves the salinity tolerance in rice at seedling stage. *Frontiers in Plant Science, 11.* https://doi.org/10.3389/fpls.2020.00833

Kumar, V., Shriram, V., Jawali, N., & Shitole, M. G. (2007). Differential response of indica rice genotypes to NaCl stress in relation to physiological and biochemical parameters. *Archives of Agronomy and Soil Science, 53*(5). https://doi.org/10.1080/03650340701576800

Läuchli, A., & Grattan, S. R. (2007). Plant growth and development under salinity stress. In *Advances in molecular breeding toward drought and salt tolerant crops* (pp. 1–32). Springer, Dordrecht.

Luo, Q., Zheng, Q., Hu, P., Liu, L., Yang, G., Li, H., Li, B., & Li, Z. (2021). Mapping QTL for agronomic traits under two levels of salt stress in a new constructed RIL wheat population. *Theoretical and Applied Genetics, 134*(1), 171–189. https://doi.org/10.1007/s00122-020-03689-8

Martinez-Beltran, J. (2005). Overview of salinity problems in the world and FAO strategies to address the problem. In *Managing saline soils and water: science, technology and social issues. Proceedings of the international salinity forum*, Riverside, California, 2005.

MDV. (2021). *The Joint FAO/IAEA Mutant Variety Database*. International Atomic Energy Agency (IAEA).

172 Cereal Crops

Michael, T. P., & Jackson, S. (2013). The first 50 plant genomes. *The Plant Genome, 6*(2). https://doi.org/10.3835/plantgenome2013.03.0001in

Miransari, M., & Smith, D. (2019). Sustainable wheat (*Triticum aestivum* L.) production in saline fields: A review. *Critical Reviews in Biotechnology, 39*(8), 999–1014. https://doi .org/10.1080/07388551.2019.1654973

Mueller, N. D., Gerber, J. S., Johnston, M., Ray, D. K., Ramankutty, N., & Foley, J. A. (2012). Closing yield gaps through nutrient and water management. *Nature, 490*(7419). https:// doi.org/10.1038/nature11420

Nam, M. H., Bang, E., Kwon, T. Y., Kim, Y., Kim, E. H., Cho, K., Park, W. J., Kim, B. G., & Yoon, I. S. (2015). Metabolite profiling of diverse rice germplasm and identification of conserved metabolic markers of rice roots in response to long-term mild salinity stress. *International Journal of Molecular Sciences, 16*(9). https://doi.org/10.3390/ ijms160921959

Navarro-León, E., López-Moreno, F. J., Atero-Calvo, S., Albacete, A., Ruiz, J. M., & Blasco, B. (2020). CAX1a tilling mutations modify the hormonal balance controlling growth and ion homeostasis in brassica rapa plants subjected to salinity. *Agronomy, 10*(11). https://doi.org/10.3390/agronomy10111699

Nayyeripasand, L., Garoosi, G. A., & Ahmadikhah, A. (2021). Genome-wide association study (GWAS) to identify salt-tolerance QTLs carrying novel candidate genes in rice during early vegetative stage. *Rice, 14*(1). https://doi.org/10.1186/s12284-020-00433-0

Qadir, M., Quillérou, E., Nangia, V., Murtaza, G., Singh, M., Thomas, R. J., Drechsel, P., & Noble, A. D. (2014). Economics of salt-induced land degradation and restoration. *Natural Resources Forum, 38*(4), 282–295. https://doi.org/10.1111/1477-8947.12054

Qin, H., Li, Y., & Huang, R. (2020). Advances and challenges in the breeding of salt-tolerant rice. In *International Journal of Molecular Sciences* (Vol. 21, Issue 21, pp. 1–15). MDPI AG. https://doi.org/10.3390/ijms21218385

Rahman, A., Ahmed, K., Butler, A., & Hoque, M. (2018). Influence of surface geology and micro-scale land use on the shallow subsurface salinity in deltaic coastal areas: A case from southwest Bangladesh. *Environmental Earth Sciences, 77*(12). https://doi.org/10 .1007/s12665-018-7594-0

Rana, M. M., Takamatsu, T., Baslam, M., Kaneko, K., Itoh, K., Harada, N., Sugiyama, T., Ohnishi, T., Kinoshita, T., Takagi, H., & Mitsui, T. (2019). Salt tolerance improvement in rice through efficient SNP marker-assisted selection coupled with speed-breeding. *International Journal of Molecular Sciences, 20*(10). https://doi.org/10.3390 /ijms20102585

Razzaq, A., Kaur, P., Akhter, N., Wani, S. H., & Saleem, F. (2021). Next-Generation Breeding Strategies for Climate-Ready Crops. In *Frontiers in Plant Science* (Vol. 12). https://doi .org/10.3389/fpls.2021.620420

Ren, Z.-H., Gao, J.-P., Li, L., Cai, X., Huang, W., Chao, D.-Y., Zhu, M., Wang, Z.-Y., Luan, S., & Lin, H. (2005). A rice quantitative trait locus for salt tolerance encodes a sodium transporter. *Nature Genetics, 37*(10), 1141–1146. https://doi.org/10.1038/ng1643

Sanoubar, R., Cellini, A., Gianfranco, G., & Spinelli, F. (2020). Osmoprotectants and antioxidative enzymes as screening tools for salinity tolerance in radish (*Raphanus sativus*). *Horticultural Plant Journal, 6*(1), 14–24. https://doi.org/10.1016/j.hpj.2019.09.001

Sharma, S., Upadhyaya, H. D., Varshney, R. K., & Gowda, C. L. L. (2013). Pre-breeding for diversification of primary gene pool and genetic enhancement of grain legumes. *Frontiers in Plant Science, 4*(AUG). https://doi.org/10.3389/fpls.2013.00309

Shi, J., Gao, H., Wang, H., Lafitte, H. R., Archibald, R. L., Yang, M., Hakimi, S. M., Mo, H., & Habben, J. E. (2017). ARGOS8 variants generated by CRISPR-Cas9 improve maize grain yield under field drought stress conditions. *Plant Biotechnology Journal, 15*(2). https://doi.org/10.1111/pbi.12603

Singh, V. K., Singh, B. D., Kumar, A., Maurya, S., Krishnan, S. G., Vinod, K. K., Singh, M. P., Ellur, R. K., Bhowmick, P. K., & Singh, A. K. (2018). Marker-assisted introgression of saltol QTL enhances seedling stage salt tolerance in the rice variety "pusa Basmati 1." *International Journal of Genomics, 2018.* https://doi.org/10.1155/2018/8319879

Sun, Y., Jiao, G., Liu, Z., Zhang, X., Li, J., Guo, X., Du, W., Du, J., Francis, F., Zhao, Y., & Xia, L. (2017). Generation of high-amylose rice through CRISPR/Cas9-mediated targeted mutagenesis of starch branching enzymes. *Frontiers in Plant Science, 8.* https://doi.org/10.3389/fpls.2017.00298

Takagi, H., Tamiru, M., Abe, A., Yoshida, K., Uemura, A., Yaegashi, H., Obara, T., Oikawa, K., Utsushi, H., Kanzaki, E., Mitsuoka, C., Natsume, S., Kosugi, S., Kanzaki, H., Matsumura, H., Urasaki, N., Kamoun, S., & Terauchi, R. (2015). MutMap accelerates breeding of a salt-tolerant rice cultivar. *Nature Biotechnology, 33*(5), 445–449. https://doi.org/10.1038/nbt.3188

Tao, R., Ding, J., Li, C., Zhu, X., Guo, W., & Zhu, M. (2021). Evaluating and screening of agro-physiological indices for salinity stress tolerance in wheat at the seedling stage. *Frontiers in Plant Science, 12.* https://doi.org/10.3389/fpls.2021.646175

Tariq, R., Ali, J., & Arif, M. (2019). Morphological screening and SalTol region based SSR markers analysis of rice (Oryza sativa) genotypes for salinity tolerance at seedling stage. *International Journal of Agriculture and Biology, 21*(1), 25–33. https://doi.org/10.17957/IJAB/15.0000

Wu, F., Yang, J., Yu, D., & Xu, P. (2020). Identification and validation a major QTL from "Sea Rice 86" seedlings conferred salt tolerance. *Agronomy, 10*(3). https://doi.org/10.3390/agronomy10030410

Xiong, H., Guo, H., Xie, Y., Zhao, L., Gu, J., Zhao, S., Li, J., & Liu, L. (2017). RNAseq analysis reveals pathways and candidate genes associated with salinity tolerance in a spaceflight-induced wheat mutant. *Scientific Reports, 7*(1). https://doi.org/10.1038/s41598-017-03024-0

Yamaguchi, T., & Blumwald, E. (2005). Developing salt-tolerant crop plants: Challenges and opportunities. In *Trends in Plant Science* (Vol. 10, Issue 12). https://doi.org/10.1016/j.tplants.2005.10.002

Yazdizadeh, M., Fahmideh, L., Mohammadi-Nejad, G., Solouki, M., & Nakhoda, B. (2020). Association analysis between agronomic traits and AFLP markers in a wide germplasm of proso millet (*Panicum miliaceum* L.) under normal and salinity stress conditions. *BMC Plant Biology, 20*(1). https://doi.org/10.1186/s12870-020-02639-2

Yildiz, M., Aycan, M., Özdemir, B., Darcin, E. S., & Koçak, E. Y. (2018). Morphological and physiological responses of commonly cultivated bread wheat (*Triticum aestivum* L.) cultivars in Turkey to salt stress. In R. Efe, M. Zincirkiran, & I. Curebal (Eds.), *Recent Researches in Science and Landscape Management* (pp. 539–553). Cambridge Scholars Publishing.

Yin, X., & J. Zhang. (2010). Recent patents on plant transgenic technology. *Recent Patents on Biotechnology, 4*(2). https://doi.org/10.2174/187220810791110688

Yue, E., Cao, H., & Liu, B. (2020). Osmir535, a potential genetic editing target for drought and salinity stress tolerance in Oryza sativa. In *Plants* (Vol. 9, Issue 10). https://doi.org/10.3390/plants9101337

Zhang, L., Ma, H., Chen, T., Pen, J., Yu, S., & Zhaoe, X. (2014). Morphological and physiological responses of cotton (*Gossypium hirsutum* l.) plants to salinity. *PLoS ONE, 9*(11). https://doi.org/10.1371/journal.pone.0112807

Zhang, X., Wang, J., Huang, J., Lan, H., Wang, C., Yin, C., Wu, Y., Tang, H., Qian, Q., Li, J., & Zhang, H. (2012). Rare allele of OsPPKL1 associated with grain length causes extra-large grain and a significant yield increase in rice. *Proceedings of the National Academy of Sciences of the United States of America, 109*(52). https://doi.org/10.1073/pnas.1219776110

Zhang, Y., Bai, Y., Wu, G., Zou, S., Chen, Y., Gao, C., & Tang, D. (2017). Simultaneous modi-
    fication of three homoeologs of TaEDR1 by genome editing enhances powdery mildew
    resistance in wheat. *Plant Journal*, *91*(4). https://doi.org/10.1111/tpj.13599
Zimny, T., Sowa, S., Tyczewska, A., & Twardowski, T. (2019). Certain new plant breeding
    techniques and their marketability in the context of EU GMO legislation – recent devel-
    opments. In *New Biotechnology* (Vol. 51, pp. 49–56). Elsevier B.V. https://doi.org/10
    .1016/j.nbt.2019.02.003
Zörb, C., Geilfus, C. M., & Dietz, K. J. (2019). Salinity and crop yield. *Plant Biology*, *21*
    Supplement 1, 31–38. https://doi.org/10.1111/plb.12884c

9 Metabolomics-Assisted Breeding for Enhancing Yield and Quality of Cereals

Qurat ul Ain Sani, Nosheen Fatima,
Qurat ul Ain Ali Hira, Rimsha Azhar,
Midhat Mahboob, Salman Nawaz
Faiza Munir, and Rabia Amir

CONTENTS

DOI: 10.1201/9781003250845-9

9.1 INTRODUCTION TO PLANT METABOLOMICS

The term "metabolome" was coined by the scientist Steven Oliver in his 1998 review article on yeast functional genomics in correlation with the term's genome, proteome, and transcriptome (Oliver et al. 1998). Various research papers were subsequently published at the end of the century, describing almost all of the prospects that metabolomics offered (Duenas et al. 2017). Metabolomics was considered an emerging field until 2010 and the youngest among the triad of systems biology (Alseekh and Fernie 2018).

Plants generate millions of metabolites at a time that are difficult to discriminate into a phenotype. Metabolomics aids in the equitable demarcation of the genotypes based on their metabolic profile (Sumner, Mendes, and Dixon 2003). Because the phenotype of an organism is linked to its metabolites, the identification of related metabolites aids in a better understanding of any genetic system (Rathahao-Paris et al. 2016). Few techniques used in metabolomics had been invented previously, and its significance was only recognized in the post-genomic era. Proteomics, genomics, and metabolomics have provided eminent tools for breeding plant crops (Saxena and Cramer 2013).

This newly established functional genomics technology has been of great value to scientists. Such technologies eventually aim to provide full coverage of the whole-genome sequence or the entire transcriptional data of an organism and are expanding the research horizons. The capability to produce data on the molecular organization of living beings has never been so important, specifically the data from multivariate and complementary sources. It has led to a paradigm shift in understanding and thinking about complicated biological questions. Contradictory expertise is being integrated to acquire new insights into cell behavior, the organization of a plant, and its response to any stress. DNA microarrays, proteomics, and other techniques have already broadened our knowledge. This shift from a reductionist to a universal approach has paved the way for new prospects. However, new challenges have arisen, mainly regarding the latest field of functional genomics, i.e., metabolomics where the particular focus is on the biochemical complement of cells and tissues (Hall 2006).

The technology is continuously evolving due to the developments in analytical chemistry, such as instrumentation, statistical methods, analytical software, analytical techniques, and computational methodologies to enhance data collection, analysis, and interpretation (Emwas et al. 2019). *Arabidopsis thaliana*, the model plant, has been comprehensively researched using surplus genomic tools and techniques, assisting functional genomics analysis. In the past few years, metabolomics studies have been extended in plants to discover their gene functions (Sharma et al. 2018). Metabolomics has also been utilized in the identification of metabolic biomarkers that help in predicting the prospects of a plant's performance (Meyer et al. 2007), such as the identification of a herbicide's mode of action by comparing herbicides of already known and unknown targets (Trenkamp et al. 2009), and in evaluating the differences between genetically modified (GM) plants and traditional cultivars (Catchpole et al. 2005).

Similar approaches intended to provide accurate explanations of any intervention; this approach also utilized the coupling of statistical methods to chemical analytics that were unique to many people examining plant metabolism. At the start of the metabolomics journey, both kinds of experiments were conducted at comparatively low output, with a single publication stating data sets having a sample size of less than 100. Among many, three studies are worth mentioning here. The first is the experiment of Fiehn and his co-workers who used gas chromatography integrated with mass spectrometry (GC-MS) for the evaluation of 326 analytes to compare a developmental variant (stomatal density and distribution, add) and the metabolic variant (digalactosyldiacylglycerol, dgd) of *A. thaliana* concerning their wild types. Expectedly, they concluded that the metabolic variant showed more differences than the developmental variant from their respective wild types (Fiehn et al. 2000).

Moreover, experiments conducted on GC-MS profiling of 27 unknown and 60 known metabolites from a variety of potato (*Solanum tuberosum*) lines expressed increased mucolytic activity in a tuber-specific manner in the wild-type tuber material provided with different exogenous concentrations of sugar. This genetically engineered material was formed to increase the tuber starch content; however, in reality, it gave rise to an opposite phenotype. Metabolic profiling demonstrated that this was the outcome of the upregulation of amino acid biosynthesis and respiration (Roessner et al. 2001).

The third study conducted was a liquid chromatography-mass spectrometry (LC-MS)-based metabolite profiling of *A. thaliana* which showed 2000 dissimilar mass signals from the leaves and roots, with many among them signifying specialized metabolites. This experiment utilized the chalcone synthase-deficient tt4 mutant to validate that subtle dissimilarities between the samples could be observed. However, demonstration of the resolving power and the sensitivity of this technique greatly broadened the array of metabolic profiling during this period. Overall, these studies greatly increased the understanding of metabolic reactions to relative disruption. Furthermore, they also provided a direct understanding of the behavior of metabolic pathways compared to previous experiments. In this vein, correlation analysis was considered to be an impactful technology for allocating metabolites to their respective biochemical pathway (Sweetlove, Nielsen, and Fernie 2017).

9.2 ANALYTICAL TECHNIQUES AND METHODOLOGIES EXPLOITED IN METABOLOMICS

Several analytical technologies have proven effective in the quantification of metabolites, such as liquid chromatography coupled with tandem mass spectrometry

(LC-MS/MS) (Huang et al. 2013) or single-stage mass spectrometry (LC-MS) (Liu et al. 2014), high or ultra-high performance liquid chromatography paired with fluorescent or UV detection (HPLC/UPLC) (Molz et al. 2014), gas chromatography integrated with mass spectrometry (Emwas et al. 2015), and nuclear magnetic resonance (NMR) spectroscopy (Kim et al. 2013).

Every analytical platform has its pros and cons. Selecting a technique is primarily dependent on the focus of the study being conducted together with the nature of the sample. On the other hand, the choice of technique or techniques is sometimes determined by its expense, accessibility, and the skills available. Also, no single analytical technique is sufficient for the complete quantification and identification of the entire metabolome of a biological sample. Therefore, metabolomic studies generally use multiple analytical techniques (Emwas et al. 2019).

Initially, in profiling transgenic plants or environmentally stressed plants, LC-MS, GC-MS, or direct-injection MS were utilized (Alseekh and Fernie 2018). Many metabolomics platforms were used to gather data on crop plants as well. Several analytical techniques and methodologies are exploited in metabolomics studies, but only a brief overview of some of the main strategies is discussed here.

9.2.1 NUCLEAR MAGNETIC RESONANCE SPECTROSCOPY

Nuclear magnetic resonance works on the principle of the absorbance and re-emission of energy by the nuclei of some of the atoms, which are swayed by an outside magnetic field. NMR quantitates and identifies a broad range of molecular compounds with detection limits in the order of nanometers (nM) or micrometers (μM). This method can be automatized and is non-destructive to molecules, producing a quick analysis that provides the opportunity to simultaneously measure numerous small molecules from the metabolome (Segers et al. 2019).

Nuclear magnetic resonance spectroscopy is also utilized for the structural study of a unique metabolite of particular concern. It provides information on organic acids and carbohydrates, is effective for isoprenoid profiling, and also detects many more compounds per test than colorimetric or enzymatic methods (Sharma et al. 2018). In addition, unlike many other metabolomics platforms, it is not restricted to tissue extract or biofluid analysis and is therefore suitable for the study of intact organs, tissues, and solid or semisolid samples using the state solid-state NMR (ssNMR) or magic angle sample spinning NMR (MAS-NMR). Furthermore, the relation between two or even three different nuclei can be studied and measured with the help of multidimensional NMR methodologies.

NMR-based plant metabolomics analysis has been carried out successfully in many crops (see Table 9.1 for a summary).

However, NMR has its disadvantages with lack of sensitivity being a significant disadvantage. In contrast to GC-MS or LC-MS, the sensitivity of NMR is 10–100 times less effective. This means that an NMR-based metabolomics analysis only provides knowledge of an average of 50–200 recognized metabolites with a concentration of less than 1 μM, whereas an LC-MS-based study usually returns information on 1000+ identified metabolites of concentrations greater than 10–100 nM.

TABLE 9.1

NMR-Based Plant Metabolomics Analysis Performed in Different Crops

Crop	Application	Advantage	References
Hypercium, ginseng, cannabis, Strychnos, Ephedra, Illex	Characterization and classification of medicinal plants	Quality control of medicinal plants, reproducible pharmacological efficacy	Kim et al. (2013)
Brassica, Catharanthus, Senecio, Arabidopsis, and tobacco	Nursing the response of plants and plant cell cultures to biotic or abiotic stress	Biotic and abiotic stress–resistant plant species	Kim et al. (2013)
Arabidopsis, maize, tomato, Senecio, Brassica, and tobacco	Segregation of wild species/transgenic plants or dissimilar genotypes	Better classification of genotypes	Kim et al. (2013)

9.2.2 HIGH-RESOLUTION MASS SPECTROMETRY (HRMS)

High-resolution mass spectrometry is an effective tool for the analysis of complicated sample matrices, distinguishing isotopic allocations and generating fragmentation patterns, thereby enabling the prediction of more accurate chemical formula and a library comparison for compound identification (Wallace and McCord 2020). HRMS is increasingly used in metabolomics as it allows the analysis of complicated mixtures providing high-quality metabolomics data (Junot et al. 2014).

Before high-resolution mass spectrometry, MS imaging studies in plants were restricted to particular organs of specific species and were generally focused on a particular class of compound. However, advances in mass spectrometry enabled detailed examination of plant metabolomes, including the first MS imaging studies at 5 µM, which uncovered a comprehensive configuration of plant metabolites. A broad range of chemical compounds was identified such as carbohydrates, phenols, phenolic choline esters, lipids, phospholipids, glycerides, and glycosides in germinating and mature oilseed rape.

HRMS was utilized in oilseed crops to map metabolites at their highly active stages, i.e., germination and maturity, concerning plant performance and seed quality. Numerous compounds were detected including methyl sinapate, sinapine, phosphatidylcholines, triacylglycerols, and cyclic spermidine conjugate. They were identified based on precise mass measurements. The high resolution and high mass accuracy generated MS images of narrow bin width ($\Delta m/z = \pm 5$ ppm), consequently inhibiting interference from adjacent peaks. The HRMS images produced by this methodology provide more detailed spatial distributions, thereby enabling the interpretation of metabolite functions.

HRMS imaging allows the detection of marker compounds for predicting and investigating the quality of plant and pathogen interaction. For example, in Fusarium head blight (FHB), which is a devastating disease of cereals, fungus-specific compounds were identified in diseased wheat grains and their distribution complemented

TABLE 9.2

HRMS-Based Plant Metabolomics Analysis Performed in Different Crops

Crop	Application	Advantage	References
Wheat	Detection of marker compound (fungus-specific compound)	Predicted the quality of plant–pathogen interaction	Bhandari et al. (2015)
Oil seed crops	Metabolites mapping at germination and maturity stage of plants	Interpreted metabolite functions	Bhandari et al. (2015)
Mullbery fruit, grape seed, hypericum species, sunflower leaves, cauliflower	Characterization of plant secondary metabolites	Detected plant natural products	—

the fluorescence labeling. The distribution of such a marker indicates its functional role. This type of information may be utilized as a reference to detect the local and systemic effects of biotic and abiotic stresses, the developmental stages along with wildtype/genotype research. HRMS-based plant metabolomics analysis has been successfully carried out in many other crops as well (see Table 9.2 for a summary).

Large-scale coupling of HRMS with traditional LC-MS techniques could help in building up an inventory of plant metabolome. Moreover, in the future, high-resolution MS will offer a novel vision in the field of plant research and development (Bhandari et al. 2015).

9.3 SEPARATION TECHNIQUES EXPLOITED IN METABOLOMICS

The theory of chromatography is that the molecules in the mixture are employed onto the exterior or into the solid, and the fluid stationary phase is splitting from each other while running with the assistance of the mobile phase. In this process of separation, many factors are effective, including the molecular characteristics that are related to the absorption, the partition of solid and liquid, affinity, and the difference between the molecular weights (Porath 1997).

In the following, we discuss five different types of separation techniques exploited in metabolomics.

9.3.1 LIQUID CHROMATOGRAPHY

Liquid chromatography has been used in plant sciences for many years (Wolfender et al. 2009). In the study of metabolomics, this particular method has application in an essential range of metabolites which may extend to MW compounds equal to 1000 Da and presents many different physical and chemical assets. The sample running is also very feasible and in general, there is no requirement of derivation. Nowadays, the massive acceptance of LC-MS is due to atmospheric pressure ionization (API) interfaces, which include electrospray ionization (ESI) and atmospheric pressure chemical ionization (APCI). Ionization for both ESI and APCI appears at atmospheric pressure. In metabolomics, ESI is mostly used but all these API techniques have been used. They all have application according to the type of analysis and the biological matrixes (Grata et al. 2008).

9.3.2 HPLC

HPLC is a very powerful technique in terms of the fast analysis of bioactive components because it allows systematic sketching of very complicated samples of plants, and it places particularly emphasis on the detection and evaluation of the constituents of the discovered complexes. A good and authentic chromatographic fingerprint must point most of its crests points in correspondence to the effective constituent and poisonous ingredients. The identification of complex bioactive phytochemical components in plant samples is enabled by the distinctive fragmentation patterns of reference standards, the type of analyzed samples, whether polar or non-polar, that are attained from their preservation time data, the online (that is UV) spectral information, the information that is presented in the literature, and the sources of the compounds. However, HPLC can only provide constrained structural information such as the UV spectrum and the provisional characterization of eluted compound peaks in samples by comparison with the standard reference compounds. LC-MS identifies unknown constituents in plant extracts using the separation abilities of HPLC and accurate structural categorization by MS (Yang et al. 2009).

9.3.3 SPME MS

Without a doubt, one of the most essential stages in analytical methods is the preparation of the samples. It is estimated that the sample preparation step takes up 60% of the time, cost, and workload of the overall process. A group of researchers developed a solvent-free technique, which is an advanced sample preparation technique called solid-phase microextraction (SPME) (Arthur and Pawliszyn 1990). It is broadly utilized with other separation techniques such as liquid chromatography and gas chromatography for academic research or routine analysis because it is reliable, versatile, less costly, and convenient.

9.3.4 SFC MS

In the development of analytical approaches, the field of metabolomics relies on two features: high-throughput analysis and broad metabolome coverage. Based on these features, supercritical fluid chromatography (SFC) is considered the most suitable alternate technique in the field of metabolomics. In comparison to the LC-based techniques, SFC is more suitable for green analytical chemistry because of its elevated ratio of CO_2 in the mobile phase which is a non-toxic and recycled fluid. Furthermore, in the last few years a major improvement has been made in the expansion of various interfaces to hyphenate SFC with the MS (Vickers 2017).

9.3.5 GC-MS AND GC-GC MS

Metabolomics is used in the identification of biomarkers that are present in biological samples in the form of small molecule metabolites. As they are diverse, their separation is quite difficult and challenging (Klavins et al. 2014). In metabolomics based on GC-MS, the metabolites are initially divided on a column of gas chromatography and then exposed to mass spectrometry. GC-GC MS is a two-dimensional gas chromatography-mass spectrometry in which two GC columns are used, generally connected through a thermal modulator. Conventionally, the second column is smaller than the first with the conditions that there are distinct stationary phases and they are typically functional at elevated temperatures. Hence, GC-GC MS gives enhanced chromatography at the highest capacity and selectivity, and the lowest detection limit for small molecule evaluation. This provides a strong method for the analytical study of metabolites in complicated biological samples. Currently, large-scale utilization of the GC-GC MS technique is limited because of difficulties in terms of data analysis. Bioinformatics has devised tools that greatly help in decreasing the time required for processing metabolomics studies (Shi et al. 2014).

9.4 METABOLOMICS-ASSISTED CEREAL BREEDING

An understanding of metabolomics is critical in crop breeding, stress tolerance, biotic stress, abiotic stress, and pathogenic resistance, and massive development and research are in progress for cereal improvements (Shulaev et al. 2008). Under stress conditions, a metabolomic platform provides understanding at the metabolomics and genetic levels in a plant. Metabolites help in the identification of different impacts of abiotic/biotic stresses and witness changes in biochemical levels (Garcia-Cela et al. 2018).

The annual yield of cereals crops is affected by biotic and abiotic stresses. Metabolomics techniques coupled with other "omics" tools overcome the effects caused by abiotic/biotic stresses in plants (Piasecka, Kachlicki, and Stobiecki 2019). For a better understanding of plant biology and the mechanisms in plants under stress, the above-mentioned methodologies help in the analysis of exogenous and endogenous metabolite activity (Sung et al. 2015). Plant metabolites are classified into three domains: primary metabolites, secondary metabolites (SMs), and hormones (Erb and Kliebenstein 2020).

Primary metabolites contribute toward cellular structure, cellular energy, and phytohormones (Verma, Ravindran, and Kumar 2016). Secondary metabolites activate phenolic compounds produced through different signaling pathways that include the malonic acid pathway, the shikimic acid pathway, isoprenoids, terpene via the malonic acid pathway, and the 2-C-methylerythritol 4-phosphate (MEP) pathway (Cheah et al. 2020). A diverse family of phytoanticipins composites is a secondary metabolite particularly in dicot plants (Hammerschmidt 1999).

9.4.1 METABOLOMICS-ASSISTED BREEDING TO IMPROVE CEREALS COMPOSITION AND YIELD

To date, metabolomics-assisted breeding in cereal crops has improved 50,000 edible plant species globally; however, few of them are supplied as food. With the rapid growth of the world's population, food production needs to be enhanced to cater to the dietary requirements of people. Breeders and researchers are focusing on specific metabolomic traits to improve the overall composition of crops (Moose, Dudley, and Rocheford 2004).

The discovery of the metabolomic quantitative trait loci (mQTLs) led to metabolomic profiling in metabolite genome-wide association studies (mGWAS) for crop improvement. Comparatively, many genes in QTL cause problems while being investigated using transgenic approaches (Kaur et al. 2015). In GWAS, various genome regions have been determined that link with the metabolomics of plants. In the maize plant, the addition of pro-vitamin A has been achieved through GWAS (Palaisa et al. 2003), kernel alignment and glucose contents (Wilson et al. 2004), and the quantity of phytate in mustard (Zhao et al. 2007).

In a complex eukaryotic system like plants, combining genomics and metabolomics data leads to the annotation of genetic factors (Goossens et al. 2003). Several potential genes have been identified as a result of these combined methods. Integrated transcriptome and metabolomic studies of *Allium fistulosum*, a monosomic for genetic material from *A. cepa*, have discovered a "hot spot" important flavonoid accretion in *A. fistulosum* (Abdelrahman et al. 2019).

Furthermore, opium-free poppies are one of the most well-known and prominent cases of metabolically modified plants (Larkin et al. 2007). To regulate the synthesis of morphine and other pharmaceutically significant compounds such as the baine, researchers used transgenic approaches and RNA interference (RNAi) silencing. Moreover, metabolic engineering has been used to improve the nutritional content of crops, such as the protein quality of maize (QPM), which is a mutant form of maize that produces more lysine and won the World Food Prize in 2000 (Prasanna et al. 2001). However, these plants saw certain unexpected modifications, such as changes in the amount of other free amino acids (Wang and Larkins 2001). The discovery of metabolic biomarkers, on the other hand, demands extensive use of univariate and multivariate static analysis of data. Rice GC-MS on non-targeted metabolomics reporting was used to establish a series of biomarkers for its developmental phases (Tarpley et al. 2005). The metabolites implicated in wild and farmed soybean salt stress adaptation mechanisms have also recently been discovered (Li et al. 2019).

9.4.2 Metabolomics-Assisted Breeding to Control Biotic Stress in Cereals

Plants have a sessile nature and are widely exposed to different stresses. The stresses induced by pathogens – viruses, bacteria, pests, and nematodes – are called biotic stresses (McDowell and Dangl 2000). Every year, around 30% of cereal yield is lost due to biotic stresses (Savary et al. 2019). The major aim of pathogens is to extract nutrients and find shelter in plants; however, to achieve these, pathogens induce disease in plants. There are four major types of pathogenic groups: necrotrophic, biotrophic, viral, and insect, based on the nutrition acquisition methods (Freeman and Beattie 2008). A plant's response to biotic stress is to develop an essential defense mechanism for protection and inhibit the invading pathogens (Figure 9.1). Pathogen-associated molecular pattern (PAMP) and effector-triggered

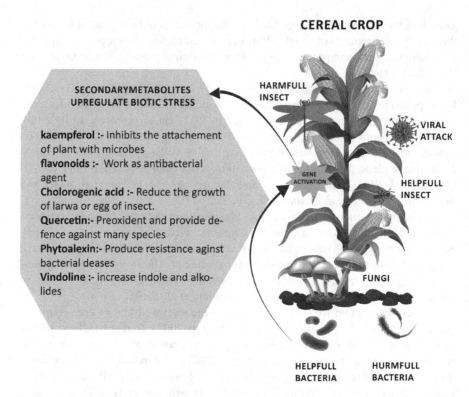

FIGURE 9.1 Microbes and insects are involved in the biosynthesis of secondary metabolites in crop protection. Among the metabolites that protect plants against them are kaempferol, flavonoids, chlorogenic acid, quercetin, phytoalexin, and vindoline.

immunity (ETI) are induced as a defense mechanism on the cellular level (Bigeard, Colcombet, and Hirt 2015).

A plant's defense system is composed of multiple layers of defense mechanisms that act as physical barriers (Pieterse et al. 2009). Likewise, physical barriers provided by plants have efficacious chemical defense molecules termed "phytoanticipins" (González-Lamothe et al. 2009). Chemical defense compounds are secondary metabolites; they represent the first layer of defense activated as antimicrobial compounds against an attack in plants. Significantly, plants activate a network of signaling associated with the activation of hormones and reactive oxygen species (ROS). The signaling cascades are associated with the activation of kinase, and further activation of defense genes via transcription factors (TFs). Thus, several antimicrobial compounds such as phenolics, phytoanticipins, phytoalexins, and secondary metabolites are released in these cascades (Jain et al. 2019).

9.4.2.1 Metabolomics Response toward Necrotrophic Pestilential

A fungal model has been used in the metabolomic study of necrotrophic pestilential. Many criteria have been set to identify the specific defense pathway of metabolomes against a pathogen. For example, Fusarium head blight diseases affect monocotyledonous plants including rice, maize, barley, triticale, and wheat (Choo 2006). Fusarium is categorized as a "necrotrophic pathogen", which produces mycotoxin chemicals that kill the plant cell and can grow and feed on the host cell. More than 100 quantitative trait loci (QTL) genes have an association with FHB resistance on chromosome 7 in barley and wheat plants (Lemmens et al. 2005). The resistance in plants is controlled by polygenes, which depend on environmental conditions, water concentration, heritability, type of soil, and quantity of nutrients in the surrounding environment.

It has also been observed that metabolites produce resistance clusters (RR) derived from chemicals such as phenylpropanoid, fatty acid, terpenoid pathway, kaempferol, and flavonoids (Bollina et al. 2010). Resistance to *Fusarium graminearum* in barley spikelets is classified as Type II resistance (Schroeder and Christensen 1963), while trichothecene non-producing (tri5-) Fusarium mutants in wheat plants are unable to spread within the inoculation spikes (Jansen et al. 2005). The combination of the resistance and the susceptibility of the barley plant and trichothecene producing and non-producing *F. graminearum strains*, is an effective model for studying the metabolic responses, leading to the determination of the tolerance rate of the barley plant toward this fungal infection (Kumaraswamy et al. 2012). This study confirms the presence of constitutive-resistance related (CRR) and induced-resistance related (IRR) metabolites. CRR compounds consist of pelargonidin 3-O-rutinoside (POR), 8E-heptadecenoic acid, vitexin, and coniferyl aldehyde.

9.4.2.2 Metabolomics Response toward Biotrophic Potential

Hemibiotrophic pestilential show natural effects on the host plant. The characteristics of life from hemibiotrophic show unaffected on the host cell (Koga et al. 2004). While the interaction between fungi and rice shows that the cell membrane loses integration through plasmolysis, necrotrophic infection produces granulated tissue and other symptoms in plants. The inconsistency of *Magnoporthe oryzae* shows its lifestyle as biotrophic compatible and fundamentally interactive. *M. oryzae* infects plants at the germination stage (Kankanala, Czymmek, and Valent 2007).

Xanthomonas oryzae pv. oryzae (Xoo) causes bacterial blight in rice plants. Studies show that rice protein changes in response to *Xoo* infection, and the catabolism of carbon is reduced. Resistance to disease, cell metabolism, and signaling transduction pathways are regulated by the Resistance (R) gene and putative receptors (PR) such as kinase. In particular, probenazole (PBZ), thaumatin-like protein (PR5), β-1,3-glucanase, and the domain of unknown function 26 (DUF26) were found to be key factors in previous studies (Mahmood et al. 2006). Studies showed that *Oryza meyeriana* (wild species) use as an interruption, has a strong resistance against *Xoo*. An upregulation of the transduction pathways of proteins, LysM receptor-like kinase, and defense proteins and downregulation of ROS enzymes (peroxidase) and modification of the cell wall (via pectin acetylesterase and expansins) have been reported. Likewise, phosphorites of TFs, epigenetic factors, kinases, and disease-resistant proteins are functionally applicable to resistance in the IRBB5 gene in *Xoo* (Hou et al. 2015).

Moreover, metabolomic studies revealed XA21 gene expression in wild-type (WT) plants. After treating plants, XA21 consists of sensitive metabolites including pigments, fatty acids, arginine, and rutin required for polyamine biosynthesis. The virulence signal acetophenone is downcast on plant XA21. Peptidoglycan recognition proteins (PGRP), likewise known as plant growth endorsing rhizobacteria, help to promote growth, root exudates, nutrients uptake, and phytohormone formation (Kandasamy et al. 2009). In 2013, the first metabolite profiling was performed in a rice cultivation infected through *Azospirillum* species. A study revealed that phenolic compounds, hydroxycinnamic derivatives, and flavonoids depend on PGPR strain interaction with plants (Chamam et al. 2013).

9.4.2.3 Metabolomics Response toward Viral Pestilential

Viral infection is spread through insects as vectors. It is involved in carbon metabolism pathways such as glyceraldehyde-3-phosphate dehydrogenase (GAPDH) as a glycolytic enzyme. These enzymes perform multi-functions in various non-metabolic processes and their production increases during viral infection. Major fluctuations have been observed in amino acids, which help to provide elements for virus-related replication, defense mechanisms, ROS accumulation and damage, and energy supply for respiration and are major responses in viral infection (Alexander and Cilia 2016). Antiviral bioactive secondary metabolites have been used in response to viral infection, with cytosinpentidemycin showing upregulation in superoxide, peroxidase, and dismutase in response to black-streaked dwarf virus infection (BSDV). Furthermore, the enhanced synthesis quality of PR proteins, such as PR10, PR5, HSP, and Bet v1 allergen has been linked to viral resistance in rice (Yu et al. 2018).

9.4.2.4 Metabolomics Response toward Insect Pestilential

The interaction between insects and plants introduces dynamic mechanisms that lead insects to invade the plant body and plants to defend themselves. Plants resist insects in two ways involving secondary metabolites as part of their defense system: herbivore morbidity through toxic excretes and rebuffed herbivores by excreting secondary metabolites which invites herbivores enemies (Lu et al. 2018). The defense mechanism strategy of plants against insects involves jasmonic acid (JA) signaling, phytohormones, detoxification, granular formation at the cell wall, and secretion of secondary metabolites (Zogli et al. 2020).

Using metabolomics, a study revealed that the rice plant responds to brown planthopper (BPH) infection changes in lipid transportation and quantitative changes in metabolism, protein transportation, and phytohormone signals induced by BPH in both resistive and susceptible cultivars (Zhang et al. 2019). Phytohormones play a crucial role in rice and hopper interaction. Salicylic acid (SA) works as evasive between rice and hopper (Peng et al. 2016), whereas JA acts differently in resistant and susceptible crops (Zhang et al. 2019). Studies show that inducing SA and JA increased the rhizosphere strain *Bacillus velezensis* (YC7010) that takes plants toward resistance to BPH infection in rice. Additionally, it has been observed that resistance in plants is also determined by flavonoid synthesis (Harun-Or-Rashid et al. 2018).

9.4.3 METABOLOMICS-ASSISTED BREEDING TO
CONTROL ABIOTIC STRESS IN CEREALS

Abiotic stresses are serious threats to plants. Abiotic stresses include cold, heavy metals, wounds, drought, salinity, UV light, air pollution, temperature, humidity, and soil erosion resulting in a global reduction in cereal crops, and leading to economic loss worldwide (Nakabayashi and Saito 2015). To improve crops, researchers are focusing on signaling pathways to regulate stress, tolerances, and resistance through functional proteins and metabolites against abiotic stresses (Hirayama and Shinozaki 2010). The aggregation of macromolecules with antioxidative action in vitro plays a vital role in minimizing the reactive oxygen species–induced stresses. The analysis of abiotic stresses under "omics" integrates metabolomics studies to identify the genes involved in metabolic processes in cereals crops (Saito 2013).

Under salinity conditions, excessive Na+ ion influx in cells is termed ion toxicity, which leads to the impairment of processes regarding the uptake of minerals and water by roots (Wu et al. 2013). Cereal crops adopt several approaches to acclimatize in salinity conditions. There are two strategies to minimize osmotic stress in plants: produce numerous compatible solutes and effectually classify Na+ ions inside vacuoles to minimize osmotic stress.

Plants utilize an escape strategy during the reproductive phase before a summer drought. Tolerance is the mechanism through which plants sustain regular homeostasis and cellular metabolism through osmosis (Wei et al. 2013). Ion toxicity–reduced nutrient uptake, imbalance in the osmotic mechanism, and chaos in metabolic conditions lead to decreases in growth and physiological activities (Meng et al. 2019).

Three aspects are considered important for the establishment of tolerance in plants: reducing damage in the plant, continuing growth at a low rate, and managing the homeostatic condition during abiotic stresses. Metabolites act as the base element for the emergence of stress tolerance in plants. Different metabolic pathways and regulatory pathways such as photosynthesis, ROS, amino acid synthesis, and tricarboxylic acid (TCA) cycle are adversely affected by salinity (Sobhanian et al. 2010).

Soil waterlogging is also a major aspect of abiotic stress that affects crop yield. Continued long periods of waterlogging create a hypoxic condition in roots and directly prevent carbon integration and the production of photosynthesis in shoots of crops (Wei et al. 2013). There are three vital stages to acclimatizing to rain stress: transduction pathways, metabolites adaption, and morphological variation (Jackson and Colmer 2005). Several metabolites produced in response to abiotic stress along with their functions are listed in Table 9.3. Acclimatization is a response to phenotypic changes acquired through environmental changes (Sanchez et al. 2012).

The severity of the stress is related to the developmental phase of the crop when it is exposed to such stress, and depends on the intensity and exposure period along with environmental conditions (Lafitte et al. 2007). Abiotic stresses are increasing daily in cereal crops, thus stress-resistant crops need to be developed with a strong coping mechanism. For this purpose, it is necessary to understand the mechanism of plant tolerance to abiotic stress. Most abiotic stresses are studied in wheat, rice, barley, maize, and cotton, for their economic significance. Information related to protein and metabolites is crucial for modulating and overcoming different stresses, helping the plant to grow and survive (Figure 9.2; Parida, Panda, and Rangani 2018).

TABLE 9.3

Functions of Different Metabolites Produced against Abiotic Stresses

Abiotic stress	Metabolites	Sources plants	Functions	References
Salinity stress				
	Polyamines	*Oryza sativa*	Polyphenol production	—
	Tropane alkaloids	*Datura innoxia*	Scopolamine production	—
	Carotenoid	*Daucus carota*	Protect plant from extensive photo light	—
Drought				
	Tremalose	*Hordeum vulgare*	Protection against osmosis	Shi, Zhu, Wan, Li, and Zheng (2015)
	Hydroxyproline	*Atriplex halimus*	Protein integrity in their structures	
	Succinate	*Vitis vinifera*	Development and metabolism activity	Sun et al. (2016)
Heavy metal stress				
	Antheraxanthin	*S. cereale*	Protect cell from light	Janik, Maksymiec, Mazur, Garstka, and Gruszecki (2010)
	Xanthophyll	*P. stratiotes*	Protect against light damage	Singh, Pandey, and Singh (2011)
	Glutamate	*T. aestivum*	Osmoprotection and signaling	Gajewska, Niewiadomska, Tokarz, Slaba, and Skłodowska (2013)
Water logging				
	Fructose	*P. sylvestris*	Osmotic regulation and production of ROS	—
	Spermidine	*P. persica*	Involved in signal transduction and cell membrane integration	—
Temperature				
	Allantoin	*S. lycopersicum*	Nitrogen transport and assimilation	Luengwilai, Saltveit, and Beckles (2012)
	2-ketoisocaproic	*S. lycopersicum*	Branched-chain amino acid catabolism	Luengwilai et al. (2012)

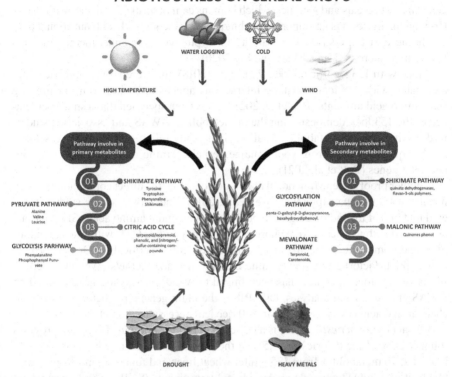

FIGURE 9.2 In response to abiotic stresses, primary and secondary metabolites are produced through various cell signalling pathways. Plants can respond to abiotic stress through these pathways by contacting essential metabolites and activating their defensive mechanisms. The following primary metabolites were produced by each pathway: Shikimate pathway (Tyrosine, Tryptophan, Phenylalanine, Shikimate), Pyruvate pathway (Alanine, valine, leucine), Citric acid pathway (Terpenoid, phenolic, nitrogen/sulfur-containing compounds), Glycolysis pathway (Phenylaniline, Phospho-phenyl Pyruvate). However, Secondary metabolites in different pathways are enlisted as follow: Shikimate pathway (Quinate dehydrogenase-flavan 3-ols polymers), Glycosylation pathway (penta-O-galloyl-β-D-glucopyranose, hexa-hydroxy-diphenyl), Malonic pathway (Quinones phenol), Mevalonate pathway (Terpenoid, Carotenoids).

9.4.4 Metabolomics-Assisted Breeding to Escalate Amino Acid Contents in Cereals

Integrated metabolomics and functional genomics techniques may help boost crop yields. Here, we will study loci and genes for the content of amino acid in major crops, which is crucial for global food security. Improving the minerals, vitamins, and protein content of important crops is critical for resolving serious public health issues associated with nutritional deficiencies. Animal proteins are more expensive than plant proteins, and a huge portion of the world's population relies on them to meet their dietary requirements. While free amino acid supply determines protein quantity, rigorous synthesis regulation and a high catabolic rate have impeded breeding progress (Galili,

Amir, and Fernie 2016). Another issue is the restricted supply of some essential amino acids in most cereals and legumes, which humans cannot form (Galili and Amir 2013). Crop biofortification is an important technique to tackle global health nutritional deficits because of the restricted availability of tryptophan, threonine, methionine, and lysine in most crops (Yang, Zhao, and Liu 2020).

Along with 153 significant loci, Wen et al. (2018) discovered 308 candidate genes associated with 528 loci. Epistatic interactions appear to be important in limiting free amino acid amounts (Pott et al. 2021). it has been specified that a small percentage of the 153 loci, demonstrating the of metabolites GWAS and association studies in detecting natural metabolic variation. This study relied on previously disclosed RNA sequencing data from young maize kernels. Amino acid variation can influence 308 genes (Pott et al. 2021).

Another proof of gene functional annotation could be the direct correlation between metabolites and gene expression (Wen et al. 2018). The authors of the study discovered that one of the indicated genes could contain essential amino acids. Six aspartate amino acid pathway genes were chosen because they linked QTL with lysine, methionine, and homoserine. The arogenate dehydratase gene is critical for Opaque2 (bZIP transcription factor) and phenylalanine buildup in maize kernels (Wen et al. 2018). In maize, tryptophan content has been linked to two decarboxylase genes. Based on GWAS and functional analyses, OsZIP18 is the main genetic predictor of branched-chain amino acid (BCAA) content in 520 rice leaf accessions (Sun et al. 2020).

Wheat (*Triticum aestivum* L.) is a key crop in terms of output. This plant provides roughly 20% of the calories and 25% of the protein consumed by humans. A study looked at 76 metabolites from 135 winter wheat lines and found six micro-quantitative trait loci (mQTLs) single nucleotide polymorphisms (SNPs). The largest wheat investigation reported so far looked at 558 metabolite levels and 10 agronomic trait QTLs to assess the genetic connection between metabolite levels and agronomic traits in wheat. In plants, the metabolome is often considered the link between the genome and the phenome. Metabolites could be utilized to predict complex agronomic traits, promoting growth and lowering the cost of breeding. LC-MS/MS gathered 1260 metabolites, 467 of which had their chemical structure determined. Compared to previous wheat metabolome investigations, this study's results showed considerable progress in metabolite detection. Polyphenols and flavonoids were included since they are important in plant biotic/abiotic stress and have several effects on the body. Amino acids, nucleic acids, phytohormones, and lipids had strong interactions. Similar findings have been reported in rice, wheat, and tomato. Phenomenamides and flavonoids were also discovered. A correlation study will be useful in the future to identify unknown metabolites and pathways. The study found 1005 methylation quantitative trait loci (mQTL) in a Wheat660K high-density genomic map–based linkage analysis. Among them were many high-resolution mQTLs along with 68 "hot spots" on chromosomes 4B and 1B, including many mQTL areas that impacted the quantities of several metabolic products (Peng et al. 2018).

9.4.5 Targeted Metabolomics in Transgenic Cereals

Targeted metabolomics is an analytical method in which a variety of metabolites are quantified by employing 13C or 15N (external or internal) different compounds recognized as labeled isotopes. Furthermore, the investigation is carried out using the quantitative or semi-quantitative approach with the help of internal standards, while metabolic changes in animal models in response to genetic alteration can be used to describe the reported phenotypes. It has proven to uncover so-called "silent phenotypes" or genes that have no visible influence on physical traits or behavior when disrupted, by combining metabolomics with other omic techniques (Wiley 2001).

A large research count has been carried out on cereals to determine the effects of quantity variation on sequence variation (Chen, Dong, et al. 2016). Several research groups on rice have employed metabolomics to study the changes in metabolites between different types and natural variations of the plant (Gong et al. 2013). Similarly, metabolomics research in maize has helped scientists to identify and select superior genotypes with improved nutritional composition (Venkatesh et al. 2016). Amino acid metabolism regulates drought stress in maize. Drought modulates photorespiration by increasing glycine and serine levels. Drought-tolerant maize grain size is connected to glycine and myo-inositol accumulation, suggesting these metabolites could be exploited to detect drought-tolerant maize. Drought-tolerant rice plants produce more allantoin, glucose, gluconic acid, and glucopyranoside than non-tolerant rice plants (Degenkolbe et al. 2013).

9.4.6 UNTARGETED METABOLOMICS IN TRANSGENIC CEREALS

Untargeted metabolomics examines all detected metabolites in a sample. It is used for fingerprinting and profiling metabolites. A metabolite fingerprinting technique is used to categorize samples based on the overall patterns of metabolite signals seen in the sample. Metabolite profiling, on the other hand, quantifies specific metabolite properties, allowing recognition of statistically varied amounts of metabolites across sample groups. Metabolomics has made it possible to capture a significant portion of the metabolome of complex biological systems in a high-throughput fashion. As a result, metabolomics offers a wide range of applications in basic science and biotechnology. Unintentional metabolic alterations are recognized as statistically significant disparities between a newly produced crop and its parent lines. When an engineering method is used, these changes are not immediately anticipated (Christ et al. 2018).

Untargeted metabolomics integrated with other high-throughput platforms, such as proteomics and transcriptomics, was utilized to analyze the kernel composition of maize and to elucidate the molecular supervision of metabolic pathways enabling the manufacture of phenolic antioxidants. Fourteen corn lines from China were analyzed using LC-MS and GC-MS, and basic and necessary macronutrients metabolites were identified (7 lipids, 5 carbohydrates, 17 amino acids, electron carriers, prosthetic groups, and 3 cofactors). Also discovered were dihydrokaempferol and costunolide, along with vitamin E, stigmasterol, and sitosterol. Transcriptomic, proteomic, and metabolic data were used to construct an integrative metabolic map (Rao et al. 2014). Corn kernel diversity could be better characterized by combining untargeted platforms with targeted approaches (Liu et al. 2021).

9.5 APPLICATIONS OF METABOLOMICS-ASSISTED CROP BREEDING

Metabolomics is the most effective way to study plant abiotic stress tolerance. It is now being utilized to hunt for novel plant metabolites. Plants respond to abiotic stress by making phytohormones. Activating a metabolic network yields a new bioactive molecule. It shows the general techniques from diagnostics through metabolomics-aided crop enhancement breeding. The following sections discuss other applications.

9.5.1 BIOMARKERS FOR TRANSGENIC CROP EVALUATION

The "substantial equivalence" between a GM crop and its conventional crop counterpart in terms of target metabolites is the basis for evaluating the risk of these GM crops. It is now possible to determine if two substances are substantially equivalent using systems biology and the application of several "omics" technologies (such as metabolomics) (Takahashi et al. 2005). For transgenic tomato, barley, soybean, maize, and rice, metabolic-based platforms have already been applied. It is still early days in the study of metabolomics, which is used to determine the significant equivalence of transgenic goods and to estimate the environmental threat. Considering these unknowns and uncertainties, we don't know whether differences in metabolite accumulation between wild-type and transgenic plants have favorable or detrimental implications on the plant, its environment, or the consumer's diet. But as metabolomics technology improves, systems for establishing significant equivalence remain promising (Hall and de Maagd 2014).

9.5.2 PREDICTOR OF HETEROSIS

In the last century, plant breeding focused on improving hybrid progeny performance (e.g., yield) over the homozygous parents. Metabolite accumulation has recently been studied as a heterotic response predictor in Arabidopsis. In Arabidopsis, combining genetic and metabolic indicators improves heterotic prediction. Carbon supply and growth are linked to ecotypes that accumulate large amounts of biomass, and starch mobilization and utilization efficiency are key factors in determining the balance between carbon supply and growth. Inbred seedlings have a different metabolic profile and biomass than hybrid Arabidopsis seedlings (Meyer et al. 2012). Based on hybrid metabolite profiles, researchers can predict the hybrid performance in maize. The metabolomes of hybrid maize seedlings differ from their parents and hybrids accumulate fewer major metabolites. Deviations from usual metabolite levels were shown to be associated with lower biomass in hybrids compared to inbred parents (Steinfath et al. 2010).

Studies of Arabidopsis and corn link heterosis to general and specialized metabolism. As a result of this early research, some substances can operate as predictors or biomarkers of hybrid performance. The capacity of these metabolites to predict traits of relevance is now being explored statistically and computationally (Lisec et al. 2011).

6. Metabolomics implementation in Plant Breeding and Genetics

Cereal crops are considered a major source of nutrition around the globe as they are enriched with minerals, carbohydrates, vitamins, and fats (Ballini, Lauter, and Wise 2013). A detailed study has been conducted on cereal crops to measure the variation in the number of metabolites present in them alongside their association in sequence variation (Chen et al. 2013). Groups conducting the research utilized the potential of metabolomics to explore the diversity of metabolites between different varieties of rice and their natural variants (Degenkolbe et al. 2013). Considering maize research groups, with the help of metabolomics, have enabled differentiating, and can subsequently choose genotypes that are superior as their composition of nutrients have been enhanced (Fang et al. 2016). Recently, the metabolomic research approach has been used for surveying chemical diversity present between distinctive rice and maize varieties as well as in their natural variants. Considering the maize crop, amino acid metabolism regulates drought stress (Jansen et al. 2005). In both barley and wheat, the pool of amino acids is speeded up by the cold stress and the induction of genes known as GABA-shunt, promoting the conversion of glutamate to GABA (Koelmel et al. 2017). It is understood that there is an accumulation of flavone-glycosides/flavones in cereal grains, protecting them from herbivores and abiotic stresses. For example, rice crops produce ample amounts of flavone-glycosides, protecting the rice crops from herbivores and abiotic stresses. Additionally, in maize, the defense system induced by herbivory, portrayed an increase in N-hydroxycinnamoyl tyramines, azelaic acid, 1,3-benzoxazine-4-ones, tryptophan, and phospholipids (Marti et al. 2013). Reports have revealed that during plant–pathogen interaction, metabolites related to resistance are accumulated. Taking the example of a tolerant wheat variety, metabolites such as coumaroylagmatine and coumaroylputrescine are accumulated during Fusarium head blight (Matsuda et al. 2012).

9.6 QTL MAPPING FOR REFINING CROP METABOLOMICS

When using quantitatively genetic approaches there is a need for applications that have high-throughput efficiency. However, due to the automation involved, reduced capital investment has been observed which has inclined researchers to adopt metabolomics technologies. Additionally, measurement of metabolites to a broader classic's extent is taken as a sign of key advancement when using the metabolomic techniques in an untargeted manner. A substantial increase in the sensitivity and accuracy of the quantification has also been observed. The use of repeatable and fine-scale, high-throughput metabolomic approaches for the collection of data on the phenotypes involved in genetic studies on a quantitative scale, represents evident progression at the technological level in the characterization of the natural genetic variation. "Genetical genomics" is a term that was coined 20 years ago as a result of a combination of genetic segregation and keeping in view the scale and scope of the genome data sets, which are also known as "genomic data sets" (Mazzucotelli et al. 2006). The availability of a high-throughput collection of metabolomics, proteomics, and transcriptomics data has proved useful for researchers in utilizing the

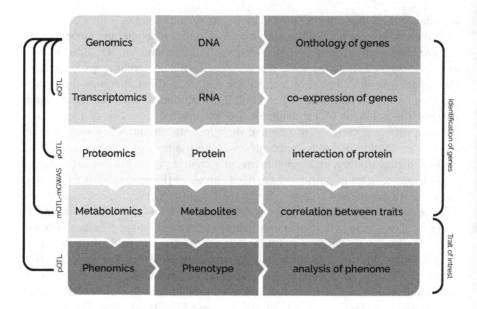

FIGURE 9.3 The central dogma of cells is expressed in the form of metabolites that further convert to phenotypic expressions. Through proper chain analysis, QTL mapping can identify metabolomics and phenomics expression in crops.

above-mentioned technologies in metabolic quantitative trait loci (mQTL), eQTL, and incarnations (pQTL) of quantitative trait locus mapping, respectively, which allow the analysis of genetic variations in the expression of mRNA, protein phenotypes, and metabolic traits (Muthamilarasan et al. 2016). When comparing the genetic effects of the variation in a molecular trait with a case study which involves morphological variation, it is crystal clear that there are more variations in the molecular trait (Obata et al. 2015). As a result, narrow localization at genomic intervals occurs in mQTLs than QTLs, which are associated with variations in the morphology. A well-planned analysis was carried out to clearly illustrate the earlier mentioned point. Twenty-seven morphological traits and 14 metabolite traits were used for the analysis, from a set of rounds of about 5000 lines of maize and 30 million SNPs. Upon detection, it was observed that there were less than 10 mQTLs for each metabolic trait while there were 30 QTLs per morphological trait on average (Figure 9.3; Okazaki and Saito 2016).

9.7 FUTURE PROSPECTS AND LIMITATIONS

Current metabolomic analytical techniques are slow in achieving comprehensive quantitative and qualitative analysis of the metabolites involved in metabolic pathways in living organisms (Wallace and McCord 2020). Limitations exist in metabolomic research (Wang et al. 2015). The instruments used for testing and the required equipment for metabolomic research are expensive, and skilled labor is needed to operate the machinery. Despite all these limitations, metabolomic technology has the flexibility to be combined with other new technologies. Mass spectrometry-based metabolomics is extremely significant and efficient and LC/MS and GC/MS provide an understandable and comprehensive basis for problem-solving in metabolite research and elucidating the metabolic pathways (Yang et al. 2019). Mass spectrometry plays a pivotal role in metabolomics research as it is considered an important and efficient analytical tool. Advances in the metabolomics research field have helped in the discovery of putative disease biomarkers, which has contributed to providing a close insight into the pathogenesis of various diseases (Zhang et al. 2015). The inability to compare and correlate the results of these studies is considered a big issue because the results of these studies are produced from the same samples procured by maverick research groups. This is the major gridlock in this field's development. Preparation so sample, matrix of sample along with the residual effects are such factors which lead to variability in data. These challenges and complexities can only be overcome by using more reliable and efficient absolute concentration determinations, compared to previously used relative metabolite concentration measurements, because the new measurement method of concentrations is unconstrained by the analytic platform, the protocol, and the method used. This method is not straightforward for MS, but it is very important.

9.8 CONCLUSION

To date, this field of metabolomics research has made exceptional progress. Although it is a newly emerging technology in the field of research, it faces some serious challenges to its applications and methods. Analytical techniques, the acquisition of data, and analysis and analytical instruments require further improvement. Considering it from the application angle, although a humungous number of predominant leading light metabolites are about variation at the genetic level or pathological and physiological changes have been procured, it is very challenging in establishing an expert system that can be worked in clinical predictive diagnosis so that diagnostic rationalization can be achieved. This approach comes with both opportunities and challenges. Perpetual furtherance and unremitting optimization of methods used in metabolic research, safety gauging of drugs, will be more efficient and accurate, interpretation and understanding of the processes of the disease will be crystal clear as well as in the field of the plant breeding and genetics, there will be a start of something new which will completely revolutionize the research aspects which will ultimately contribute not only toward the human health but also toward every aspect of the living beings and to things which are in close association with them.

REFERENCES

Abdelrahman, Mostafa, Sho Hirata, Yuji Sawada, Masami Yokota Hirai, Shusei Sato, Hideki Hirakawa, Yoko Mine, Keisuke Tanaka, and Masayoshi Shigyo. 2019. 'Widely targeted metabolome and transcriptome landscapes of *Allium fistulosum–A. cepa* chromosome addition lines revealed a flavonoid hot spot on chromosome 5A', *Scientific Reports*, 9: 1–15.
Alexander, Mariko M, and Michelle Cilia. 2016. 'A molecular tug-of-war: global plant proteome changes during viral infection', *Current Plant Biology*, 5: 13–24.

Alseekh, Saleh, and Alisdair R Fernie. 2018. 'Metabolomics 20 years on: what have we learned and what hurdles remain?', *The Plant Journal*, 94: 933–42.

Arthur, Catherine L, and Janusz Pawliszyn. 1990. 'Solid phase microextraction with thermal desorption using fused silica optical fibers', *Analytical Chemistry*, 62: 2145–48.

Ballini, Elsa, Nick Lauter, and Roger Wise. 2013. 'Prospects for advancing defense to cereal rusts through genetical genomics', *Frontiers in Plant Science*, 4: 117.

Bhandari, Dhaka Ram, Qing Wang, Wolfgang Friedt, Bernhard Spengler, Sven Gottwald, and Andreas Römpp. 2015. 'High resolution mass spectrometry imaging of plant tissues: towards a plant metabolite atlas', *Analyst*, 140: 7696–709.

Bigeard, Jean, Jean Colcombet, and Heribert Hirt. 2015. 'Signaling mechanisms in pattern-triggered immunity (PTI)', *Molecular Plant*, 8: 521–39.

Bollina, Venkatesh, G Kenchappa Kumaraswamy, Ajjamada C Kushalappa, Thin Miew Choo, Yves Dion, Sylvie Rioux, Denis Faubert, and Habiballah Hamzehzarghani. 2010. 'Mass spectrometry-based metabolomics application to identify quantitative resistance-related metabolites in barley against Fusarium head blight', *Molecular Plant Pathology*, 11: 769–82.

Catchpole, Gareth S, Manfred Beckmann, David P Enot, Madhav Mondhe, Britta Zywicki, Janet Taylor, Nigel Hardy, Aileen Smith, Ross D King, and Douglas B Kell. 2005. 'Hierarchical metabolomics demonstrates substantial compositional similarity between genetically modified and conventional potato crops', *Proceedings of the National Academy of Sciences*, 102: 14458–62.

Chamam, Amel, Hervé Sanguin, Floriant Bellvert, Guillaume Meiffren, Gilles Comte, Florence Wisniewski-Dyé, Cédric Bertrand, and Claire Prigent-Combaret. 2013. 'Plant secondary metabolite profiling evidences strain-dependent effect in the *Azospirillum–Oryza sativa* association', *Phytochemistry*, 87: 65–77.

Cheah, Boon Huat, Hou-Ho Lin, Han-Ju Chien, Chung-Ta Liao, Li-Yu D Liu, Chien-Chen Lai, Ya-Fen Lin, and Wen-Po Chuang. 2020. 'SWAtH-MS-based quantitative proteomics reveals a uniquely intricate defense response in *Cnaphalocrocis medinalis*-resistant rice', *Scientific Reports*, 10: 1–11.

Chen, Wei, Liang Gong, Zilong Guo, Wensheng Wang, Hongyan Zhang, Xianqing Liu, Sibin Yu, Lizhong Xiong, and Jie Luo. 2013. 'A novel integrated method for large-scale detection, identification, and quantification of widely targeted metabolites: application in the study of rice metabolomics', *Molecular Plant*, 6: 1769–80.

Chen, Xian, Yan Dong, Chulang Yu, XianPing Fang, Zhiping Deng, Chengqi Yan, and Jianping Chen. 2016. 'Analysis of the proteins secreted from the *Oryza meyeriana* suspension-cultured cells induced by *Xanthomonas oryzae* pv. *oryzae*', *PloS One*, 11: e0154793.

Choo, Thin Meiw. 2006. 'Breeding barley for resistance to fusarium head blight and mycotoxin accumulation', *Plant Breeding Reviews*, 26: 125.

Christ, Bastien, Tomáš Pluskal, Sylvain Aubry, and Jing-Ke Weng. 2018. 'Contribution of untargeted metabolomics for future assessment of biotech crops', *Trends in Plant Science*, 23: 1047–56.

Degenkolbe, Thomas, Phuc T Do, Joachim Kopka, Ellen Zuther, Dirk K Hincha, and Karin I Köhl. 2013. 'Identification of drought tolerance markers in a diverse population of rice cultivars by expression and metabolite profiling', *PLoS One*, 8: e63637.

Duenas, Maria Emilia, Adam T Klein, Liza E Alexander, Marna D Yandeau-Nelson, Basil J Nikolau, and Young Jin Lee. 2017. 'High spatial resolution mass spectrometry imaging reveals the genetically programmed, developmental modification of the distribution of thylakoid membrane lipids among individual cells of maize leaf', *The Plant Journal*, 89: 825–38.

Emwas, Abdul-Hamid M, Zeyad A Al-Talla, Yang Yang, and Najeh M Kharbatia. 2015. 'Gas chromatography–mass spectrometry of biofluids and extracts', in *Metabonomics* (Springer).

Emwas, Abdul-Hamid, Raja Roy, Ryan T McKay, Leonardo Tenori, Edoardo Saccenti, GA Gowda, Daniel Raftery, Fatimah Alahmari, Lukasz Jaremko, and Mariusz Jaremko. 2019. 'NMR spectroscopy for metabolomics research', *Metabolites*, 9: 123.

Erb, Matthias, and Daniel J Kliebenstein. 2020. 'Plant secondary metabolites as defenses, regulators, and primary metabolites: the blurred functional trichotomy', *Plant Physiology*, 184: 39–52.

Fang, Heng, Aihua Zhang, Jingbo Yu, Liang Wang, Chang Liu, Xiaohang Zhou, Hui Sun, Qi Song, and Xijun Wang. 2016. 'Insight into the metabolic mechanism of scoparone on biomarkers for inhibiting Yanghuang syndrome', *Scientific Reports*, 6: 1–9.

Fiehn, Oliver, Joachim Kopka, Peter Dörmann, Thomas Altmann, Richard N Trethewey, and Lothar Willmitzer. 2000. 'Metabolite profiling for plant functional genomics', *Nature Biotechnology*, 18: 1157–61.

Freeman, Brian C, and Gwyn A Beattie. 2008. 'An overview of plant defenses against pathogens and herbivores', *The Plant Health Instructor*.

Gajewska, E., Niewiadomska, E., Tokarz, K., Słaba, M., & Skłodowska, M. (2013). Nickel-induced changes in carbon metabolism in wheat shoots. *Journal of Plant Physiology*, *170*(4), 369–377.

Galili, Gad, and Rachel Amir. 2013. 'Fortifying plants with the essential amino acids lysine and methionine to improve nutritional quality', *Plant Biotechnology Journal*, 11: 211–22.

Galili, Gad, Rachel Amir, and Alisdair R Fernie. 2016. 'The regulation of essential amino acid synthesis and accumulation in plants', *Annual Review of Plant Biology*, 67: 153–78.

Garcia-Cela, Esther, Elisavet Kiaitsi, Angel Medina, Michael Sulyok, Rudolf Krska, and Naresh, Magan. 2018. 'Interacting environmental stress factors affects targeted metabolomic profiles in stored natural wheat and that inoculated with *F. graminearum*', *Toxins*, 10: 56.

Gong, Liang, Wei Chen, Yanqiang Gao, Xianqing Liu, Hongyan Zhang, Caiguo Xu, Sibin Yu, Qifa Zhang, and Jie Luo. 2013. 'Genetic analysis of the metabolome exemplified using a rice population', *Proceedings of the National Academy of Sciences*, 110: 20320–25.

González-Lamothe, Rocío, Gabriel Mitchell, Mariza Gattuso, Moussa S Diarra, François Malouin, and Kamal, Bouarab. 2009. 'Plant antimicrobial agents and their effects on plant and human pathogens', *International Journal of Molecular Sciences*, 10: 3400–19.

Goossens, Alain, Suvi T Häkkinen, Into Laakso, Tuulikki Seppänen-Laakso, Stefania Biondi, Valerie De Sutter, Freya Lammertyn, Anna Maria Nuutila, Hans Söderlund, and Marc Zabeau. 2003. 'A functional genomics approach toward the understanding of secondary metabolism in plant cells', *Proceedings of the National Academy of Sciences*, 100: 8595–600.

Grata, Elia, Julien Boccard, Davy Guillarme, Gaetan Glauser, Pierre-Alain Carrupt, Edward E Farmer, Jean-Luc Wolfender, and Serge Rudaz. 2008. 'UPLC–TOF-MS for plant metabolomics: a sequential approach for wound marker analysis in *Arabidopsis thaliana*', *Journal of Chromatography B*, 871: 261–70.

Hall, Robert D. 2006. 'Plant metabolomics: from holistic hope, to hype, to hot topic', *New Phytologist*, 169: 453–68.

Hall, Robert D, and Ruud A de Maagd. 2014. 'Plant metabolomics is not ripe for environmental risk assessment', *Trends in Biotechnology*, 32: 391–92.

Hammerschmidt, Ray. 1999. 'Phytoalexins: what have we learned after 60 years?', *Annual Review of Phytopathology*, 37: 285–306.

Harun-Or-Rashid, Md, Hyun-Jin Kim, Seon-In Yeom, Hyeon-Ah Yu, Md Manir, Surk-Sik Moon, Yang Jae Kang, and Young Ryun Chung. 2018. 'Bacillus velezensis YC7010 enhances plant defenses against brown planthopper through transcriptomic and metabolic changes in rice', *Frontiers in Plant Science*, 9: 1904.

Hirayama, Takashi, and Kazuo Shinozaki. 2010. 'Research on plant abiotic stress responses in the post-genome era: past, present and future', *The Plant Journal*, 61: 1041–52.

Hou, Yuxuan, Jiehua Qiu, Xiaohong Tong, Xiangjin Wei, Babi R Nallamilli, Weihuai Wu, Shiwen Huang, and Jian Zhang. 2015. 'A comprehensive quantitative phosphoproteome analysis of rice in response to bacterial blight', *BMC Plant Biology*, 15: 1–15.

Huang, Yin, Yuan Tian, Geng Li, Yuanyuan Li, Xinjuan Yin, Can Peng, Fengguo Xu, and Zunjian Zhang. 2013. 'Discovery of safety biomarkers for realgar in rat urine using UFLC-IT-TOF/MS and 1 H NMR based metabolomics', *Analytical and Bioanalytical Chemistry*, 405: 4811–22.

Jackson, MB, and TD Colmer. 2005. 'Response and adaptation by plants to flooding stress', *Annals of Botany*, 96: 501–05.

Jain, Priyanka, Himanshu Dubey, Pankaj Kumar Singh, Amolkumar U Solanke, Ashok K Singh, and TR Sharma. 2019. 'Deciphering signalling network in broad spectrum near isogenic lines of rice resistant to *Magnaporthe oryzae*', *Scientific Reports*, 9: 1–13.

Janik, E., Maksymiec, W., Mazur, R., Garstka, M., & Gruszecki, W. I. (2010). Structural and functional modifications of the major light-harvesting complex II in cadmium-or copper-treated Secale cereale. *Plant and Cell Physiology*, 51(8), 1330–1340.

Jansen, Carin, Diter Von Wettstein, Wilhelm Schäfer, Karl-Heinz Kogel, Angelika Felk, and Frank J Maier. 2005. 'Infection patterns in barley and wheat spikes inoculated with wild-type and trichodiene synthase gene disrupted Fusarium graminearum', *Proceedings of the National Academy of Sciences*, 102: 16892–97.

Junot, Christophe, François Fenaille, Benoit Colsch, and François Bécher. 2014. 'High resolution mass spectrometry based techniques at the crossroads of metabolic pathways', *Mass Spectrometry Reviews*, 33: 471–500.

Kandasamy, Saveetha, Karthiba Loganathan, Raveendran Muthuraj, Saravanakumar Duraisamy, Suresh Seetharaman, Raguchander Thiruvengadam, Balasubramanian Ponnusamy, and Samiyappan Ramasamy. 2009. 'Understanding the molecular basis of plant growth promotional effect of *Pseudomonas fluorescens* on rice through protein profiling', *Proteome Science*, 7: 1–8.

Kankanala, Prasanna, Kirk Czymmek, and Barbara Valent. 2007. 'Roles for rice membrane dynamics and plasmodesmata during biotrophic invasion by the blast fungus', *The Plant Cell*, 19: 706–24.

Kaur, Shubhneet, Parmjit S Panesar, Manab B Bera, and Varinder Kaur. 2015. 'Simple sequence repeat markers in genetic divergence and marker-assisted selection of rice cultivars: a review', *Critical Reviews in Food Science and Nutrition*, 55: 41–49.

Kim, Ji Won, Sung Ha Ryu, Siwon Kim, Hae Won Lee, Mi-sun Lim, Sook Jin Seong, Suhkmann Kim, Young-Ran Yoon, and Kyu-Bong Kim. 2013. 'Pattern recognition analysis for hepatotoxicity induced by acetaminophen using plasma and urinary 1H NMR-based metabolomics in humans', *Analytical Chemistry*, 85: 11326–34.

Klavins, Kristaps, Hedda Drexler, Stephan Hann, and Gunda Koellensperger. 2014. 'Quantitative metabolite profiling utilizing parallel column analysis for simultaneous reversed-phase and hydrophilic interaction liquid chromatography separations combined with tandem mass spectrometry', *Analytical Chemistry*, 86: 4145–50.

Koelmel, Jeremy P, Candice Z Ulmer, Christina M Jones, Richard A Yost, and John A Bowden. 2017. 'Common cases of improper lipid annotation using high-resolution tandem mass spectrometry data and corresponding limitations in biological interpretation', *Biochimica et biophysica acta*, 1862: 766.

Koga, Hironori, Koji Dohi, Osamu Nakayachi, and Masashi Mori. 2004. 'A novel inocula-tion method of Magnaporthe grisea for cytological observation of the infection process using intact leaf sheaths of rice plants', *Physiological Molecular Plant Pathology*, 64: 67–72.

Kumaraswamy, GK, AC Kushalappa, TM Choo, Y Dion, and S Rioux. 2012. 'Differential metabolic response of barley genotypes, varying in resistance, to trichothecene-pro-ducing and-nonproducing (tri5–) isolates of Fusarium graminearum', *Plant Pathology*, 61: 509–21.

Lafitte, HR, Guan Yongsheng, Shi Yan, and ZK Li. 2007. 'Whole plant responses, key pro-cesses, and adaptation to drought stress: the case of rice', *Journal of Experimental Botany*, 58: 169–75.

Larkin, Philip J, James AC Miller, Robert S Allen, Julie A Chitty, Wayne L Gerlach, Susanne Frick, Toni M Kutchan, and Anthony J Fist. 2007. 'Increasing morphinan alkaloid pro-duction by over-expressing codeinone reductase in transgenic *Papaver somniferum*', *Plant Biotechnology Journal*, 5: 26–37.

Lemmens, Marc, Uwe Scholz, Franz Berthiller, Chiara Dall'Asta, Andrea Koutnik, Rainer Schuhmacher, Gerhard Adam, Hermann Buerstmayr, Ákos Mesterházy, and Rudolf Krska . 2005. 'The ability to detoxify the mycotoxin deoxynivalenol colocalizes with a major quantitative trait locus for Fusarium head blight resistance in wheat', *Molecular Plant-Microbe Interactions*, 18: 1318–24.

Li, J., Zhu, S. F., Zhao, X. L., Liu, Y. X., Wan, M. H., Guo, H., ... Tang, W. F. (2015). Metabolomic profiles illuminate the efficacy of Chinese herbal Da-Cheng-Qi decoc-tion on acute pancreatitis in rats. *Pancreatology*, 15(4), 337–343.

Li, Mingxia, Jie Xu, Rui Guo, Yuan Liu, Shiyao Wang, He Wang, Abd Ullah, and Lianxuan Shi. 2019. 'Identifying the metabolomics and physiological differences among Soja in the early flowering stage', *Plant Physiology and Biochemistry*, 139: 82–91.

Lisec, Jan, Lilla Römisch-Margl, Zoran Nikoloski, Hans-Peter Piepho, Patrick Giavalisco, Joachim Selbig, Alfons Gierl, and Lothar Willmitzer. 2011. 'Corn hybrids display lower metabolite variability and complex metabolite inheritance patterns', *The Plant Journal*, 68: 326–36.

Liu, Mei-Ling, Peng Zheng, Zhao Liu, Yi Xu, Jun Mu, Jing Guo, Ting Huang, Hua-Qing Meng, and Peng Xie. 2014. 'GC-MS based metabolomics identification of possible novel biomarkers for schizophrenia in peripheral blood mononuclear cells', *Molecular Biosystems*, 10: 2398–406.

Liu, Rui, Zheng-Xue Bao, Pei-Ji Zhao, and Guo-Hong Li. 2021. 'Advances in the study of metabolomics and metabolites in some species interactions', *Molecules*, 26: 3311.

Lu, Hai-ping, Ting Luo, Hao-wei Fu, Long Wang, Yuan-yuan Tan, Jian-zhong Huang, Qing Wang, Gong-yin Ye, Angharad MR Gatehouse, and Yong-gen Lou . 2018. 'Resistance of rice to insect pests mediated by suppression of serotonin biosynthesis', *Nature Plants*, 4: 338–44.

Luengwilai, K., Saltveit, M., & Beckles, D. M. (2012). Metabolite content of harvested Micro-Tom tomato (Solanum Lycopersicum L.) fruit is altered by chilling and protective heat-shock treatments as shown by GC–MS metabolic profiling. *Postharvest Biology and Technology*, 63(1), 116–122.

Mahmood, Tariq, Asad Jan, Makoto Kakishima, and Setsuko Komatsu. 2006. 'Proteomic analysis of bacterial-blight defense-responsive proteins in rice leaf blades', *Proteomics*, 6: 6053–65.

Marti, Guillaume, Matthias Erb, Julien Boccard, Gaetan Glauser, Gwladys R Doyen, Neil Villard, Christelle AM Robert, Ted CJ Turlings, Serge Rudaz, and Jean-Luc Wolfender. 2013. 'Metabolomics reveals herbivore-induced metabolites of resistance and suscepti-bility in maize leaves and roots', *Plant, Cell & Environment*, 36: 621–39.

Matsuda, Fumio, Yozo Okazaki, Akira Oikawa, Miyako Kusano, Ryo Nakabayashi, Jun Kikuchi, Jun-Ichi Yonemaru, Kaworu Ebana, Masahiro Yano, and Kazuki Saito. 2012. 'Dissection of genotype–phenotype associations in rice grains using metabolome quantitative trait loci analysis', *The Plant Journal*, 70: 624–36.

Mazzucotelli, Elisabetta, Alfredo Tartari, Luigi Cattivelli, and Giuseppe Forlani. 2006. 'Metabolism of γ-aminobutyric acid during cold acclimation and freezing and its relationship to frost tolerance in barley and wheat', *Journal of Experimental Botany*, 57: 3755–66.

McDowell, John M, and Jeffery L Dangl . 2000. 'Signal transduction in the plant immune response', *Trends in Biochemical Sciences*, 25: 79–82.

Meng, Qingfeng, Ravi Gupta, Cheol Woo Min, Soon Wook Kwon, Yiming Wang, Byoung Il Je, Yu-Jin Kim, Jong-Seong Jeon, Ganesh Kumar Agrawal, and Randeep Rakwal. 2019. 'Proteomics of Rice—Magnaporthe oryzae interaction: what have we learned so far?', *Frontiers in Plant Science*, 10: 1383.

Meyer, Rhonda C, Matthias Steinfath, Jan Lisec, Martina Becher, Hanna Witucka-Wall, Ottó Törjék, Oliver Fiehn, Änne Eckardt, Lothar Willmitzer, and Joachim Selbig. 2007. 'The metabolic signature related to high plant growth rate in Arabidopsis thaliana', *Proceedings of the National Academy of Sciences*, 104: 4759–64.

Meyer, Rhonda C, Hanna Witucka-Wall, Martina Becher, Anna Blacha, Anastassia Boudichevskaia, Peter Dörmann, Oliver Fiehn, Svetlana Friedel, Maria Von Korff, and Jan Lisec. 2012. 'Heterosis manifestation during early Arabidopsis seedling development is characterized by intermediate gene expression and enhanced metabolic activity in the hybrids', *The Plant Journal*, 71: 669–83.

Molz, Patrícia, Joel Henrique Ellwanger, Carla Eliete Iochims dos Santos, Johnny Ferraz Dias, Deivis de Campos, Valeriano Antonio Corbellini, Daniel Prá, Marisa Terezinha Lopes Putzke, and Silvia Isabel Rech Franke. 2014. 'A metabolomics approach to evaluate the effects of shiitake mushroom (Lentinula edodes) treatment in undernourished young rats', *Nuclear Instruments and Methods in Physics Research Section B: Beam Interactions with Materials and Atoms*, 318: 194–97.

Moose, Stephen P, John W Dudley, and Torbert R Rocheford. 2004. 'Maize selection passes the century mark: a unique resource for 21st century genomics', *Trends in Plant Science*, 9: 358–64.

Muthamilarasan, Mehanathan, Annvi Dhaka, Rattan Yadav, and Manoj Prasad. 2016. 'Exploration of millet models for developing nutrient rich graminaceous crops', *Plant Science*, 242: 89–97.

Nakabayashi, Ryo, and Kazuki Saito. 2015. 'Integrated metabolomics for abiotic stress responses in plants', *Current Opinion in Plant Biology*, 24: 10–16.

Obata, Toshihiro, Sandra Witt, Jan Lisec, Natalia Palacios-Rojas, Igor Florez-Sarasa, Salima Yousfi, Jose Luis Araus, Jill E Cairns, and Alisdair R Fernie. 2015. 'Metabolite profiles of maize leaves in drought, heat, and combined stress field trials reveal the relationship between metabolism and grain yield', *Plant Physiology*, 169: 2665–83.

Okazaki, Yozo, and Kazuki Saito. 2016. 'Integrated metabolomics and phytochemical genomics approaches for studies on rice', *GigaScience*, 5: s13742-016-0116-7.

Oliver, Stephen G, Michael K Winson, Douglas B Kell, and Frank Baganz. 1998. 'Systematic functional analysis of the yeast genome', *Trends in Biotechnology*, 16: 373–78.

Palaisa, Kelly A, Michele Morgante, Mark Williams, and Antoni Rafalski. 2003. 'Contrasting effects of selection on sequence diversity and linkage disequilibrium at two phytoene synthase loci', *The Plant Cell*, 15: 1795–806.

Parida, Asish K, Ashok Panda, and Jaykumar Rangani. 2018. 'Metabolomics-guided elucidation of abiotic stress tolerance mechanisms in plants', in *Plant Metabolites and Regulation under Environmental Stress* (Elsevier).

Peng, Lei, Yan Zhao, Huiying Wang, Jiajiao Zhang, Chengpan Song, Xinxin Shangguan, Lili Zhu, and Guangcun He. 2016. 'Comparative metabolomics of the interaction between rice and the brown planthopper', *Metabolomics* 12: 1–15.

Peng, Yanchun, Hongbo Liu, Jie Chen, Taotao Shi, Chi Zhang, Dongfa Sun, Zhonghu He, Yuanfeng Hao, and Wei Chen. 2018. 'Genome-wide association studies of free amino acid levels by six multi-locus models in bread wheat', *Frontiers in Plant Science*, 9: 1196.

Piasecka, Anna, Piotr Kachlicki, and Maciej Stobiecki. 2019. 'Analytical methods for detection of plant metabolomes changes in response to biotic and abiotic stresses', *International Journal of Molecular Sciences*, 20: 379.

Pieterse, Corné MJ, Antonio Leon-Reyes, Sjoerd Van der Ent, and Saskia CM Van Wees. 2009. 'Networking by small-molecule hormones in plant immunity', *Nature Chemical Biology*, 5: 308–16.

Porath, Jerker. 1997. 'From gel filtration to adsorptive size exclusion', *Journal of Protein Chemistry*, 16: 463–68.

Pott, Delphine M, Sara Durán-Soria, Sonia Osorio, and José G Vallarino. 2021. 'Combining metabolomic and transcriptomic approaches to assess and improve crop quality traits', *CABI Agriculture and Bioscience*, 2: 1–20.

Prasanna, BM, SK Vasal, B Kassahun, and NN Singh. 2001. 'Quality maize protein', *Current Science*, 81: 1308–19.

Rao, Jun, Fang Cheng, Chaoyang Hu, Sheng Quan, Hong Lin, Jing Wang, Guihua Chen, Xiangxiang Zhao, Danny Alexander, and Lining Guo. 2014. 'Metabolic map of mature maize kernels', *Metabolomics*, 10: 775–87.

Rathahao-Paris, Estelle, Sandra Alves, Christophe Junot, and Jean-Claude Tabet. 2016. 'High resolution mass spectrometry for structural identification of metabolites in metabolomics', *Metabolomics*, 12: 1–15.

Roessner, Ute, Alexander Luedemann, Doreen Brust, Oliver Fiehn, Thomas Linke, Lothar Willmitzer, and Alisdair R Fernie. 2001. 'Metabolic profiling allows comprehensive phenotyping of genetically or environmentally modified plant systems', *The Plant Cell*, 13: 11–29.

Saito, Kazuki. 2013. 'Phytochemical genomics—a new trend', *Current Opinion in Plant Biology*, 16: 373–80.

Sanchez, Diego H, Franziska Schwabe, Alexander Erban, Michael K Udvardi, Joachim Kopka. 2012. 'Comparative metabolomics of drought acclimation in model and forage legumes', *Plant Cell, and Environment*, 35: 136–49.

Savary, Serge, Laetitia Willocquet, Sarah Jane Pethybridge, Paul Esker, Neil McRoberts, Andy Nelson. 2019. 'The global burden of pathogens and pests on major food crops', *Nature Ecology, and Evolution*, 3: 430–39.

Saxena, Ashish, and Christopher S Cramer. 2013. 'Metabolomics: a potential tool for breeding nutraceutical vegetables', *Advances in Crop Science and Technology*, 1: 2.

Schroeder, HW, and JJ Christensen. 1963. 'Factors affecting resistance of wheat to scab caused by Gibberella zeae', *Phytopathology*, 53: 831–38.

Segers, Karen, Sven Declerck, Debby Mangelings, Yvan Vander Heyden, and Ann Van Eeckhaut. 2019. 'Analytical techniques for metabolomic studies: a review', *Bioanalysis*, 11: 2297–318.

Sharma, Kapil, Supriya Sarma, Abhishek Bohra, Abhijit Mitra, Naveen K Sharma, and Anirudh Kumar. 2018. 'Plant metabolomics: an emerging technology for crop improvement', in *New visions in plant science*, 1st edn. IntechOpen, London: 65–79.

Shi, Xue, Xiaoli Wei, Imhoi Koo, Robin H Schmidt, Xinmin Yin, Seong Ho Kim, Andrew Vaughn, Craig J McClain, Gavin E Arteel, and Xiang Zhang. 2014. 'Metabolomic analysis of the effects of chronic arsenic exposure in a mouse model of diet-induced fatty liver disease', *Journal of Proteome Research*, 13: 547–54.

Shulaev, Vladimir, Diego Cortes, Gad Miller, and Ron Mittler. 2008. 'Metabolomics for plant stress response', *Physiologia Plantarum*, 132: 199–208.

Singh, J. S., Pandey, V. C., & Singh, D. P. (2011). Efficient soil microorganisms: A new dimension for sustainable agriculture and environmental development. *Agriculture, Ecosystems and Environment*, *140*(3–4), 339–353.

Sobhanian, Hamid, Nasrin Motamed, Ferdous Rastgar Jazii, Takuji Nakamura, and Setsuko Komatsu. 2010. 'Salt stress induced differential proteome and metabolome response in the shoots of Aeluropus lagopoides (Poaceae), a halophyte C4 plant', *Journal of Proteome Research*, 9: 2882–97.

Steinfath, Matthias, Tanja Gärtner, Jan Lisec, Rhonda C Meyer, Thomas Altmann, Lothar Willmitzer, and Joachim Selbig. 2010. 'Prediction of hybrid biomass in Arabidopsis thaliana by selected parental SNP and metabolic markers', *Theoretical and Applied Genetics*, 120: 239–47.

Sumner, Lloyd W, Pedro Mendes, and Richard A Dixon. 2003. 'Plant metabolomics: large-scale phytochemistry in the functional genomics era', *Phytochemistry*, 62: 817–36.

Sun, Yangyang, Yuheng Shi, Guige Liu, Fang Yao, Yuanyuan Zhang, Chenkun Yang, Hao Guo, Xianqing Liu, Cheng Jin, and Jie Luo. 2020. 'Natural variation in the OsbZIP18 promoter contributes to branched-chain amino acid levels in rice', *New Phytologist*, 228: 1548–58.

Sung, Jwakyung, Suyeon Lee, Yejin Lee, Sangkeun Ha, Beomheon Song, Taewan Kim, Brian M Waters, and Hari B Krishnan. 2015. 'Metabolomic profiling from leaves and roots of tomato (*Solanum lycopersicum* L.) plants grown under nitrogen, phosphorus or potassium-deficient condition', *Plant Science*, 241: 55–64.

Sweetlove, Lee J, Jens Nielsen, and Alisdair R Fernie. 2017. 'Engineering central metabolism–a grand challenge for plant biologists', *The Plant Journal*, 90: 749–63.

Takahashi, Hideyuki, Yuji Hotta, Mitsunori Hayashi, Maki Kawai-Yamada, Setsuko Komatsu, and Hirofumi Uchimiya. 2005. 'High throughput metabolome and proteome analysis of transgenic rice plants (Oryza sativa L.)', *Plant Biotechnology Journal*, 22: 47–50.

Tarpley, Lee, Anthony L Duran, Tesfamichael H Kebrom, and Lloyd W Sumner. 2005. 'Biomarker metabolites capturing the metabolite variance present in a rice plant developmental period', *BMC Plant Biology*, 5: 1–12.

Trenkamp, Sandra, Peter Eckes, Marco Busch, and Alisdair R Fernie. 2009. 'Temporally resolved GC-MS-based metabolic profiling of herbicide treated plants treated reveals that changes in polar primary metabolites alone can distinguish herbicides of differing mode of action', *Metabolomics*, 5: 277–91.

Venkatesh, Tyamagondlu V, Alexander W Chassy, Oliver Fiehn, Sherry Flint-Garcia, Qin Zeng, Kirsten Skogerson, and George G Harrigan. 2016. 'Metabolomic assessment of key maize resources: GC-MS and NMR profiling of grain from B73 hybrids of the nested association mapping (NAM) founders and of geographically diverse landraces', *Journal of Agricultural and Food Chemistry*, 64: 2162–72.

Verma, Vivek, Pratibha Ravindran, and Prakash P Kumar. 2016. 'Plant hormone-mediated regulation of stress responses', *BMC Plant Biology*, 16: 1–10.

Vickers, Neil J. 2017. 'Animal communication: when i'm calling you, will you answer too?', *Current Biology*, 27: R713–R15.

Wallace, M Ariel Geer, and James P McCord. 2020. 'High-resolution mass spectrometry', in *Breathborne Biomarkers and the Human Volatilome* (Elsevier).

Wang, Biao, Yingdang Ren, Chuantao Lu, and Xifeng Wang. 2015. 'iTRAQ-based quantitative proteomics analysis of rice leaves infected by Rice stripe virus reveals several proteins involved in symptom formation', *Virology Journal*, 12: 1–21.

Wang, Xuelu, and Brian A Larkins. 2001. 'Genetic analysis of amino acid accumulation in opaque-2 maize endosperm', *Plant Physiology*, 125: 1766–77.

Wei, Wenliang, Donghua Li, Linhai Wang, Xia Ding, Yanxin Zhang, Yuan Gao, and Xiurong Zhang. 2013. 'Morpho-anatomical and physiological responses to waterlogging of sesame (*Sesamum indicum* L.)', *Plant Science*, 208: 102–11.

Wen, Weiwei, Min Jin, Kun Li, Haijun Liu, Yingjie Xiao, Mingchao Zhao, Saleh Alseekh, Wenqiang Li, Francisco de Abreu e Lima, and Yariv Brotman. 2018. 'An integrated multi-layered analysis of the metabolic networks of different tissues uncovers key genetic components of primary metabolism in maize', *The Plant Journal*, 93: 1116–28.

Wiley, John. 2001. Sons Inc. (New York, NY).

Wilson, Larissa M, Sherry R Whitt, Ana M Ibáéñez, Torbert R Rocheford, Major M Goodman, and Edward S Buckler IV. 2004. 'Dissection of maize kernel composition and starch production by candidate gene association', *The Plant Cell*, 16: 2719–33.

Wolfender, Jean-Luc, Gaetan Glauser, Julien Boccard, and Serge Rudaz. 2009. 'MS-based plant metabolomic approaches for biomarker discovery', *Natural Product Communications*, 4: 1934578X0900401019.

Wu, Dezhi, Shengguan Cai, Mingxian Chen, Lingzhen Ye, Zhonghua Chen, Haitao Zhang, Fei Dai, Feibo Wu, and Guoping Zhang. 2013. 'Tissue metabolic responses to salt stress in wild and cultivated barley', *PLoS One*, 8: e55431.

Yang, Min, Jianghao Sun, Zhiqiang Lu, Guangtong Chen, Shuhong Guan, Xuan Liu, Baohong Jiang, Min Ye, and De-An Guo. 2009. 'Phytochemical analysis of traditional Chinese medicine using liquid chromatography coupled with mass spectrometry', *Journal of Chromatography A*, 1216: 2045–62.

Yang, Qiang, Ai-hua Zhang, Jian-hua Miao, Hui Sun, Ying Han, Guang-li Yan, Fang-fang Wu, and Xi-jun Wang. 2019. 'Metabolomics biotechnology, applications, and future trends: a systematic review', *Rsc Advances*, 9: 37245–57.

Yang, Qingqing, Dongsheng Zhao, and Qiaoquan Liu. 2020. 'Connections between amino acid metabolisms in plants: lysine as an example', *Frontiers in Plant Science*, 11: 928.

Yu, Lu, Wenli Wang, Song Zeng, Zhuo Chen, Anming Yang, Jing Shi, Xiaozhen Zhao, Baoan Song. 2018. 'Label-free quantitative proteomics analysis of cytosinpeptidemycin responses in southern rice black-streaked dwarf virus-infected rice', *Pesticide Biochemistry and Physiology*, 147: 20–26.

Zhang, Aihua, Huiyu Wang, Hui Sun, Yue Zhang, Na An, Guangli Yan, Xiangcai Meng, and Xijun Wang. 2015. 'Metabolomics strategy reveals therapeutical assessment of limonin on nonbacterial prostatitis', *Food & Function*, 6: 3540–49.

Zhang, Xiaoyun, Fuyou Yin, Suqin Xiao, Chunmiao Jiang, Tengqiong Yu, Ling Chen, Xue Ke, Qiaofang Zhong, Zaiquan Cheng, and Weijiao Li. 2019. 'Proteomic analysis of the rice (Oryza officinalis) provides clues on molecular tagging of proteins for brown planthopper resistance', *BMC Plant Biology*, 19: 1–11.

Zhao, Jianjun, Maria-Joao Paulo, Diaan Jamar, Ping Lou, Fred Van Eeuwijk, Guusje Bonnema, Dick Vreugdenhil, and Maarten Koornneef. 2007. 'Association mapping of leaf traits, flowering time, and phytate content in Brassica rapa', *Genome*, 50: 963–73.

Zogli, Prince, Lise Pingault, Sajjan Grover, and Joe Louis. 2020. 'Ento (o)mics: the intersection of 'omic' approaches to decipher plant defense against sap-sucking insect pests', *Current Opinion in Plant Biology*, 56: 153–61.

10 Metabolic Responses in Plants under Abiotic Stresses

Ahmad Ali, Kiran Khurshid, Namrah Ahmad,
Nida Mushtaq, Rabia Amir, and Faiza Munir

CONTENTS

DOI: 10.1201/9781003250845-10

10.1 INTRODUCTION

Climate changes due to deforestation and fossil fuel usage trigger enhancement of greenhouse gasses and carbon dioxide levels. These changes might be intensified, as the average global temperature rise and decrease in soil moisture and participation are 6.4°C and 20%, respectively (Field et al. 2014; Schiermeier 2008). Flooding, heat waves, drought and other abiotic stresses can have a great influence on the world economy due to metabolic alteration, reducing yield and productivity. Due to a recorded 188 events of heavy storms, drought and flooding, economic losses in the United States were above \$1 trillion from 1980 to 2014 (https://www.ncdc. noaa.gov/cdo-web/). Agriculture productivity enhancement is also required for the growing world population. By 2050 there will be 2.3 billion additional people, and to feed them a 70% increase in productivity is required (Tilman et al. 2011; Wani and Sah 2014). Identification of plant response to various abiotic stresses is an essential aspect of plant biotechnology (Qin et al. 2011). In combination, these abiotic stresses can have additive, negative and severe consequences on productivity (Mittler 2006; Rizhsky et al. 2002). Therefore, further research is required to identify the effects of abiotic stresses on plant metabolism and to apply biotechnological tools to obtain more tolerant plants (Lobell and Gourdji 2012; Prasad et al. 2009). Metabolic pathways alteration, transcriptional and proteomic reprograming are the responses of plants toward abiotic stresses, where many transcriptional factors, metabolic enzymes, chaperones, detoxification enzymes and protein kinases are involved (Maruyama et al. 2014; Nakashima et al. 2014).

Continuously occurring evolutionary processes promoted the diversification of plant metabolism, as 200,000 plant secondary metabolites have been discovered with multiple functions. Metabolism regulates physiological, defense and signaling responses under adverse environmental conditions. Concentration, storage, biosynthesis and transport of metabolites (primary and secondary) are affected by abiotic stresses. Carbohydrate, amine and amino acid metabolic pathways are involved in metabolic adjustment against these stresses. Activation of prompt metabolic responses restores energetic and chemical imbalance disturbed by abiotic stresses and is pivotal to survival and acclimation. Time-series analyses have shown that transcriptional activities respond to different stresses slower than metabolic activities. Thus, comprehensive and integrative investigations are needed to study these synchronous responses and to correlate them with corresponding abiotic stresses. Integration of data collected from metabolic research in response to abiotic stresses is revealing about the tolerance traits that can be transferred to economical crops. Metabolomics is based on approaches like mass spectrometry, and liquid and gas chromatography which help in depicting the changes of metabolites in response to stresses. Certainly, current knowledge about metabolite regulation toward different stresses like drought, salinity, oxidative, heat, flooding and ozone in several plant species is largely contributed by metabolomics (Krasensky and Jonak 2012; Obata and Fernie 2012). Another potent approach is phytohormone engineering to produce tolerant and high-yielding crops. Phytohormones are important for plant metabolism, as they act as chemical messengers, regulate cellular processes and signal transduction pathways (Kazan 2015; Voß et al. 2014).

In this chapter, we discuss identified metabolic responses, state-of-the-art developments on the metabolic responses to several abiotic stresses and their results in the form of enhanced crop yield and productivity, along with prospects.

10.2 PLANT GROWTH RESPONSES UNDER ABIOTIC STRESSES

Plants respond to abiotic stresses such as temperature, salinity and drought (Matich et al. 2019) through complex and multigenic networks of biochemical, molecular and physiological responses (Gill et al. 2016). Physiological responses include mechanical restraints, altered activities of macromolecules, oxidative stress leading cellular damage (Bechtold and Field 2018) and a decrease in osmotic potential with changed cellular environment (Singh et al. 2020). Salinity stress disrupts ionic and osmotic homeostasis by altering ion distribution and water potential affecting homeostasis at the cellular level (Liu et al. 2019) that leads to molecular damage and cellular-level growth arrest (Cramer et al. 2011). Similar responses are associated with heat, cold, salinity, drought and mineral deficiency (Ahanger et al. 2016). Salt, drought, and osmotic and ionic stress affect the water potential at low levels by regulating turgor and active transport of solutes for hydrogen ion gradient control (Ahanger et al. 2017), thus creating ionic balance (Huang 2018). Fluctuation in ionic balance serves as the primary sensor for secondary signaling pathways, which regulate defense and tolerance mechanisms (Barlow et al. 2015). Metabolic changes under heat stress are species and habitat specific for plants (Wang et al. 2019). High temperature denatures enzymes (Hatfield and Prueger 2015) and hinders metabolic or synthetic processes like pigment biosynthesis, light-dependent reactions and metabolism of carbon (Fahad et al. 2017) causing oxidative burst (Ohama et al. 2017), electron flow imbalance and partial impairment of pigment (Hemantaranjan et al. 2014). Drought stress limits nutrients' availability to plants for growth mechanisms that influence productivity (Ahanger et al. 2016). Drought reduces transpiration due to faulty membrane transporters, which lower the mineral nutrient transportation rate causing an imbalance in nutrient distribution as a secondary response (Ahanger et al. 2017). Under heat stress, the conductance of water from roots to aerial parts of the plants is compromised (Hatfield and Prueger 2015). Inorganic ion uptake by roots is inhibited when there is water deficiency and insufficient nutrient provision to leaves and shoots (Huang et al. 2016). Thus, the plant responds by early closure of guard cells, which decreases transpiration (Fahad et al. 2017).

10.3 REGULATION OF METABOLIC PROCESSES: OSMOLYTE ACCUMULATION

At the cellular and organismic level, plants are known to adopt a variety of responses of the physiological and biochemical nature in order to tackle stress conditions (Nahar et al. 2016). A widely found stress resistance technique in plants is the alteration of the status of metabolites. Such metabolites include galactinol, raffinose, trehalose and gluconate, which are carbohydrates. Metabolites that are amino acids include glutamate, asparagine and proline, while there are polyamines like spermine, spermidine and putrescine (Parida et al. 2018). These metabolites are compatible solutes or osmoprotectants acting as minute, nontoxic molecules, electrically neutral at molar concentrations, and accumulate in plants under abiotic stress environments as a defensive mechanism (Nahar et al. 2016). Some of the protective roles osmoprotectants play under stress conditions are maintenance of cellular tension by restoration of osmotic equilibrium, hunting reactive oxygen species (ROS) to reintroduce redox metabolism and balancing of proteins and cellular entities to be developed (Vahdati and Leslie 2013). Plants that regulated increased levels of osmolyte biosynthesis/metabolite genes gave better tolerance to stress (Marwein et al. 2019). It is important to know that some metabolic variations are common in all kinds of stresses like drought, salt and temperature, while other variations are specific. For instance, proline accumulation occurs upon all other kinds of stresses except high-temperature stresses. Proline and raffinose concentration increase rapidly in response to drought, cold and salt stresses, while the main carbohydrate process varies quickly in a complex fashion that is time dependent (Krasensky and Jonak 2012).

Accumulation of amino acids has been seen in numerous studies on plants under stress conditions. This is either due to amino acid synthesis or from increased breakdown of proteins due to a stressful environment. While the overall effect of an increase in amino acid signifies cell damage in some species, an increase in amino acids in some species has a beneficial effect during stress adaptation (Hayat et al. 2012). Proline can act as cryoprotectant, signaling molecule, molecular structure stabilizer and ROS scavenger under stress that results in dehydration, involving freezing, salinity, heavy metals and drought, as a molecular chaperone and an osmoprotectant (Nahar et al. 2016). During the occurrence of stress there is proline production, while during recovery from stress there is proline breakdown (Singh et al. 2015). Proline is synthesized in cytosol or chloroplast from glutamate, which is broken into glutamate-semialdehyde (GSA) by Δ-1-pyrroline-5-carboxylate synthetase (P5CS). Pyrroline-5-carboxylate (P5C) is formed from GSA, which is then converted into proline by reduction of proline dehydrogenase (ProDH). There is an increased accommodation of proline-rich cell wall linker proteins in plasma membrane due to proline overproduction. The production of such a metabolite is increased while its breakdown is decreased at the same time during osmotic or salt stress conditions (Zagorchev et al. 2014). Proline maintenance is done by various protein kinases that are crucial for drought, salt and cold tolerance. Recent investigation reveals that there is a promoting role of MAPK-rooted signaling in the proline accommodation under stress conditions. The levels of accommodation of non-protein amino acid γ-aminobutyric acid (GABA) increase under salt stresses (Krasensky and Jonak 2012).

Polyamines (PAs) have been involved in reducing oxidative stress and protecting membranes (Krasensky and Jonak 2012). During dehydration, stress glycine betaines (GBs) are the most present quaternary ammonium compounds in plants (Nahar et al. 2016). The notable roles of GB are removal of oxidative damage, protection of photosystem II, and stabilization of membranes (Chen and Murata 2011). During dehydration or freezing, trehalose substitutes water by forming hydrogen bonds with macromolecules or membranes (Nahar et al. 2016). Carbohydrates such as fructans and starch are accumulated as storage materials that can be used through times of reduced energy availability in plants. The plant increases the levels of some metabolites while decreasing the levels of others to regulate normal functioning when exposed to stress environments (Krasensky and Jonak 2012). Osmolyte regulation under abiotic stress is shown in Figure 10.1.

10.4 PLANT METABOLOMICS INVOLVES PLANT RESPONSES AT DIFFERENT LEVELS

Plants reconfigure their metabolic system to gain tolerance against abiotic stresses. Under abiotic stress conditions, plant developmental processes and metabolism is affected in terms of restraint of metabolic chemicals, lack of substrate, overabundance need for specific metabolites or a blend of these components (Obata and Fernie 2012).

10.4.1 CARBOHYDRATE/SUGAR, AMINO ACID AND FATTY ACID METABOLISM

Sugars such as mono-oligosaccharides, di-oligosaccharides, raffinose and stachyose are involved in the regulation of many physiological processes and immune responses in plants. Sugar-associated immunity is termed as "sweet immunity" or "sugar enhanced immunity" (Sami et al. 2016). Application of sugar spray activates sugar-triggered immunity to protect plants under high disease stress (Trouvelot et al. 2014). For example, in crop plants, negative impacts of salt stress can be overcome by using sugar molecules, as they can act as osmolytes and signaling molecules. Furthermore, an increase in the intensity of sugar molecules like glucose, sucrose and fructose during salt stresses shows a varied stress tolerance (Rosa et al. 2009). Sugars such as glucose can prompt stomatal closure and prevent tissue dehydration and increase tolerance to drought stress. At the cellular level, cold stress is harmful (Yuanyuan et al. 2009), therefore in many plants sucrose, raffinose, glucose and other sugar molecules provide resistance against cold stress by interacting with the lipid bilayer, e.g., exogenous use of sucrose in *Arabidopsis thaliana* increases resistance to cold stress (Fàbregas and Fernie 2019).

Conversely, in many plant species, amino acid accumulation has been reported under stress conditions (Zeier 2013). The relationship between an amino acid and plant immunity against pathogens has been known for a long time, as proline is an exemplary model since it is realized that its accumulation is related to drought tolerance. Similarly, γ-aminobutyric acid (GABA) levels are upregulated under drought stress (Zinta et al. 2018). Past investigations indicated that during the reactions of stress, upregulated aromatic amino acids were utilized for the synthesis of secondary metabolites and play a crucial role in response to abiotic stresses (Less and Galili 2008). However, it was demonstrated that respiration pathways utilize these amino acids (particularly branched-chain amino acids) under drought stress to minimize its effect (Zinta et al. 2018).

Moreover, fatty acids and lipids play important roles in plant biological processes, which is especially significant for those procedures that underlie the plant resistance reaction. For instance, fatty acids like oxylipins, diacylglycerol, phenolics, phosphatidylglycerol (PG) and omega 3 desaturase fatty acids play important roles in abiotic stresses (Lim et al. 2017; Upchurch 2008). Lipid-dependent signaling also plays

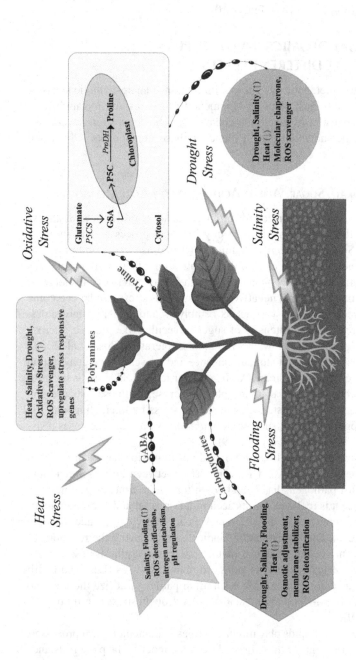

FIGURE 10.1 Osmolyte regulation under abiotic stresses. Proline acts as a molecular chaperone and ROS scavenger under abiotic stresses. During drought, salinity and heat stress, the level of proline is upregulated. However, during heat stress there is no notable proline accumulation. Proline is synthesized in cytosol or chloroplast from glutamate by GSA and P5C reductase. Accumulation of GABA is upregulated in response to salinity and flooding stress. GABA plays an important role in ROS detoxification, nitrogen metabolism and pH regulation. Drought, heat, salinity and flooding stress results in upregulation of carbohydrates. Carbohydrates are essentially involved in osmotic adjustment, ROS detoxification and in stabilizing membranes. Accumulation of polyamines significantly increases in response to salinity, heat, drought and oxidative stress. Polyamines function in ROS detoxification and in upregulation of stress-responsive genes (Suprasanna et al. 2016). GABA, γ-aminobutyric acid; ROS, reactive oxygen species; P5CS, Δ-1-pyrroline-5-carboxylate synthetase; GSA, glutamate-semialdehyde; P5C, pyrroline-5-carboxylate; ProDH, proline dehydrogenase.

an important role to make plants adaptive to various abiotic stresses by activating transcription cascade and the gene expressions leading to stress adaptation (Hou et al. 2016a). Drought and salt stress onset involve the accumulation of phosphatidic acid by targeting many proteins. For instance, phosphatidic acid has accounted for advanced stomata closing by blocking protein phosphatase ABI1 (McLoughlin and Testerink 2013).

Similarly, under chilling stress sphingolipids and their metabolites are associated with an assortment of cell and environmental reactions. Phosphorylated ceramides and phytosphingosine-1-phosphate contribute to overcoming such stress conditions. Sphingosine phosphorylated derivatives are additionally related to ABA-dependent stomatal closing and the adaptive responses under water-deficit conditions (Michaelson et al. 2016).

10.4.2 ROLE OF PHYTOHORMONES

A small amount of each phytohormone can enhance a plant's adaptability in abiotic stress conditions (Kaur et al. 2019). Phytohormones include jasmonates (JAs), auxin (Aux), gibberellins (GAs), cytokinin (CKs), salicylic acid (SA), ethylene (ET), brassinosteroids (BRs) and abscisic acid (ABA) (Wani et al. 2016). Among these phytohormones, the endogenous and exogenous application of ABA leads to cold, salinity, drought and heat stress tolerance by manipulating many physiological responses such as stomatal closing, seed dormancy, and production of storage proteins and lipids (Vishwakarma et al. 2017). It has been estimated that *Cucumis sativus* L. (cucumber) seedlings grown under cold stress result in spillage of cell contents and water, and subsequently cell shrinkage. Therefore, exogenously applied ABA to these seedlings fundamentally reduces their stresses (Rikin and Richmond 1976). Interestingly, auxin-related genes and auxin production play an important role under different stresses, i.e., osmotic, drought and salt stress. For instance, osmotic stress is a causative factor of oxidative damage caused by ROS and can interrupt normal cellular functions. Therefore, auxin combines stress signals of other phytohormones to hunt ROS and normalize root development, which is a stress-induced morphological response (Sharma et al. 2015). Cytokinins have both negative and positive consequences under environmental stress conditions (Zwack and Rashotte 2015). Various examinations conducted on plant taxa discovered that the phytohormone cytokinin is downregulated under prolonged stress. Whereas, numerous investigations have detailed both short-range and continued rise in cytokinin levels, especially because of severe stress conditions (Zwack and Rashotte 2015). These stress conditions include high light, altered photoperiod and temperature, salinity, osmotic, drought and mineral deficiency (Cortleven omega 2019). Ethylene is a hormone that exists in gaseous form and has different functions in regulating plant metabolic processes and stress conditions (Khan et al. 2017). However, the onset of stress tolerance depends on the varying concentration of ethylene under stress conditions (Dubois et al. 2018). Ethylene likewise significantly enhances plant strength to heat stress. Different biotic or abiotic stresses increase ET concentration, which builds plant endurance probabilities under these unfavorable conditions (Wani et al. 2016). Like other hormones, GAs also play important roles in several plant functions

(Vishal and Kumar 2018) and under different abiotic stresses (e.g., osmotic, salinity and chilling stress). Hence, under different abiotic stresses, gibberellin signaling is downregulated, which results in the reduction of plant growth because GA reduction is associated with accumulation of DELLA protein (a transcriptional regulator). Such changes lead to a reduction in plant growth and enhance stress tolerance (Colebrook et al. 2014). Additionally, BRs can emphatically impact plant reactions to abiotic stresses, for example, heat, cold, water deficiency, salt, pesticides and metal stress (Ahammed et al. 2020). Beside other hormones, the JAs trigger the induction of several defense genes in response to salt, heavy metal, drought, nutrient deficiency, low temperature and other abiotic stresses (Ali and Baek 2020). Salicylic acid is an inducer of several genes, some of which encode for HSPs (heat shock proteins) and chaperones (Ahmad et al. 2019). Specifically, the importance of SA has been increased in plant immunology, as it can increase plant tolerance against several abiotic stresses through salicylic acid-induced control of plant metabolic procedures in plants such as *Arabidopsis thaliana*. For instance, exogenously applied SA regulates the ICS1 (isochorismate synthase 1) gene, which in turn increases drought tolerance (Khan et al. 2015). In higher plants, abiotic stress resistance induced by salicylic acid might be established with proteins related with stress, e.g., PR proteins, late embryogenesis abundant proteins (LAE), some enzymes acting as antioxidants, 14-3-3 proteins, RS1-like proteins which are salt stress root proteins, two glutamine synthase isoforms and several other stress-related proteins (Kang et al. 2014).

10.4.2.1 Crosstalk between Phytohormones under Abiotic Stresses

Under different abiotic stresses, several plant hormones interact to fine-tune defense mechanisms. Crosstalks are the merging points that usually take place among signal transduction pathways. And then a subsequent defense signaling response is structured by this crosstalk, as shown in Figure 10.2 (Wani et al. 2016).

10.4.2.2 Metabolic Engineering of Phytohormones

The engineering of phytohormones enables biotechnologists to increase stress adaptation. In transgenic approaches, the enzymes, genes and other signal transduction pathways responsible for phytohormones biosynthesis are targeted (Ciura and Kruk 2018). Identification, characterization and integration of genes responsible for phytohormones biosynthesis will be helpful in transgenic crop development (Sharma et al. 2019). The objective genes for plant hormone catabolism and anabolism are listed in Table 10.1.

10.4.2.3 Metabolic Responses in Plants Correlated with Circadian Rhythms

Plants have an internal clock, called circadian rhythms, which foresees natural signs, for example, temperature, humidity, light and plant metabolism (Espinoza et al. 2010). The transcriptome examinations proved that the circadian clock regulates one-third of all the genes in Arabidopsis or 30% of a plant's genome transcript is regulated by circadian rhythms.

This is particularly true for sugar-responsive genes because products of photosynthesis are usually stored in the leaves during the daytime, generally as starch, and reused for the sucrose synthesis and exported at nighttime indicating that many

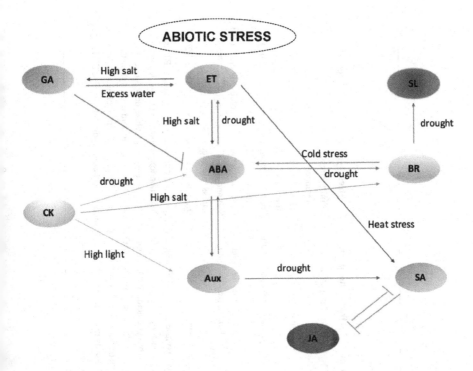

FIGURE 10.2 Crosstalk between phytohormones under abiotic stress. Crosstalk of these phytohormones allows plants to coordinate complex processes required under abiotic stresses. These meeting points are essentially important to trigger defense responses, as shown above (Wani et al. 2016). ET, ethylene; GA, gibberellic acid; SA, salicylic acid; CK, cytokinin; ABA, abscisic acid; BR, brassinosteroid; Aux, auxin; JA, jasmonic acid; SL, strigolactones.

sugar-responsive genes are regulated by circadian rhythms (Bläsing et al. 2005). Similarly, under constant light conditions, the quantity of numerous metabolites of the Calvin–Benson cycle fluctuate in bean plants (Farré and Weise 2012). Circadian rhythms are also considered important for lipid and fatty acid biosynthesis because the lipid and fatty acid biosynthesis enzymes are transcriptionally controlled by circadian rhythms (Harmer et al. 2000). Fatty acid, especially oleic acid (18:1) and linolenic acid (18:3), concentrations fluctuate with circadian rhythms. In *Arabidopsis thaliana*, oleic acid (18:1) concentration was decreased with a subsequent increase of linolenic acid (18:3) under 14-hour dark and 10-hour light conditions. When the same plants were exposed to light conditions, the (18:3) concentration was increased with a detriment of (18:1). However, amino acid synthesis is also regulated by circadian rhythms. It has been proposed that amino acid fluctuations might be due to diurnal expressions of protein synthesis (Yang and Midmore 2005). For instance, in *Arabidopsis thaliana*, the majority of the aminos have elevated levels toward the start of the night frame in the leaves of Arabidopsis, which is due to diurnal cycles (Farré and Weise 2012). Circadian rhythms can regulate hormone biosynthesis in many plants. For example, auxin levels in rosette leaves of Arabidopsis fluctuate by

TABLE 10.1
Objective Genes for Plant Hormone Catabolism and Anabolism

Phytohormone	Genes involved in anabolism	Genes involved in catabolism	Transgenic plant	Stress type	Reference
BR	DWF4/DWARF4	OsGSK1	*Arabidopsis thaliana*	Salinity stress tolerance	—
ET	ACC synthase	ACC oxidase, ACD	Tomato	Salt and cold tolerance	Klay et al. (2014)
AUX	IAA, tms1, YUC6, NIT	GH3	*Arabidopsis* and potato	Enhanced drought tolerance	Kim et al. (2013)
GA	GA20ox, GA3ox	GA2ox	*Arabidopsis thaliana*	Drought tolerance	Colebrook et al. (2014)
ABA	ZEP, NCED, SDR, AAO, LOS5	?	*Arabidopsis thaliana*	Salt and mannitol stress tolerance	Singh et al. (2015)
Cytokinins	ARRS	CKX	Rice	Drought tolerance	Reguera et al. (2013)

Abbreviations: AAO, aldehyde oxidase; DWF4/DWARF4, gene encoding steroid 22α-hydroxylase (CYP90B1GH3), CKX, cytokinin oxidase/dehydrogenase; IPT, isopentenyl transferase; UGT, UDP-glycosyl-transferase; 13-LOX, 13-li-poxygenase; tms1, agrobacterial gene encoding tryptophan monooxygenase; NIT, nitriles converting indole-3-acetonitrile to IAA; ZEP, zeaxanthin epoxidase; ACC, 1-aminocyclopropane-1-carboxylic acid; ACD, 1-aminocyclopropane-1-carboxylase deaminase; GA2OX, GA 20-oxidase; iaaM, bacterial gene encoding tryptophan monoxidase; LOS5, molybdenum cofactor required for AAO activity; JMT, jasmonic acid carboxyl methyltransferase, NCED, 9-cis-epoxycartenoid dioxygenase; YUC, gene encoding enzyme converting tryptamine to N-hydroxyl-tryptamine; SDR, short-chain alcohol dehydrogenase/reductase; GH3, proteins responsible for conjugation of Aux to amino acids; ARRS, Arabidopsis response factors; CKX, cytokinin oxidase/dehydrogenase; OsGSK1, oryzae sativa glycogen synthase kinase-3 like gene 1; GA2ox, GA 2-oxidase; GA3oX, GA 3-oxidase.

circadian rhythms (Staiger 2002). In sorghum, gibberellin (GA) biosynthesis is also controlled by circadian rhythms (Lee et al. 1998). In barley, wheat and rye, ethylene biosynthesis is regulated with the circadian clock (Espinoza et al. 2010).

10.4.2.4 Regulation of Circadian Rhythms by Metabolites

Transcription factors encoded by clock genes mutually regulate each other. It was reported that in *Arabidopsis thaliana* CCA1 (clock associated 1), LHY (late elongated hypocotyl) and TOC1 (timing of CAB expression) transcription factors regulate several circadian rhythm outputs (Kim et al. 2019; Staiger 2002). However, in Arabidopsis, photosynthesis is regulated by the circadian clock periodically and this periodic photosynthesis in turn regulates the circadian clock as a feedback mechanism. This was proved by inhibition of photosynthesis, which elongates circadian rhythms (Haydon et al. 2017). In *Arabidopsis thaliana*, sugar can trigger the circadian clock by regulating its genes early in the photoperiod and thus expressing a "metabolic dawn" (Haydon et al. 2013).

10.5 METABOLIC ALTERATIONS IN RESPONSE TO DIFFERENT ABIOTIC STRESSES

10.5.1 Drought Stress

A plant produces ABA in response to drought stress conditions to adapt itself for survival. ABA production triggers stomatal closure, metabolic alteration and expresses drought-related genes, as illustrated in Figure 10.3 (Seki et al. 2002; Yamaguchi-Shinozaki and Shinozaki 2006). For drought stress there are two transcriptional regulatory systems, ABA-dependent and ABA-independent, demonstrated by molecular analysis. In Arabidopsis, several main signaling molecules have been identified and analyzed, which showed involvement in signal transduction, for example, transcription factors ABF2/ABA-AREB1 (Furihata et al. 2006) and DREB2A (Sakuma et al. 2006), and calcium-dependent protein kinases CPK6 and CPK3 (Mori et al. 2006), and protein kinases such as RPK1 (Osakabe et al. 2006) and SRK2C (Umezawa et al. 2004). Receptors for ABA are reported such as CHLH and FCA RNA binding protein (Shen et al. 2006). ABA's role in drought stress is confirmed through the silencing *AtNCED3* gene, which codes for 9-cis-epoxycarotenoid dioxygenase enzyme. 9-cis-Epoxycarotenoid dioxygenase cleaves C40 carotenoids into Xanthoxin C15, which is ABA precursor in its biosynthesis. Mutants having *nced3* showed a decrease in the accumulation of ABA and reduced tolerance toward drought (Iuchi et al. 2001; Schwartz et al. 1997). Sugars, trehalose, amino acids, sucrose, amines and sugar alcohols accumulate in several species under drought stress, demonstrated by physiological studies, which act as antioxidants, osmolytes and scavengers for survival and are associated with maintaining and protecting the structure and function of cellular components (Bartels and Sunkar 2005). Metabolic pathways are engineered in plants through genetic engineering by introducing drought-related genes and showed high tolerance (Umezawa et al. 2006a). Thus, metabolism of ABA and osmolytes are associated with drought stress.

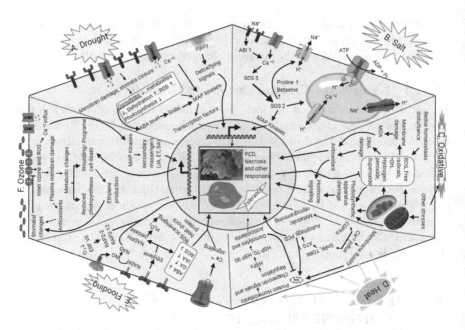

FIGURE 10.3 Metabolic alterations in response to different abiotic stresses. (a) Drought signals perception causes membrane damage, Ca^{+2} influx and cell process alterations, which lead to the activation of the MAP kinase cascade that triggered transcription regulation for tolerance (Seki et al. 2002; Yamaguchi-Shinozaki and Shinozaki 2006). (b) Maintenance of ion homeostasis under salinity stress. The salt overly sensitive (SOS) pathway is activated by the salinity stress. The increase in cytosolic Ca^{+2} is sensed by SOS3 which activates SOS2. The activated SOS3–SOS2 protein complex phosphorylates the plasma membrane Na^+/H^+ antiporter, SOS1 (Yang and Guo 2018b). (c) ROS damage membrane, DNA and photosynthesis, activating hormonal signaling and antioxidant accumulation that cause cell death/PCD and ROS scavenging, respectively. (d) HSP activation under heat stress to induce antioxidants and osmolyte accumulation. Ca^{++} influx triggers autophagy, PCD and metabolic reprogramming by employing kinases, e.g., SnRK1, TORK, CDPKs and ATG1 (Fahad et al. 2017). (e) Flooding stress decreases availability of oxygen (O_2) and induces production of ethylene (C_2H_4) and nitric oxide (NO) leading to the activation of the downstream signal transduction pathway (Lekshmy et al. 2015). (f) Ozone entry upregulates ROS and damages membrane that cause metabolism disturbance, which lead to PCD, reduced photosynthesis, ethylene production and stomatal closure. On the other hand, Ca^{+2} influx also triggers PCD (Morgan et al. 2004; Saxena et al. 2019). Ca^{+2}, calcium ion; ROS, reactive oxygen species; ABA, abscisic acid; SnRK, sucrose non-fermenting 1 related protein kinases; MAP kinases, mitogen-activated protein kinases; ABI1, abscisic acid insensitive 1; SOS, salt overly sensitive pathway; ATP, adenosine triphosphate; ADP, adenosine diphosphate; Pi, inorganic phosphate; H^+, hydrogen ion; Na^+, sodium ion; ATPase, adenosine triphosphatase; MDA, malondialdehyde; CDPKs, calcium-dependent protein kinases; TORK, serine/threonine protein kinase; ATG1, autophagy-related genes; HSFs, heat stress transcription factors; HSP, heat shock protein; NO, nitric oxide; GA, gibberellic acid; IAA, indole acetic acid; NADH/NAD, nicotinamide adenine dinucleotide; NADPH, nicotinamide adenine dinucleotide phosphate; O_2, oxygen; ERF VII, group VII ethylene response factors; RAP2.2/RAP2.12, ethylene response transcription factor; H_2O_2, hydrogen peroxide; JA, jasmonic acid; ET, ethylene; SA, salicylic acid; PCD, programmed cell death.

10.5.1.1 Osmolyte Accumulation

Sugars, for example sucrose, sorbitol, trehalose and raffinose family oligosaccharides (RFO), sugar alcohol (mannitol), amines such as polyamines and glycine betaine, and amino acids (proline) accumulate and act as osmolytes, stabilize proteins, and maintain structures and turgidity of cell under drought conditions (Bartels and Sunkar 2005). Many genes are identified as drought-inducible with distinct functions that contribute to these osmolyte metabolisms, which might be used to enhance osmolyte accumulation.

Sugar (galactinol, fructan, RFO, trehalose) and sugar alcohol (D-ononitol and mannitol) accumulation are enhanced against drought stress upon insertion of genes involved in their metabolism (Bartels and Sunkar 2005; Cook et al. 2004; Garg et al. 2002; Kasuga et al. 1999; Taji et al. 2002; Umezawa et al. 2006b). Galactinol synthase (AtGolS2) gene overexpression helps in the accumulation of raffinose and galactinol (Taji et al. 2002). *DREB1A/CBF3* overexpression enhances tolerance toward cold and drought stress as it gathers more raffinose and galactose as compared to the wild variety (Avonce et al. 2004; Taji et al. 2002; Valliyodan and Nguyen 2006). These results concluded that raffinose and galactinol play the role of osmoprotectants in drought conditions. D-ononitol and sugar alcohol protect membranes and enzymes from hydrogen radicals in drought conditions, as studied in tobacco upon myo-inositol O-methyltransferase gene insertion (Vinocur and Altman 2005).

In the plant defense mechanism, amino acid proline (pro) plays a diverse role, as it works as a stabilizer of subcellular structure, energy sink, free radical scavenger, a signal related to stress and controller of osmotic modulation (Nanjo et al. 1999). Pro accumulation is correlated with drought tolerance shown by *P5CS* overexpression or by *ProDH* antisense suppression in several plants (Bartels and Sunkar 2005). Polyamines (APs) such as spermine, putrescine and spermidine are required for plant growth, developmental processes and responses toward abiotic stress in several plants. PA accumulation is regulated by *Datura ADC* gene overexpression in response to drought stress in rice (Capell et al. 2004). Spermidine synthase, *LTI78*, *RD29A*, *COR78* and *DREB2B* upregulated in drought stress enhance the accumulation of spermidine (Kasukabe et al. 2004). This evidence suggests that PAs play important roles in responses toward drought stress. An amine (Glycine betaine) maintains the water balance between the environment and cell wall, stabilizing macromolecule activity and structure to protect the plant (Sakamoto and Murata 2002).

10.5.2 SALINITY STRESS

An intensely dangerous environmental stress factor is saline soil, which has serious impacts on crop growth, seed germination and productivity resulting in serious agricultural yield losses (Muscolo et al. 2015). High soil salinity is spread to about approximately 20% of the area that is cultivated with 6.5% of the world's total area. Salt stress is entirely caused due to excessive levels of both Na^+ and Cl^- ions in the soil. Three stress pathways induced due to salinity stress are ionic stress, osmotic stress and secondary stress (Yang and Guo 2018a). Ionic stress is caused due to increased levels of toxic ions within plant cells. Salts that are taken up by plants reach the shoots by long-distance transport via transpiration streams accumulated in the leaves. Metabolic pathways are adversely affected due to high levels of Na+ ions in the cytoplasm, which limit the uptake of other ions into the plant cell. For instance, the significance of potassium (K+) is known for many catalytic enzyme activities. Osmotic stress occurs due to high concentrations of salt in the soil. Due to increased soluble salt in the soil, there is a decrease in water potential at the surface of the root, which ultimately affects plant uptake of water, resulting in water deficiency (Parihar et al. 2015). The occurrence of osmotic and ionic stress in plants causes other toxic compounds to accumulate and cause nutritional imbalances, leading to secondary stresses in plants. Due to the generation of reactive oxygen species, the salinity stress is accompanied by oxidative stress (Isayenkov and Maathuis 2019).

The ability to detect both the hyperosmotic component and ionic Na+ component of stress is developed in plants to exhibit an effective response to salt stress. The salt overly sensitive pathway, abbreviated as SOS pathway, is a key regulator of ion homeostasis (Hill et al. 2013). Plasma membrane receptors perceive increased levels of Na+ and high osmolarity, separately, leading to enhancement in levels of cytosolic Ca+2 (Forni et al. 2017). This increase is sensed by SOS3 kinase, thereby activating SOS2. The plasma membrane Na+/H+ antiporter and the protein complex phosphorylate SOS1 from the activated SOS3–SOS2 is illustrated in Figure 10.3 (Carillo et al. 2011). In responding to excess ions, maintenance of cellular ion homeostasis is a significant adaptive trait of salt-tolerant plants. There is an increase of K+ and a decrease of Na+ to obtain a suitable K+:Na+ ratio in the cytoplasm; this prevents damage to the cell and saves from nutritional deficiency. Mechanisms that reduce Na+ are focused on restricting the Na+ ion uptake, increasing the efflux of Na+ ions and compartmentalizing Na+ ions in the vacuole. In compartmentalization, the toxic ions gather into compartments called vacuoles, whereas gathering of organic osmolytes in the cytosol occurs to ensure osmotic balance and safeguarding against oxidative stresses. The detoxification of cytosol is performed by creating a gradient of H+ across the membrane, which is electrochemical and is utilized to cause secondary compartmentalization of Na+ inside vacuoles or either its removal from the cell. So, for this purpose vacuolar H+-ATPase has a key role to play. As a matter of supported fact, H+ pumping is induced in plants owing to the salt stresses just to power the exchanger activities of Na+/H+, illustrated in Figure 10.3 (Rouphael et al. 2018). ABA pathways as well as osmotic stress signaling are both triggered by salt stress. Salinity stress increases the ABA concentration in plants and activates sucrose non-fermenting 1-related protein kinase 2 (SnRK2) kinase activities (He et al. 2018; Yang and Guo 2018b). Plants try to regulate their metabolic pathways to get used to the environment when under high-salt stress conditions, which is all about developmental responses like leaf senescence, root architecture, life cycle and growth (Isayenkov and Maathuis 2019). There is participation of mitogen-activated protein kinase (MAPK) cascades in salt-stress-centered responses in plants (Yang and Guo 2018b). Salt stress causes formation of reactive oxygen species, which cause signal transduction and responses in the plant cell. Regulation of ROS homeostasis and antioxidant defense responses are done by MAPK cascades, which form a significant detoxification signaling pathway. ROS stress is mitigated due to salt stress by enzymatic and non-enzymatic systems. Enzymatic scavengers are catalase (CAT), guaiacol peroxidase (GPOX), guaiacol peroxidase (GPX) and glutathione peroxidase (GR) (Wu et al. 2013). Carotenoids, alkaloids, ascorbic acid, flavonoids, tocopherol and pheno-lic compounds are among the non-enzymatic scavengers. Certain small molecules trigger downstream salt-stress responses by functioning as signals. To counteract NaCl stress, nitric oxide (NO) leads to an increase in the activity of H+-ATPase and H+-PPase. The SOS1 antiporter activity is upregulated by stimulation of the activity of MPK6, which is done by phosphatidic acid (PA)-meditated signaling pathways (Yang and Guo 2018b). Proline, trehalose, glucose, fructose, mannitol, glycine beta-ine, polyamine and sugar are among the several osmolytes that are accumulated in plant species in both leaves and shoots in response to salt stress. While these metabolites mainly act to adjust osmotic balance and detoxification, they may also regulate salt response by directly binding to and activating or inactivating enzymes (Deinlein et al. 2014).

10.5.3 HEAT STRESS

Growth and development of plants are highly dependent on temperature; each spe-cies of plant has developed a specific range for temperature endurance (Melillo et al. 2014). At the metabolic level, heat stress has reportedly reduced the photosynthesis rate (Ihsan et al. 2016) due to protein or enzyme denaturation (Chen et al. 2017c), e.g., PS II-associated proteins, manganese stabilizing protein (Sun et al. 2016). Increased assimilation of intercellular carbon dioxide and less photosynthetic net rate cause stomata to close (Hemantaranjan et al. 2014). ROS overproduction declines ATP and NADPH production and disbands thylakoid membranes of chloroplast by degrad-ing chlorophyll (Chen et al. 2017b). Prolonged duration of heat exposure degrades proteins and enzymes involved in chloroplast biosynthetic processes (Sun and Guo 2016). High temperature alters membrane structural properties, i.e., extra fluidity is due to disintegration of weak H-bonds in the proteins (Shi et al. 2016), which signals to activate thermo-tolerance and thermo-avoidance responses by plants (Hemantaranjan et al. 2014). This extra fluid property of lipid membranes increases their permeability, leading to leakage of ions or electrolytes through the membrane (Bahuguna et al. 2017). High temperature not only causes membrane damage at a cellular level but also denatures proteins, disturbs biosynthetic pathways, inactivates enzymes and increases photorespiration (Lawas et al. 2018). Thus, all these physi-ological changes in plants activate ROS or free radicals (Fahad et al. 2017). ROS induces apoptosis or programmed cell death (PCD) (Kerchev et al. 2020) by activat-ing hypersensitive reaction (HR) inducible genes and helps in activating heat toler-ance mechanisms by expressing proteins, e.g. HSPs (Hemantaranjan et al. 2014).

Plants induce certain stress-responsive metabolites to mediate adverse effects of heat damage, e.g., biogenic volatile organic compounds (isoprene units) (Abdelrahman et al. 2017), and high accumulation of about 11 amino acids includ-ing tryptophan, tyrosine and phenylalanine (Asai et al. 2017). Positive regulation in isoprene synthesis and its accumulation in leaves elevate ATP concentration under short-term heat stress. Biosynthesis of phenolics is downregulated, as these phenolic compounds comprise many intermediate compounds of flavonoid and anthocyanin pathway. Thus, a decrease in phenolic compounds degrades both these pathways. Plants under heat stress utilize maximum energy of cells since isoprene and flavonoid synthesis is an energy-demanding processes. Therefore, under extreme heat stress, there is substrate competition in pathways (isoprene, TCA, flavonoid) and plants will try to direct substrates to assimilate and allocate it to sites that will make their immediate availability possible under heat shock (Austen et al. 2019). An increase in the concentration of α-tocopherol or vitamin E has been reported under abiotic stimuli like heat stress. Photooxidation increases due to excessive light and elevated temperature on the leaf surface. α-Tocopherol acts as a photoprotectant to curb photooxidative stress caused by multiple stresses (Muñoz and Munné-Bosch 2019). Apocarotenoid signaling molecules ACS1 formation by the metabolism of β-carotenes through enzymatic cleavage by carotenoid cleavage dioxygenase (CCD4) enzyme. Heat stress stimulates CsCCD4a, which is a homolog of CCD4 enzyme, which in turn upregulates apo-carotenoids (Hou et al. 2016a; Hou et al. 2016b).

Increased membrane fluidity activates kinase and lipases (phospholipases) (Muñoz and Munné-Bosch 2019). Sucrose non-fermenting related protein kinase-1 (SnRK1) and target of rapamycin (TOR) kinase activate under abiotic stresses and are involved in metabolic reprogramming. SnRK1 triggered by heat stress enhances catabolism to acquire maximum energy. Both of these kinases activate ABA synthesis under stress, by FUS3 (ABA synthesis promoter) stabilization and phosphorylation by SnRK1 (Rodriguez et al. 2019). ABA is involved in developmental and growth regulation under stress conditions. TOR kinases downregulate the autophagy process by phosphorylating TOP46, which suppresses autophagy gene expression. Antagonistic to TOR kinases, overexpression of SnRK1 kinase increases autophagy under heat stress by phosphorylating ATG1 complex, which is an autophagy initiator, as shown in Figure 10.3 (Chen et al. 2017a).

Since heat stress denatures most cellular proteins, HSPs actively prevent aggregation of misfolded proteins through ubiquitin proteasomal degradation (Sedaghatmehr et al. 2019). HSPs and molecular chaperones are actively regulated by signaling molecules for hormonal regulation, PCD and autophagy (Trivedi et al. 2016). HSPs positively correlate with the production of antioxidants, e.g., isoprene volatile organic compounds (Austen et al. 2019), redox state regulation, and osmolytes, as shown in Figure 10.3. For example, ascorbate peroxidase (or APX) protects against oxidative damage. Glutathione and thioredoxin regulate reduction and oxidation. Glycine, betaine and trehalose are important osmolytes activated through Hsp60, Hsp70, Hsp90 and Hsp100, and assist in the acquisition of tolerance against heat stress (Roy et al. 2019). Heat shock transcription factors (HSFs), specifically HSFA1s, are integral and vital in function of these transcriptional networks. HSFA1 is the positive regulator of DREB, specifically DREB2A, which triggers HsfA3 by the assistance of other regulatory factors [i.e., coactivator complexes such as DNA polymerase II, subunit B3-1 (DPB3-1)] that directly induce the expression of heat response genes by producing chaperones and enzymes (Ohama et al. 2017).

10.5.4 FLOODING STRESS

Crop growth is considerably affected by flooding and is a key stress factor for plants; it leads to deteriorated yield and has adverse impacts on crop production (Tewari and Mishra 2018). Flooding is responsible for affecting 16% of agricultural products worldwide. Flooding, or waterlogging, is caused by both heavy rainfall and poor soil drainage. The environment is rendered hypoxic (meaning respiration is limited) as soon as flooding results. Roots and other microorganisms deplete the soil of oxygen and after this it becomes anoxic (meaning the respiration stops). Tolerance to hypoxia or anoxia depends on the species, external conditions and organs affected in the stage of development, and ranges from hours, days or weeks. The energy formation pathways are instantaneously shifted from aerobic respiration to anaerobic fermentation at a low oxygen level, which suppresses many developmental stages in plants (Zhou et al. 2020). The span and intensity of flooding is mainly dependent on the rate of influx of water, but it also depends on the absorption capability of the soil and the rate of water flow near the root zone (Rehem et al. 2012). In this way, the first limitation imposed on plants is the shortage of oxygen to facilitate aerobic respiration of plant parts that are submerged. When the flooding continues for an extended period of time, then another issue appears in the form of the reduced oxidation-reduction potential of soil. This reduced redox potential is the reason why toxic compounds like sulfides, ethanol, acetaldehyde, formic acid, lactic acid, soluble Fe and Mn, and acetic acid appear. This clearly shows that the flooding stress causes lack of oxygen and exposure to toxic chemicals to the plants (Striker 2012). The flooded soil introduces harsh stresses on plants because it does not help the plant fulfill its basic needs like O_2, CO_2 and sunlight due to excess water levels. These declining levels of oxygen, carbon dioxide and sunlight take a toll on the rate of photosynthesis and aerobic respiration lowering the availability of carbohydrates, which ultimately hinders growth and development of plants under stress (Voesenek and Bailey-Serres 2013). Low oxygen levels are sensed by group VII ethylene-responsive transcription factors (ERFs), i.e., RAP 2.2 and RAP 2.12, to regulate metabolic adaptation to flooding, as illustrated in Figure 10.3. Due to hypoxia, there is an increased redox potential among the soil and the plant leading to the formation of reactive oxygen species (Field et al. 2014; Tewari and Mishra 2018).

Stress due to flooding induces physiological, anatomical and morphological changes in plants for protection against potential harms (Mangena 2018). The first victim of the stress, i.e. the root, undergoes notable phenotypic changes and lower growth rates when compared to shoots, in waterlogging situations (Panozzo et al. 2019). The response that occurs mostly is the formation of adventitious roots (Zhang et al. 2017). There are either no cuticles or those present are very thin which gives it large surface area per unit root biomass due to numerous root hairs and small diameter of root hairs (Ayi et al. 2016). When the flooding stays for longer periods, the water depth elevates and the overall architecture of the plant changes in an attempt to exhibit an adaptive strategy. In order to deal with complete submergence, two changes occur: low oxygen escape syndrome (LOES) and low oxygen quiescence syndrome (LOQS). LOES enables the plant to quickly reach the water surface to allow gaseous exchange between the atmosphere and plant tissues through certain structural adaptations like fast growth of plant stems, petioles and internodes when plants are submerged. It is evident that LOES helps the plant to deal with submergence temporarily, but it certainly does not help in the long-term. For long-term adaptation, LOQS is an effective strategy. In LOQS, the plant reduces the respiration process in response to hypoxic stress (Zhou et al. 2020). Closing of stomata, transpiration rate and photosynthesis are monitored according to the flood tolerance capabilities of each plant species (Mollard et al. 2010). Some plants facilitate gas exchange under submergence by retaining a gas film that develops on the hydrophobic cuticle; this leads to improved photosynthetic rates (Loreti et al. 2018). A phytohormone that plays a vital role in coping with waterlogging stress is ethylene, and its accommodation in plants when waterlogging occurs is very common (Zhou et al. 2020). In several species there has been a reported enhancement in both expression and activity of 1-aminocyclopropane-1-carboxylic acid (ACC) synthase and an enzyme that synthesizes ethylene, i.e., ACC oxidase (Sasidharan and Voesenek 2015). An ethylene-dependent mechanism aids plants like Arabidopsis, tomato and rice in exhibiting leaf hyponasty during flooding to protect them (Yu and Huang 2017). Anaerobic respiration, fermentative metabolism and glycolysis are sought in plants when there is less oxygen concentration during flooding to maintain cell functionality (Tewari and Mishra 2018). So, some anaerobic genes are maintained at a greater expression under flooding conditions to upregulate fermentation, as it is the only process of energy generation in flooding conditions (Kim et al. 2020).

10.5.5 OZONE

Ozone (O_3) in the troposphere is a toxic pollutant that generates ROS and effects photosynthesis, metabolic pathways and defense reactions of plants. Concentration of ozone, canopy architecture, growth stage of the plant, rate of entry into the plant and environmental conditions determine the extent of effects by O_3 (Fuhrer 2009). Enhancement in O_3 concentration is proportional to assimilate availability, reductions in photosynthesis, more oxidative impairment and increase leaf senescence.

10.5.6 EFFECT ON PLANT BIOCHEMICAL PROCESSES

Ozone produces ROS, which directly damage CO_2 fixation by damaging stomatal functioning or RuBisCO activity and indirectly triggered leaf senescence, as shown in Figure 10.3 (Morgan et al. 2004). In whole photosynthesis process, chlorophyll content is reduced, and carboxylation contribution is decreased due to -SH group oxidation of RuBisCO, under high ozone concentration (Feng and Kobayashi 2009; Köllner and Krause 2000; Sandermann Jr. et al. 1998). Overall assimilate production and export from leaves is reduced. Furthermore, ozone exposure effects the effective leaf part, thus limiting interception of light. After entering the stomatal pore, ozone reacts with extracellular fluid compounds such as isoprene and ethylene, which lead to the formation of secondary oxidants (hydroxy-hydroperoxides, primary ozonides), which might react with cell membrane proteins. Radical scavengers such as polyamines and ascorbic acid prevent this reaction up to some degree (Lefohn 1991). Reaction of O_3 with phenylpropanoid or with ethylene results in the accumulation of formate, acetate and formaldehyde in affected tissue (Wellburn et al. 1994). O_3 interaction with ethylene and other hydrocarbons is proposed to be involved in injury mechanisms (Wellburn and Wellburn 1996). The effect of increasing ozone is not straightforward on crop quality, as with the loss in grain yield, the quality of grain increases. In high O_3 concentration, starch concentration decreases in wheat and grain protein concentration increases, which are accompanied by decreasing protein yield (Grandjean Grimm and Fuhrer 1992; Piikki et al. 2008; Pleijel and Wallin 1996). Protein and amino acid content changes due to O_3 effect the nitrogen metabolism cycle (Rowland et al. 1988). In potato, ascorbic acid concentration enhances, while starch and sugar content decreases with high ozone concentration (Vorne et al. 2002).

10.5.7 OZONE EFFECTS ON PRIMARY AND SECONDARY METABOLISM

Primary metabolism reduction induced by O_3 is correlated with capacity of CO_2 fixation at the cellular level, established on analysis of RuBisCO activity, transcript and protein level. Moreover, molecular study in rice and wheat have shown similar alterations in RuBisCO content, photosynthetic machinery components and other Calvin–Benson cycle enzymes such as ATP synthase, aldolase, NADP-glyceraldehyde3-phosphate dehydrogenase, RuBisCO activase, oxygen-evolving subunit of photosystem II and phosphoglycerate kinase (Agrawal et al. 2002; Sarkar et al. 2010). O_3's negative effect on cytosolic ATP-citrate lyase transcript abundance and mitochondrial alternative oxidase 2b was analyzed in soybean (Gillespie et al. 2012). These changes reorganize metabolism of mitochondria to maintain respiration rates needed for cellular damage repair and detoxification of O_3.

Accumulation of flavonoids, cell wall-bound polyphenols (lignin) and polyphenolic compounds are induced by O_3 (Langebartels et al. 1997). Primary carbon metabolism is linked with secondary ones by the phenylpropanoid pathway. Condensation of phosphoenolpyruvate (PEP) from glycolysis and erythrose 4-phosphate from the pathway of pentose phosphate is the first step for phenylalanine. In poplar trees and Scots pine, shikimate dehydrogenase activity increased when sanitize with ozone (Léger et al. 1997; Luethy-Krause et al. 1990). Furthermore, cinnamoylCoA is formed for phenylalanine, which is a precursor to produce catechin, lignin and stilbenes. Submission of ozone increases phenylalanine ammonia lyase activity (Kangasjärvi et al. 1994; Langebartels et al. 1997; Léger et al. 1997). Polyamine (putrescine) biosynthesis increases with ozone, which acts as an oxygen radical scavenger (Kangasjärvi et al. 1994; Langebartels et al. 1997). Ethylene, a wounding compound, is stimulated by ozone, which might trigger expression of defense genes, acting as a second messenger (Heath and Taylor 1997; Kangasjärvi et al. 1994; Sandermann Jr. et al. 1998). The total of anthocyanins, tannins and phenolic increases with ozone concentration elevation in plants (Pellegrini et al. 2011; Shakeri et al. 2016). Phenolic and carotenoid content increases up to 17% and 52.8%, respectively, when treated with high O_3 concentration (Bortolin et al. 2016). Furthermore, the ABA level also elevated 11 times more than the control cells upon O_3 treatment (Sun et al. 2012). Thus, several primary and secondary metabolites alter their production with exposure to ozone.

10.5.8 OZONE TRIGGERS OXIDATIVE STRESS

Oxidative stress is often determined by enhanced ozone level which effect plant via ROS-mediated form. Weaken photosynthesis, leaf injury, decreased shoots and roots growth, and low yield are triggered by harmful level of ozone as illustrated in Figure 10.3. Plants have both tolerance and avoidance approaches for survival in damage-level ozone conditions, such as activating various repair mechanisms and an antioxidant defense system, and stomata closure. Ozone reacts with ethylene and alkenes, and generates ROS in apoplast fluid, which impairs membrane lipid, increases membrane permeability, decreases membrane potential, collapses H^+-pump and enhances uptake of Ca^{+2}. ROS can also damage nucleic acid, proteins and carbohydrates. Ozone's harmful effects on growth of plants is reduce by the reaction of NO in the surrounding atmosphere and surplus ozone. Various abiotic stresses perturbed the ROS generation and degradation equilibrium which result in an considerable increase in the ROS level (Saxena et al. 2019).

10.5.9 OXIDATIVE STRESS

Oxidative damage occurs as a burst of ROS under other stresses (Kordrostami et al. 2019). ROS at low concentrations act as secondary signal molecules, but at high levels cause drastic damage to cellular components (Sewelam et al. 2016). Disturbed redox homeostasis cause oxidative damage (Negrão et al. 2017). ROS or free radicals damage cell membranes by lipid peroxidation, which results in the production of malondialdehyde or MDA (Pandey et al. 2017), free radicals and hydrogen peroxide (Zandalinas and Mittler 2018), damaging lipid membrane proteins, pigments and DNA (Anjum et al. 2016), as shown in Figure 10.3. Enzymes involved in the oxidative damage repair express in higher concentrations that indicate oxidative stress intensity in the cell, e.g., glutathione reductase (GR), catalase (CAT), superoxide dismutase (SOD) and ascorbate peroxidase (APX) (Berkowitz et al. 2016). Under stress, electrons from incomplete oxygen reduction (Huché-Thélier et al. 2016) are taken up by superoxide, H_2O_2, and OH^o (Kordrostami et al. 2017). Oxygen substitution due to excess NADPH and delay in $NADP^+$ reduction generates ROS (Ighodaro and Akinloye 2018). NADPH is a product in several physiological reactions, e.g., the citric acid cycle (Sewelam et al. 2016). An increase in the $NADPH:NADP^+$ ratio triggers H^+ potential reduction to balance this ratio (Sies 2017). ROS levels become variable within cellular compartments like mitochondria, nucleus, chloroplast, endoplasmic reticulum and cytosol due to continuous scavenging and generation (Sies et al. 2017), which helps intercellular and cell-to-cell signaling (Noctor et al. 2018), as shown in Figure 10.3. Photosynthesis decreases due to adverse changes in controlling factors indicating damage of photosynthetic apparatus (Gururani et al. 2015), and a massive burst of ROS activates autophagy and PCD.

Low molecular weight compounds having low to high antioxidant properties are abundant in plant cells, e.g., compounds like amino acids, mono- and di-carbohydrates, pigments, peptides, carotenoids, and flavonoids regulate basic anabolic processes, and complex and specific pathways of secondary molecules synthesis (Noctor et al. 2015).

Glutathione and ascorbate fundamentally function in antioxidant processing of ROS because the involvement of peroxidases enables rapid reaction with H_2O_2 (Yousuf et al. 2012) and potential reductase and connected enzymatic systems regenerate oxidized forms by acting upon reactants generated by dehydrogenases during photosynthesis (Foyer and Noctor 2011). Glutathione and ascorbates have supremacy over sacrificial antioxidants, which readily reduce and accumulate in cellular organelles by allowing repeated redox cycles of ROS for redox state regulation (Queval et al. 2011). Oxidized forms of glutathione and ascorbates gather in apoplast and vacuoles due to the absence of redox cycling mechanisms (Noctor et al. 2016). GR is one of the flavoenzymes present universally in organisms (Ighodaro and Akinloye 2018). Glutathione levels are also indicators of oxidative stress in higher plants (Noctor et al. 2015).

Ascorbate, or ascorbic acid, is an abundant carboxylic acid and metabolic agent constituting about 10% of total carbohydrate content. Ascorbic acid acts as a ROS scavenger under stress and scavenges singlet oxygen O_2^- and OH^- Ascorbates have a crucial role in photoprotection by acting as a cofactor of enzymes, i.e., "violaxanthin-deepoxidase" enzyme in xanthophyll cycles (Hossain et al. 2017) Ascorbate, being a cofactor of peroxidase enzymes, removes radicals, and regenerates tocopherols and xanthophylls, which assist in ROS quenching mechanisms (Rahantaniaina et al. 2017). Glutathione is indirectly involved in ascorbate regeneration chemically from dehydroascorbate (DHA) that is an oxidized form of ascorbate and enzymatically by the assistance of dehydroascorbate reductase (DHAR) during the metabolism of hydrogen peroxide (Foyer and Noctor 2011). Under stress, equilibrium shifts toward ROS accumulation either by hyperactivation of ROS or a decline in the capacity of antioxidants both resulting in PCD for reprogramming stress tolerance (Shumbe et al. 2017).

Carotenoids are highly lipophilic, tetraterpenes constituting a 40-carbon chain in molecular assembly linked with cellular or organelle membranes. These pigments act as UV protectants in photosynthesis (Arvayo-Enríquez et al. 2013), photoprotectants under light stress, and strong ROS receptors under oxidative stress to protect fats and pigments. The carotenoid quenching mechanism scavenge ROS through carotenoid-dependent pathways (Fischer et al. 2013) when oxidative burst surpasses the limit of carotenoids; these singlet O_2 molecules start lipid peroxidation (Kromdijk et al. 2016). Carotenoids act as lipid metabolites due to slight antioxidative properties when scavenging these singlet O_2 molecules (Foyer et al. 2017). Other than singlet oxygen molecules, superoxide produced via the reduction of oxygen is sometimes disputed in hydrogen peroxide by superoxide dismutase (Saini et al. 2018). Different cellular organelles of plant cells contain different types of SODs, i.e., prokaryotic SODs containing ferrous and manganese, and eukaryotic SODs containing copper and zinc (Farmer and Mueller 2013). Other oxidases like GOX inside peroxisomes (Foyer and Noctor 2016) and heme III peroxidases reduce superoxide; these enzymes are abundantly encoded by more than 70 genes in the model plant, Arabidopsis (Ramel et al. 2012). Quantitatively, crucial units of hydrogen peroxide generation from superoxide are ETC (photosynthetic and respiratory), GOX in photorespiration and membrane-associated NADPH oxidases, but the contributions of these processes and enzymes vary among plant species (Exposito-Rodriguez et al. 2017). Ferritin proteins with transition metals as prosthetic groups subsidize generation of hydrogen peroxides as well as inhibits the excessive production of OH° radicals, which have an otherwise significant role in cell wall metabolism (O'Brien et al. 2012).

10.6 CONCLUSION

Abiotic stress on plants is a crucial factor of climate change, which is an intricate event with an extensive array of variable impacts on the environment. Continuous exposure to such abiotic stresses causes distorted metabolic processes and impairs biomolecule concentrations. Due to inherent defense systems in plants, feasibility to endure multistress tolerance is increased in crop plants. Plants have inherent physical, morphological and molecular abilities at the cellular level to respond to stress. Such generalized and conserved cellular defense responses are membrane lipid desaturation, ROS scavenging, chaperone induction and increased concentration of compatible solutes. These responses are coupled with a multitude of abiotic stresses as consequences of membrane damage, ROS generation, protein inactivation and dehydration of tissues. These responses are mostly under the regulation of complex upstream signaling molecules, e.g., phytohormones ABA, polyamines (PAs), ROS, hydrogen sulfide, calcium ions, and pigment molecules. Similarly, downstream factors regulating gene expression include transcription factors as well. Some fundamental coordinators in defense responses at the systemic level are the phytohormones ABA, ET, JA and SA, which synchronize these intricate stress responses as the hormonal signal. Such complex pathways for stress signaling in plants have a conserved feature of MAPKs relay common to other kingdoms (fungi and metazoans) as well. Though there are other conserved systems in plants like utilization of Ca^{+2}, lipid molecules and ROS as secondary messengers, plant species are diverse in perception, generation and transduction of these molecules.

For a quantitative and comprehensive analysis of biomolecules in plants through metabolomics, when correlated with expression of their genotypes toward abiotic stresses, has helped to reveal the chemical composition of these molecules like osmoprotectants connected to biochemical and physiological alteration. These studies show metabolism adjustment, coordination of metabolic pathways and enzyme activities of a plant toward different abiotic stresses. This knowledge will uncover potential metabolites that can be introduced to economical crops and help to make strategies to enhance the adaptation of plants in different stress conditions. Strategies include not only a set of reactions but also include interconnected biochemical pathways and networks that trigger metabolite accumulation. This task is demanding as it considers those reactions in metabolic responses that are dependent on stress dose and change with the course of time. Furthermore, not only are there differences at species and genus levels but also at the tissue level, showing genetic background correlation. However, advancements in metabolomics tools will speed the improvement and design of breeding projects that will open a new era for crops having the ability to survive abiotic stresses.

REFERENCES

Abdelrahman M, El-Sayed M, Jogaiah S, Burritt DJ, Tran L-SP (2017) The "STAY-GREEN" trait and phytohormone signaling networks in plants under heat stress. *Plant Cell Rep* 36:1009–1025.

Agrawal GK, Rakwal R, Yonekura M, Kubo A, Saji H (2002) Proteome analysis of differentially displayed proteins as a tool for investigating ozone stress in rice (*Oryza sativa L.*) seedlings. *Proteomics* 2:947–959.

Ahammed GJ, Li X, Liu A, Chen S (2020) Brassinosteroids in plant tolerance to abiotic stress. *J Plant Growth Regul* 39:1451–1464.

Ahanger MA, MoradTalab N, Fathi Abd-Allah E, Ahmad P, Hajiboland R (2016) Plant growth under drought stress: Significance of mineral nutrients. In: *Water Stress and Crop Plants: A Sustainable Approach*, John Wiley & Sons, vol. 2: Pp. 649–668.

Ahanger MA, Tomar NS, Tittal M, Argal S, Agarwal R (2017) Plant growth under water/salt stress: ROS production; antioxidants and significance of added potassium under such conditions. *Physiol Mol Biol Plants* 23:731–744.

Ahmad F, Singh A, Kamal A (2019) Salicylic acid–mediated defense mechanisms to abiotic stress tolerance. In: *Plant Signaling Molecules*. Elsevier, pp. 355–369.

Ali M, Baek K-H (2020) Jasmonic acid signaling pathway in response to abiotic stresses in plants. *Int J Mol Sci* 21:621.

Anjum NA et al. (2016) Catalase and ascorbate peroxidase—representative H2O2-detoxifying heme enzymes in plants. *Environ Sci Pollut Res* 23:19002–19029.

Arvayo-Enríquez H, Mondaca-Fernández I, Gortárez-Moroyoqui P, López-Cervantes J, Rodríguez-Ramírez R (2013) Carotenoids extraction and quantification: A review *Anal Methods* 5:2916–2924.

Asai T, Matsukawa T, Kajiyama S (2017) Metabolomic analysis of primary metabolites in citrus leaf during defense responses. *J Biosci Bioeng* 123:376–381.

Austen N, Lake JA, Phoenix G, Cameron DD (2019) The regulation of plant secondary metabolism in response to abiotic stress: Interactions between heat shock and elevated CO2. *Front Plant Sci* 10:1463.

Avonce N, Leyman B, Mascorro-Gallardo JO, Van Dijck P, Thevelein JM, Iturriaga G (2004) The *Arabidopsis trehalose-6-P synthase AtTPS1* gene is a regulator of glucose, abscisic acid, and stress signaling. *Plant Physiol* 136:3649–3659.

Ayi Q, Zeng B, Liu J, Li S, van Bodegom PM, Cornelissen JH (2016) Oxygen absorption by adventitious roots promotes the survival of completely submerged terrestrial plants. *Ann Bot* 118:675–683.

Bahuguna RN, Solis CA, Shi W, Jagadish KS (2017) Post-flowering night respiration and altered sink activity account for high night temperature-induced grain yield and quality loss in rice (*Oryza sativa L.*). *Physiol Plant* 159:59–73.

Barlow K, Christy B, O'leary G, Riffkin P, Nuttall J (2015) Simulating the impact of extreme heat and frost events on wheat crop production: A review. *Field Crops Res* 171:109–119.

Bartels D, Sunkar R (2005) Drought and salt tolerance in plants. *Crit Rev Plant Sci* 24:23–58.

Bechtold U, Field B (2018) Molecular mechanisms controlling plant growth during abiotic stress. *J Exp Bot* 69:2753–2758.

Berkowitz O, De Clercq I, Van Breusegem F, Whelan J (2016) Interaction between hormonal and mitochondrial signalling during growth, development and in plant defence responses. *Plant Cell Environ* 39:1127–1139.

Bläsing OE et al. (2005) Sugars and circadian regulation make major contributions to the global regulation of diurnal gene expression in Arabidopsis. *Plant Cell* 17:3257–3281.

Bortolin RC et al. (2016) Chronic ozone exposure alters the secondary metabolite profile, antioxidant potential, anti-inflammatory property, and quality of red pepper fruit from *Capsicum baccatum*. *Ecotoxicol Environ Saf* 129:16–24.

Capell T, Bassie L, Christou P (2004) Modulation of the polyamine biosynthetic pathway in transgenic rice confers tolerance to drought stress. *Proc Natl Acad Sci* 101:9909–9914.

Carillo P, Annunziata MG, Pontecorvo G, Fuggi A, Woodrow P (2011) Salinity stress and salt tolerance. *Abiotic Stress in Plants–Mechanisms and Adaptations* 1:21–38.

Chen L et al. (2017a) The AMP-activated protein kinase KIN10 is involved in the regulation of autophagy in *Arabidopsis*. *Front Plant Sci* 8:1201.

Chen TH, Murata N (2011) Glycinebetaine protects plants against abiotic stress: Mechanisms and biotechnological applications. *Plant Cell Environ* 34:1–20.

Chen Y-E et al. (2017b) Responses of photosystem II and antioxidative systems to high light and high temperature co-stress in wheat. *Environ Exp Bot* 135:45–55.

Chen Y, Wang H, Hu W, Wang S, Snider JL, Zhou Z (2017c) Co-occurring elevated tempera-ture and waterlogging stresses disrupt cellulose synthesis by altering the expression and activity of carbohydrate balance-associated enzymes during fiber development in cotton. *Environ Exp Bot* 135:106–117.

Ciura J, Kruk J (2018) Phytohormones as targets for improving plant productivity and stress tolerance. *J Plant Physiol* 229:32–40.

Colebrook EH, Thomas SG, Phillips AL, Hedden P (2014) The role of gibberellin signalling in plant responses to abiotic stress. *J Exp Biol* 217:67–75.

Cook D, Fowler S, Fiehn O, Thomashow MF (2004) A prominent role for the *CBF* cold response pathway in configuring the low-temperature metabolome of Arabidopsis. *Proc Natl Acad Sci* 101:15243–15248.

Cortleven A, Leuendorf JE, Frank M, Pezzetta D, Bolt S, Schmülling T (2019) Cytokinin action in response to abiotic and biotic stresses in plants. *Plant Cell Environ* 42:998–1018.

Cramer GR, Urano K, Delrot S, Pezzotti M, Shinozaki K (2011) Effects of abiotic stress on plants: A systems biology perspective. *BMC Plant Biol* 11:1–14.

Das K, Roychoudhury A (2014) Reactive oxygen species (ROS) and response of antioxi-dants as ROS-scavengers during environmental stress in plants. *Front Environ Sci* 2 doi:10.3389/fenvs.2014.00053.

Deinlein U, Stephan AB, Horie T, Luo W, Xu G, Schroeder JI (2014) Plant salt-tolerance mechanisms. *Trends Plant Sci* 19:371–379.

Dubois M, Van den Broeck L, Inzé D (2018) The pivotal role of ethylene in plant growth. *Trends Plant Sci* 23:311–323.

Espinoza C et al. (2010) Interaction with diurnal and circadian regulation results in dynamic metabolic and transcriptional changes during cold acclimation in Arabidopsis. *PloS One* 5:e14101.

Exposito-Rodriguez M, Laissue PP, Yvon-Durocher G, Smirnoff N, Mullineaux PM (2017) Photosynthesis-dependent H_2O_2 transfer from chloroplasts to nuclei provides a high-light signalling mechanism. *Nat Commun* 8:1–11.

Fàbregas N, Fernie AR (2019) The metabolic response to drought. *J Exp Bot* 70:1077–1085.

Fahad S et al. (2017) Crop production under drought and heat stress: Plant responses and management options. *Front Plant Sci* 8:1147.

Farmer EE, Mueller MJ (2013) ROS-mediated lipid peroxidation and RES-activated signal-ing. *Annu Rev Plant Biol* 64:429–450.

Farré EM, Weise SE (2012) The interactions between the circadian clock and primary metab-olism. *Curr Opin Plant Biol* 15:293–300.

Feng Z, Kobayashi K (2009) Assessing the impacts of current and future concentrations of surface ozone on crop yield with meta-analysis. *Atmos Environ* 43:1510–1519.

Field CB et al. (2014) Summary for policymakers. In: *Climate Change 2014: Impacts, Adaptation, and Vulnerability. Part A: Global and Sectoral Aspects. Contribution of Working Group II to the Fifth Assessment Report of the Intergovernmental Panel on Climate Change*. Cambridge University Press, pp. 1–32.

Fischer BB, Hideg E, Krieger-Liszkay A (2013) Production, detection, and signaling of sin-glet oxygen in photosynthetic organisms. *Antioxid Redox Signal* 18:2145–2162.

Forni C, Duca D, Glick BR (2017) Mechanisms of plant response to salt and drought stress and their alteration by rhizobacteria. *Plant Soil* 410:335–356.

Foyer CH, Noctor G (2011) Ascorbate and glutathione: The heart of the redox hub. *Plant Physiol* 155:2–18.

Foyer CH, Noctor G (2016) Stress-triggered redox signalling: What's in pROSpect? *Plant Cell Environ* 39:951–964.

Foyer CH, Ruban AV, Noctor G (2017) Viewing oxidative stress through the lens of oxidative signalling rather than damage. *Biochem J* 474:877–883 doi:10.1042/bcj20160814.

Fuhrer J (2009) Ozone risk for crops and pastures in present and future climates. *Naturwissenschaften* 96:173–194.

Furihata T, Maruyama K, Fujita Y, Umezawa T, Yoshida R, Shinozaki K, Yamaguchi-Shinozaki K (2006) Abscisic acid-dependent multisite phosphorylation regulates the activity of a transcription activator *AREB1*. *Proc Natl Acad Sci* 103:1988–1993.

Garg AK, Kim J-K, Owens TG, Ranwala AP, Do Choi Y, Kochian LV, Wu RJ (2002) Trehalose accumulation in rice plants confers high tolerance levels to different abiotic stresses. *Proc Natl Acad Sci* 99:15898–15903.

Gill SS, Anjum NA, Gill R, Tuteja N (2016) Abiotic Stress Signaling in Plants–An Overview. In: *Abiotic Stress Response in Plants*, pp. 1–12. doi:10.1002/9783527694570.ch1.

Gillespie KM et al. (2012) Greater antioxidant and respiratory metabolism in field-grown soybean exposed to elevated O3 under both ambient and elevated CO2. *Plant Cell Environ* 35:169–184.

Grandjean Grimm A, Fuhrer J (1992) The response of spring wheat (*Triticum aestivum L.*) to ozone at higher elevations: I. Measurement of ozone and carbon dioxide fluxes in open-top field chambers. *New Phytol* 121:201–210.

Gururani MA, Venkatesh J, Tran LSP (2015) Regulation of photosynthesis during abiotic stress-induced photoinhibition. *Mol Plant* 8:1304–1320.

Harmer SL et al. (2000) Orchestrated transcription of key pathways in *Arabidopsis* by the circadian clock. *Science* 290:2110–2113.

Hatfield JL, Prueger JH (2015) Temperature extremes: Effect on plant growth and development. *Weather Clim Extrem* 10:4–10.

Hayat S, Hayat Q, Alyemeni MN, Wani AS, Pichtel J, Ahmad A (2012) Role of proline under changing environments: A review. *Plant Signal Behav* 7:1456–1466.

Haydon MJ, Mielczarek O, Frank A, Román Á, Webb AA (2017) Sucrose and ethylene signaling interact to modulate the circadian clock. *Plant Physiol* 175:947–958.

Haydon MJ, Mielczarek O, Robertson FC, Hubbard KE, Webb AA (2013) Photosynthetic entrainment of the *Arabidopsis thaliana* circadian clock. *Nature* 502:689–692.

He M, He C-Q, Ding N-Z (2018) Abiotic stresses: General defenses of land plants and chances for engineering multistress tolerance. *Front Plant Sci* 9:1771.

Heath R, Taylor G (1997) Physiological processes and plant responses to ozone exposure. In: *Forest Decline and Ozone*. Springer, pp. 317–368.

Hemantaranjan A, Bhanu AN, Singh M, Yadav D, Patel P, Singh R, Katiyar D (2014) Heat stress responses and thermotolerance. *Adv Plants Agric Res* 1:1–10.

Hill CB, Jha D, Bacic A, Tester M, Roessner U (2013) Characterization of ion contents and metabolic responses to salt stress of different *Arabidopsis AtHKT1; 1* genotypes and their parental strains. *Mol Plant* 6:350–368.

Hossain MA, Munné-Bosch S, Burritt DJ, Diaz-Vivancos P, Fujita M, Lorence A (2017) *Ascorbic Acid in Plant Growth, Development and Stress Tolerance*. Springer.

Hou Q, Ufer G, Bartels D (2016a) Lipid signalling in plant responses to abiotic stress. *Plant Cell Environ* 39:1029–1048.

Hou X, Rivers J, León P, McQuinn RP, Pogson BJ (2016b) Synthesis and function of apocarotenoid signals in plants. *Trends Plant Sci* 21:792–803.

Huang R-d (2018) Research progress on plant tolerance to soil salinity and alkalinity in sorghum. *J Integr Agric* 17:739–746.

Huang S, Van Aken O, Schwarzländer M, Belt K, Millar AH (2016) The roles of mitochondrial reactive oxygen species in cellular signaling and stress response in plants. *Plant Physiol* 171:1551–1559.

Huché-Thélier L, Crespel L, Le Gourrierec J, Morel P, Sakr S, Leduc N (2016) Light signaling and plant responses to blue and UV radiations—Perspectives for applications in horticulture. *Environ Exp Bot* 121:22–38.

Ighodaro O, Akinloye O (2018) First line defence antioxidants-superoxide dismutase (SOD), catalase (CAT) and glutathione peroxidase (GPX): Their fundamental role in the entire antioxidant defence grid. *Alexandria J Med* 54:287–293.

Ihsan MZ, El-Nakhlawy FS, Ismail SM, Fahad S (2016) Wheat phenological development and growth studies as affected by drought and late season high temperature stress under arid environment. *Front Plant Sci* 7:795.

Isayenkov SV, Maathuis FJ (2019) Plant salinity stress: Many unanswered questions remain. *Front Plant Sci* 10:80.

Iuchi S et al. (2001) Regulation of drought tolerance by gene manipulation of *9-cis-epoxycarotenoid dioxygenase*, a key enzyme in abscisic acid biosynthesis in *Arabidopsis*. *Plant J* 27:325–333.

Kang G, Li G, Guo T (2014) Molecular mechanism of salicylic acid-induced abiotic stress tolerance in higher plants. *Acta Physiol Plant* 36:2287–2297.

Kangasjärvi J, Talvinen J, Utriainen M, Karjalainen R (1994) Plant defence systems induced by ozone. *Plant Cell Environ* 17:783–794.

Kasuga M, Liu Q, Miura S, Yamaguchi-Shinozaki K, Shinozaki K (1999) Improving plant drought, salt, and freezing tolerance by gene transfer of a single stress-inducible transcription factor. *Nat Biotechnol* 17:287–291.

Kasukabe Y, He L, Nada K, Misawa S, Ihara I, Tachibana S (2004) Overexpression of *spermidine synthase* enhances tolerance to multiple environmental stresses and up-regulates the expression of various stress-regulated genes in transgenic *Arabidopsis thaliana*. *Plant Cell Physiol* 45:712–722.

Kaur P et al. (2019) Phytohormones in improving abiotic stress tolerance in plants. In: *Plant Tolerance to Environmental Stress: Role of Phytoprotectants*.CRC Press, 81–102.

Kazan K (2015) Diverse roles of jasmonates and ethylene in abiotic stress tolerance. *Trends Plant Sci* 20:219–229.

Kerchev P, van der Meer T, Sujeeth N, Verlee A, Stevens CV, Van Breusegem F, Gechev T (2020) Molecular priming as an approach to induce tolerance against abiotic and oxidative stresses in crop plants. *Biotechnol Adv* 40:107503.

Khan MIR, Fatma M, Per TS, Anjum NA, Khan NA (2015) Salicylic acid-induced abiotic stress tolerance and underlying mechanisms in plants. *Front Plant Sci* 6:462.

Khan NA, Khan MIR, Ferrante A, Poor P (2017) Ethylene: A key regulatory molecule in plants. *Front Plant Sci* 8:1782.

Kim S-C, Nusinow DA, Sorkin ML, Pruneda-Paz J, Wang X (2019) Interaction and regulation between lipid mediator phosphatidic acid and circadian clock regulators. *Plant Cell* 31:399–416.

Kim, S. M., Hsu, A., Lee, Y. H., Dresselhaus, M., Palacios, T., Kim, K. K., & Kong, J. (2013). The effect of copper pre-cleaning on graphene synthesis. *Nanotechnology*, 24(36), 365602.

Kim Y, Shahzad R, Lee I-J (2020) Regulation of flood stress in plants. In: *Plant Life Under Changing Environment*. Elsevier, pp. 157–173.

Klay, I., Pirrello, J., Riahi, L., Bernadac, A., Cherif, A., Bouzayen, M., & Bouzid, S. (2014). Ethylene response factor Sl-ERF. B. 3 is responsive to abiotic stresses and mediates salt and cold stress response regulation in tomato. *The Scientific World Journal*, https://doi.org/10.1155/2014/167681.

Köllner B, Krause G (2000) Changes in carbohydrates, leaf pigments and yield in potatoes induced by different ozone exposure regimes. *Agric Ecosyst Environ* 78:149–158.

Kordrostami M, Rabiei B, Ebadi AA (2019) Oxidative Stress in Plants: Production, Metabolism, and Biological Roles of Reactive Oxygen Species. In: *Handbook of Plant and Crop Stress*, Fourth Edition. CRC Press, pp. 85–92.

Kordrostami M, Rabiei B, Kumleh HH (2017) Biochemical, physiological and molecular evaluation of rice cultivars differing in salt tolerance at the seedling stage. *Physiol Mol Biol Plants* 23:529–544.

Krasensky J, Jonak C (2012) Drought, salt, and temperature stress-induced metabolic rearrangements and regulatory networks. *J Exp Bot* 63:1593–1608.

Kromdijk J, Głowacka K, Leonelli L, Gabilly ST, Iwai M, Niyogi KK, Long SP (2016) Improving photosynthesis and crop productivity by accelerating recovery from photoprotection. *Science* 354:857–861.

Langebartels C, Ernst D, Heller W, Lütz C, Payer H-D, Sandermann H (1997) Ozone responses of trees: Results from controlled chamber exposures at the GSF phytotron. In: *Forest Decline and Ozone*. Springer, pp. 163–200.

Lawas LMF, Zuther E, Jagadish SK, Hincha DK (2018) Molecular mechanisms of combined heat and drought stress resilience in cereals. *Curr Opin Plant Biol* 45:212–217.

Lee I-J, Foster KR, Morgan PW (1998) Photoperiod control of gibberellin levels and flowering in sorghum. *Plant Physiol* 116:1003–1011.

Lefohn AS (1991) *Surface-Level Ozone Exposures and Their Effects on Vegetation*. CRC Press.

Léger E, Deschaseaux A, Pireaux J, Cabané M, Weber E, Afif D, Dizenegremel P (1997) Relationship between primary and secondary metabolism in ozone treated poplar trees. Changes in enzyme acticities and gene expression. In: *Proceedings of the IUFRO Meeting "Somatic Cell Genetics and Molecular Genetics of Trees"*.

Lekshmy S, Jha SK, Sairam RK (2015) Physiological and molecular mechanisms of flooding tolerance in plants. In: *Elucidation of Abiotic Stress Signaling in Plants*. Springer, pp. 227–242.

Less H, Galili G (2008) Principal transcriptional programs regulating plant amino acid metabolism in response to abiotic stresses. *Plant Physiol* 147:316–330.

Li H, Liang Z, Ding G, Shi L, Xu F, Cai H (2016) A natural light/dark cycle regulation of carbon-nitrogen metabolism and gene expression in rice shoots. *Front Plant Sci* 7:1318.

Lim G-H, Singhal R, Kachroo A, Kachroo P (2017) Fatty acid–and lipid-mediated signaling in plant defense. *Annu Rev Phytopathol* 55:505–536.

Liu X, Ma D, Zhang Z, Wang S, Du S, Deng X, Yin L (2019) Plant lipid remodeling in response to abiotic stresses. *Environ Exp Bot* 165:174–184.

Lobell DB, Gourdji SM (2012) The influence of climate change on global crop productivity. *Plant Physiol* 160:1686–1697.

Loreti E, Valeri MC, Novi G, Perata P (2018) Gene regulation and survival under hypoxia requires starch availability and metabolism. *Plant Physiol* 176:1286–1298.

Luethy-Krause B, Pfenninger I, Landolt W (1990) Effects of ozone on organic acids in needles of Norway spruce and Scots pine. *Trees* 4:198–204.

Mangena P (2018) Water stress: Morphological and anatomical changes in soybean (*Glycine max* L.) plants. In: *Plant, Abiotic Stress and Responses to Climate Changes*. InTech Open, pp. 9–31.

Maruyama K et al. (2014) Integrated analysis of the effects of cold and dehydration on rice metabolites, phytohormones, and gene transcripts. *Plant Physiol* 164:1759–1771.

Marwein R et al. (2019) Genetic engineering/Genome editing approaches to modulate signaling processes in abiotic stress tolerance. In: *Plant Signaling Molecules*. Elsevier, pp. 63–82.

Matich EK, Soria NGC, Aga DS, Atilla-Gokcumen GE (2019) Applications of metabolomics in assessing ecological effects of emerging contaminants and pollutants on plants.
J Hazard Mater 373:527–535.

McLoughlin F, Testerink C (2013) Phosphatidic acid, a versatile water-stress signal in roots.
Front Plant Sci 4:525.

Melillo JM, Richmond T, Yohe G (2014) *Climate Change Impacts in the United States*, vol
52, U.S. Global Change Research Program

Michaelson LV, Napier JA, Molino D, Faure J-D (2016) Plant sphingolipids: Their importance
in cellular organization and adaption. *BBA-Mol Cell Biol L* 1861:1329–1335.

Mittler R (2006) Abiotic stress, the field environment and stress combination. *Trends Plant
Sci* 11:15–19.

Mollard FP, Striker GG, Ploschuk EL, Insausti P (2010) Subtle topographical differences
along a floodplain promote different plant strategies among *Paspalum dilatatum* subspecies and populations. *Austral Ecol* 35:189–196.

Morgan PB, Bernacchi CJ, Ort DR, Long SP (2004) An in vivo analysis of the effect of
season-long open-air elevation of ozone to anticipated 2050 levels on photosynthesis in
soybean. *Plant Physiol* 135:2348–2357.

Mori IC et al. (2006) CDPKs CPK6 and CPK3 function in ABA regulation of guard cell
S-type anion-and Ca 2+-permeable channels and stomatal closure. *PLoS Biol* 4:e327.

Muñoz P, Munné-Bosch S (2019) Vitamin E in plants: Biosynthesis, transport, and function.
Trends Plant Sci 24:1040–1051.

Muscolo A, Junker A, Klukas C, Weigelt-Fischer K, Riewe D, Altmann T (2015) Phenotypic
and metabolic responses to drought and salinity of four contrasting lentil accessions.
J Exp Bot 66:5467–5480.

Nahar K, Hasanuzzaman M, Fujita M (2016) Roles of osmolytes in plant adaptation to
drought and salinity. In: *Osmolytes and Plants Acclimation to Changing Environment:
Emerging Omics Technologies*. Springer, pp. 37–68.

Nakashima K, Yamaguchi-Shinozaki K, Shinozaki K (2014) The transcriptional regulatory
network in the drought response and its crosstalk in abiotic stress responses including
drought, cold, and heat. *Front Plant Sci* 5:170.

Nanjo T et al. (1999) Biological functions of proline in morphogenesis and osmotolerance
revealed in antisense transgenic *Arabidopsis thaliana*. *Plant J* 18:185–193.

Negrão S, Schmöckel S, Tester M (2017) Evaluating physiological responses of plants to
salinity stress. *Ann Bot* 119:1–11.

Noctor G, Lelarge-Trouverie C, Mhamdi A (2015) The metabolomics of oxidative stress.
Phytochemistry 112:33–53.

Noctor G, Mhamdi A, Foyer CH (2016) Oxidative stress and antioxidative systems: Recipes
for successful data collection and interpretation. *Plant Cell Environ* 39:1140–1160.

Noctor G, Reichheld J-P, Foyer CH (2018) ROS-related redox regulation and signaling in
plants. In: *Semin Cell Dev Biol*. Elsevier, pp 3–12.

O'Brien JA, Daudi A, Butt VS, Bolwell GP (2012) Reactive oxygen species and their role in
plant defence and cell wall metabolism. *Planta* 236:765–779.

Obata T, Fernie AR (2012) The use of metabolomics to dissect plant responses to abiotic
stresses. *Cell Mol Life Sci* 69:3225–3243.

Ohama N, Sato H, Shinozaki K, Yamaguchi-Shinozaki K (2017) Transcriptional regulatory
network of plant heat stress response. *Trends Plant Sci* 22:53–65.

Osakabe Y, Mizuno S, Maruyama K, Seki M, Shinozaki K, Shinozaki K (2006) An LRR
receptor kinase, RPK1, is a key membrane-bound regulator of abscisic acid early signaling in *Arabidopsis*. In: *Plant Cell Physiol*. Oxford University Press, pp. S103–S103.

Pandey P, Irulappan V, Bagavathiannan MV, Senthil-Kumar M (2017) Impact of combined
abiotic and biotic stresses on plant growth and avenues for crop improvement by
exploiting physio-morphological traits. *Front Plant Sci* 8:537.

Panozzo A, Dal Cortivo C, Ferrari M, Vicelli B, Varotto S, Vamerali T (2019) Morphological changes and expressions of *AOX1A*, *CYP81D8*, and putative *PFP* genes in a large set of commercial maize hybrids under extreme waterlogging. *Front Plant Sci* 10:62.

Parida AK, Panda A, Rangani J (2018) Metabolomics-guided elucidation of abiotic stress tolerance mechanisms in plants. In: *Plant Metabolites and Regulation Under Environmental Stress*. Elsevier, pp. 89–131.

Parihar P, Singh S, Singh R, Singh VP, Prasad SM (2015) Effect of salinity stress on plants and its tolerance strategies: A review. *Environ Sci Pollut Res* 22:4056–4075.

Pellegrini E, Carucci MG, Campanella A, Lorenzini G, Nali C (2011) Ozone stress in *Melissa officinalis* plants assessed by photosynthetic function. *Environ Exp Bot* 73:94–101.

Piikki K, De Temmerman L, Ojanperä K, Danielsson H, Pleijel H (2008) The grain quality of spring wheat (*Triticum aestivum L.*) in relation to elevated ozone uptake and carbon dioxide exposure. *Euro J Agron* 28:245–254.

Pleijel H, Wallin G (1996) Effects of the open-top chamber on air turbulence and light–the possible consequences for ozone uptake by cereals. In *Critical Levels for Ozone in Europe: Testing and Finalizing the Concepts*. pp. 303–307.

Prasad PV, Vu JC, Boote KJ, Allen LH (2009) Enhancement in leaf photosynthesis and upregulation of Rubisco in the C4 sorghum plant at elevated growth carbon dioxide and temperature occur at early stages of leaf ontogeny. *Funct Plant Biol* 36:761–769.

Qin F, Shinozaki K, Yamaguchi-Shinozaki K (2011) Achievements and challenges in understanding plant abiotic stress responses and tolerance. *Plant Cell Physiol* 52:1569–1582.

Queval G, Jaillard D, Zechmann B, Noctor G (2011) Increased intracellular H2O2 availability preferentially drives glutathione accumulation in vacuoles and chloroplasts. *Plant Cell Environ* 34:21–32.

Rahantaniaina M-S, Li S, Chatel-Innocenti G, Tuzet A, Issakidis-Bourguet E, Mhamdi A, Noctor G (2017) Cytosolic and chloroplastic DHARs cooperate in oxidative stress-driven activation of the salicylic acid pathway. *Plant Physiol* 174:956–971.

Ramel F, Birtic S, Ginies C, Soubigou-Taconnat L, Triantaphylidès C, Havaux M (2012) Carotenoid oxidation products are stress signals that mediate gene responses to singlet oxygen in plants. *Proc Natl Acad Sci USA* 109:5535–5540.

Reguera, M., Peleg, Z., Abdel-Tawab, Y. M., Tumimbang, E. B., Delatorre, C. A., & Blumwald, E. (2013). Stress-induced cytokinin synthesis increases drought tolerance through the coordinated regulation of carbon and nitrogen assimilation in rice. *Plant Physiology*, *163*(4), 1609–1622.

Rehem BC, Bertolde FZ, de Almeida A-AF (2012) *Regulation of Gene Expression in Response to Abiotic Stress in Plants*. InTech.

Rikin A, Richmond AE (1976) Amelioration of chilling injuries in cucumber seedlings by abscisic acid. *Physiol Plant* 38:95–97.

Rizhsky L, Liang H, Mittler R (2002) The combined effect of drought stress and heat shock on gene expression in tobacco. *Plant Physiol* 130:1143–1151.

Rodriguez M, Parola R, Andreola S, Pereyra C, Martínez-Noël G (2019) TOR and SnRK1 signaling pathways in plant response to abiotic stresses: Do they always act according to the "yin-yang" model? *Plant Sci* 288:110220.

Rosa M, Prado C, Podazza G, Interdonato R, González JA, Hilal M, Prado FE (2009) Soluble sugars: Metabolism, sensing and abiotic stress: A complex network in the life of plants. *Plant Signal Behav* 4:388–393.

Rouphael Y et al. (2018) Physiological and metabolic responses triggered by omeprazole improve tomato plant tolerance to NaCl stress. *Front Plant Sci* 9:249.

Rowland A, Borland A, Lea P (1988) *Changes in Amino-Acids, Amines and Proteins in Response to Air Pollutants*. Elsevier Applied Science Publishers Ltd.

Roy S, Mishra M, Dhankher OP, Singla-Pareek SL, Pareek A (2019) Molecular chaperones: Key players of abiotic stress response in plants. In: *Genetic Enhancement of Crops for Tolerance to Abiotic Stress: Mechanisms and Approaches, Vol. I.* Springer International Publishing, pp 125–165. doi:10.1007/978-3-319-91956-0_6.

Saini P et al. (2018) Reactive oxygen species (ROS): A way to stress survival in plants. In: *Abiotic Stress-Mediated Sensing and Signaling in Plants: An Omics Perspective.* Springer Singapore, pp 127–153. doi:10.1007/978-981-10-7479-0_4.

Sakamoto A, Murata N (2002) The role of glycine betaine in the protection of plants from stress: Clues from transgenic plants. *Plant Cell Environ* 25:163–171.

Sakuma Y, Maruyama K, Osakabe Y, Qin F, Seki M, Shinozaki K, Yamaguchi-Shinozaki K (2006) Functional analysis of an Arabidopsis transcription factor, *DREB2A*, involved in drought-responsive gene expression. *Plant Cell* 18:1292–1309.

Sami F, Yusuf M, Faizan M, Faraz A, Hayat S (2016) Role of sugars under abiotic stress. *Plant Physiol Biochem* 109:54–61.

Sandermann Jr H, Ernst D, Heller W, Langebartels C (1998) Ozone: An abiotic elicitor of plant defence reactions. *Trends Plant Sci* 3:47–50.

Sarkar A et al. (2010) Investigating the impact of elevated levels of ozone on tropical wheat using integrated phenotypical, physiological, biochemical, and proteomics approaches. *J Proteome Res* 9:4565–4584.

Sasidharan R, Voesenek LA (2015) Ethylene-mediated acclimations to flooding stress. *Plant Physiol* 169:3–12.

Saxena P, Srivastava A, Tyagi M, Kaur S (2019) Impact of tropospheric ozone on plant metabolism—a review. *Pollut Res* 38:175–180.

Schiermeier Q (2008) Water: A long dry summer. *Nature* 452:270.

Schwartz SH, Tan BC, Gage DA, Zeevaart JA, McCarty DR (1997) Specific oxidative cleavage of carotenoids by *VP14* of maize. *Science* 276:1872–1874.

Sedaghatmehr M, Thirumalaikumar VP, Kamranfar I, Marmagne A, Masclaux-Daubresse C, Balazadeh S (2019) A regulatory role of autophagy for resetting the memory of heat stress in plants. *Plant Cell Environ* 42:1054–1064 doi:10.1111/pce.13426.

Seki M et al. (2002) Monitoring the expression profiles of 7000 *Arabidopsis* genes under drought, cold and high-salinity stresses using a full-length cDNA microarray. *Plant J* 31:279–292.

Sewelam N, Kazan K, Schenk PM (2016) Global plant stress signaling: Reactive oxygen species at the cross-road. *Front Plant Sci* 7:187 doi:10.3389/fpls.2016.00187.

Shakeri A, Sahebkar A, Javadi B (2016) *Melissa officinalis* L.–A review of its traditional uses, phytochemistry and pharmacology. *J Ethnopharmacol* 188:204–228.

Sharma A et al. (2019) Phytohormones regulate accumulation of osmolytes under abiotic stress. *Biomolecules* 9:285.

Sharma E, Sharma R, Borah P, Jain M, Khurana JP (2015) Emerging roles of auxin in abiotic stress responses. In: *Elucidation of Abiotic Stress Signaling in Plants.* Springer, pp. 299–328.

Shen Y-Y et al. (2006) The Mg-chelatase H subunit is an abscisic acid receptor. *Nature* 443:823–826.

Shi W, Yin X, Struik PC, Xie F, Schmidt RC, Jagadish KS (2016) Grain yield and quality responses of tropical hybrid rice to high night-time temperature. *Field Crops Res* 190:18–25.

Shumbe L, d'Alessandro S, Shao N, Chevalier A, Ksas B, Bock R, Havaux M (2017) METHYLENE BLUE SENSITIVITY 1 (MBS1) is required for acclimation of Arabidopsis to singlet oxygen and acts downstream of β -cyclocitral. *Plant Cell Environ* 40:216–226.

Sies H (2017) Hydrogen peroxide as a central redox signaling molecule in physiological oxidative stress: Oxidative eustress. *Redox Biol* 11:613–619.

Sies H, Berndt C, Jones DP (2017) Oxidative stress. *Annu Rev Biochem* 86:715–748 doi:10.114 6/annurev-biochem-061516-045037.

Singh M, Kumar J, Singh S, Singh VP, Prasad SM (2015) Roles of osmoprotectants in improving salinity and drought tolerance in plants: A review. *Rev Environ Sci Biol* 14:407–426.

Singh RK, Deshmukh R, Muthamilarasan M, Rani R, Prasad M (2020) Versatile roles of aquaporin in physiological processes and stress tolerance in plants. *Plant Physiol Biochem* 149:178–189.

Staiger D (2002) Circadian rhythms in Arabidopsis: Time for nuclear proteins. *Planta* 214:334–344.

Striker GG (2012) Flooding stress on plants: Anatomical, morphological and physiological responses. *Botany* 1:3–28.

Sun A-Z, Guo F-Q (2016) Chloroplast retrograde regulation of heat stress responses in plants. *Front Plant Sci* 7:398.

Sun L, Su H, Zhu Y, Xu M (2012) Involvement of abscisic acid in ozone-induced puerarin production of *Pueraria thomsnii Benth.* suspension cell cultures. *Plant Cell Rep* 31:179–185.

Sun Y, Liu X, Zhai H, Gao H, Yao Y, Du Y (2016) Responses of photosystem II photochemistry and the alternative oxidase pathway to heat stress in grape leaves. *Acta Physiol Plant* 38:232.

Suprasanna P, Nikalje G, Rai A (2016) Osmolyte accumulation and implications in plant abiotic stress tolerance. In: *Osmolytes and Plants Acclimation to Changing Environment: Emerging Omics Technologies.* Springer, pp. 1–12.

Taji T et al. (2002) Important roles of drought-and cold-inducible genes for *galactinol synthase* in stress tolerance in *Arabidopsis thaliana*. *Plant J* 29:417–426.

Tewari S, Mishra A (2018) Flooding stress in plants and approaches to overcome. In: *Plant Metabolites and Regulation Under Environmental Stress.* Elsevier, pp. 355–366.

Tilman D, Balzer C, Hill J, Befort BL (2011) Global food demand and the sustainable intensification of agriculture. *Proc Natl Acad Sci* 108:20260–20264.

Trivedi DK, Huda KMK, Gill SS, Tuteja N (2016) Molecular chaperone: Structure, function, and role in plant abiotic stress tolerance. In: *Abiotic Stress Response in Plants.* Wiley International Publisher, 131–134.

Trouvelot S et al. (2014) Carbohydrates in plant immunity and plant protection: Roles and potential application as foliar sprays. *Front Plant Sci* 5:592.

Umezawa T, Fujita M, Fujita Y, Yamaguchi-Shinozaki K, Shinozaki K (2006a) Engineering drought tolerance in plants: Discovering and tailoring genes to unlock the future. *Curr Opin Biotech* 17:113–122.

Umezawa T et al. (2006b) *CYP707A3*, a major *ABA 8'-hydroxylase* involved in dehydration and rehydration response in *Arabidopsis thaliana*. *Plant J* 46:171–182.

Umezawa T, Yoshida R, Maruyama K, Yamaguchi-Shinozaki K, Shinozaki K (2004) *SRK2C*, a SNF1-related protein kinase 2, improves drought tolerance by controlling stress-responsive gene expression in *Arabidopsis thaliana*. *Proc Natl Acad Sci* 101:17306–17311.

Upchurch RG (2008) Fatty acid unsaturation, mobilization, and regulation in the response of plants to stress. *Biotechnol Lett* 30:967–977.

Vahdati K, Leslie C (2013) *Abiotic Stress: Plant Responses and Applications in Agriculture.* BoD–Books on Demand.

Valliyodan B, Nguyen HT (2006) Understanding regulatory networks and engineering for enhanced drought tolerance in plants. *Curr Opin Plant Biol* 9:189–195.

Vinocur B, Altman A (2005) Recent advances in engineering plant tolerance to abiotic stress: Achievements and limitations. *Curr Opin Biotech* 16:123–132.

Vishal B, Kumar PP (2018) Regulation of seed germination and abiotic stresses by gibberellins and abscisic acid. *Front Plant Sci* 9:838.

Vishwakarma K et al. (2017) Abscisic acid signaling and abiotic stress tolerance in plants: A review on current knowledge and future prospects. *Front Plant Sci* 8:161.

Voesenek L, Bailey-Serres J (2013) Flooding tolerance: O2 sensing and survival strategies. *Curr Opin Plant Biol* 16:647–653.

Vorne V et al. (2002) Effects of elevated carbon dioxide and ozone on potato tuber quality in the European multiple-site experiment 'CHIP-project'. *Eur J Agron* 17:369–381.

Voß U, Bishopp A, Farcot E, Bennett MJ (2014) Modelling hormonal response and development. *Trends Plant Sci* 19:311–319.

Wang Y et al. (2019) Research progress on heat stress of rice at flowering stage. *Rice Sci* 26:1–10.

Wani SH, Kumar V, Shriram V, Sah SK (2016) Phytohormones and their metabolic engineering for abiotic stress tolerance in crop plants. *Crop J* 4:162–176.

Wani SH, Sah S (2014) Biotechnology and abiotic stress tolerance in rice. *J Rice Res* 2:e105.

Wellburn A, Paul N, Mehlhorn H (1994) The relative implications of ozone formation both in the stratosphere and the troposphere. *Proc R Soc B* 102:33–47.

Wellburn F, Wellburn A (1996) Variable patterns of antioxidant protection but similar ethene emission differences in several ozone-sensitive and ozone-tolerant plant selections. *Plant Cell Environ* 19:754–760.

Wu D et al. (2013) Tissue metabolic responses to salt stress in wild and cultivated barley. *PLoS One* 8:e55431.

Xia X-J, Zhou Y-H, Shi K, Zhou J, Foyer CH, Yu J-Q (2015) Interplay between reactive oxygen species and hormones in the control of plant development and stress tolerance. *J Exp Bot* 66:2839–2856.

Yamaguchi-Shinozaki K, Shinozaki K (2006) Transcriptional regulatory networks in cellular responses and tolerance to dehydration and cold stresses. *Annu Rev Plant Biol* 57:781–803.

Yang Y, Guo Y (2018a) Elucidating the molecular mechanisms mediating plant salt-stress responses. *New Phytol* 217:523–539.

Yang Y, Guo Y (2018b) Unraveling salt stress signaling in plants. *J Integr Plant Biol* 60:796–804.

Yang Z, Midmore DJ (2005) A model for the circadian oscillations in expression and activity of nitrate reductase in higher plants. *Ann Bot* 96:1019–1026.

Yousuf PY, Hakeem KUR, Chandna R, Ahmad P (2012) Role of glutathione reductase in plant abiotic stress. In: *Abiotic Stress Responses in Plants*. Springer, pp. 149–158.

Yu Y, Huang R (2017) Integration of ethylene and light signaling affects hypocotyl growth in *Arabidopsis*. *Front Plant Sci* 8:57.

Yuanyuan M, Yali Z, Jiang L, Hongbo S (2009) Roles of plant soluble sugars and their responses to plant cold stress. *Afr J Biotechnol* 8:2004–2010.

Zagorchev L, Kamenova P, Odjakova M (2014) The role of plant cell wall proteins in response to salt stress. *Sci World J*, https://doi.org/10.1155/2014/764089.

Zandalinas SI, Mittler R (2018) ROS-induced ROS release in plant and animal cells. *Free Radic Biol Med* 122:21–27.

Zeier J (2013) New insights into the regulation of plant immunity by amino acid metabolic pathways. *Plant Cell Environ* 36:2085–2103.

Zhang Q, Huber H, Beljaars SJ, Birnbaum D, de Best S, de Kroon H, Visser EJ (2017) Benefits of flooding-induced aquatic adventitious roots depend on the duration of submergence: Linking plant performance to root functioning. *Ann Bot* 120:171–180.

Zhou W, Chen F, Meng Y, Chandrasekaran U, Luo X, Yang W, Shu K (2020) Plant waterlogging/flooding stress responses: From seed germination to maturation. *Plant Physiol Bioch* 148:228–236.

Zinta G et al. (2018) Dynamics of metabolic responses to periods of combined heat and drought in *Arabidopsis thaliana* under ambient and elevated atmospheric CO2. *J Exp Bot* 69:2159–2170.

Zwack PJ, Rashotte AM (2015) Interactions between cytokinin signalling and abiotic stress responses. *J Exp Bot* 66:4863–4871.

11 Climate Change and Cereal Modeling
Impacts and Adaptability

Arzoo Ahad, Sami Ullah Jan,
Khola Rafique, Sameera Zafar,
Murtaz Aziz Ahmad, Faiza Abbas, and Alvina Gul

CONTENTS

DOI: 10.1201/9781003250845-11

11.1 INTRODUCTION

Cereals are important staple crops in the human diet offering vital health benefits, nutritive value, and production and have great significance for the world's food security (M. Iqbal et al., 2019). *Triticum aestivum* L., *Oryza sativa*, and *Zea mays* are the main stable cereals production-wise for a major part of the world's population (Fatima et al., 2020). They are rich in minerals and nutrients, fibers, vitamin B complex, Vitamin E, thiamine, and iron. Cereals also contain high levels of carbohydrates, which make them an excellent source of energy, as they are very densely packed and in a form that our bodies can easily use. Therefore, cereals are among the most planted crops in the world (Challinor et al., 2014; M. Farooq et al., 2009). However, global threats, including drought, heat, salt, fungi, bacteria, viruses, and vegetation, are imposing serious risks to all cereals (Zaidi et al., 2016). The top three cereals are under threat by climate change, thus compromising global food security (Ahsan et al., 2020; J. Wang et al., 2018). Climate modification is a complex process, mainly mediated by anthropogenic means, which results in temperature rise, increased storms, and droughts. This process occurs naturally but human activities have sped up the process manyfold. Living organisms, especially plants, are incapable of acclimating quickly to the fluctuating environment. Figure 11.1 shows an increasing trend in global CO_2 emissions in billion tonnes, indicating a change in atmosphere.

In recent years, the productivity of cereal crops has declined because of climate change (Challinor et al., 2014; M. Farooq et al., 2009). A report disclosed that during 2020, wheat and rice yields were expected to fall up to 17% if the conditions continue (Dubey et al., 2021). A decline in the production of a major crop of a region leads to food insecurity and economic instability. The situation can worsen and lead to widespread famine. The situation is already very dire, as recent data (Figure 11.2) shows

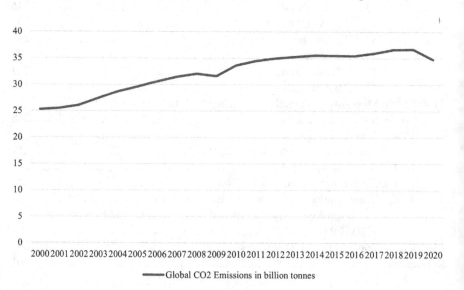

FIGURE 11.1 Global CO_2 emissions in billion tons.

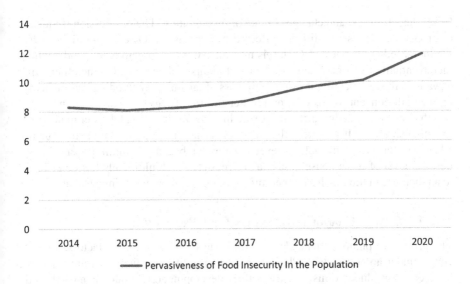

FIGURE 11.2 Pervasiveness of severe food insecurity in overall populace.

an upward trend in food insecurity (FAOSTAT, 2021). As time passes, the emissions of greenhouse gases are only increasing, leading to more aggressive consequences of climate modification (Roser, 2020). Therefore, special arrangements through modern technology are essential to get enough yield from these crops while work is being done to reduce the effects of climate change.

There have been developments in these areas, especially hybrid seeds and genetically modified organism (GMO) crops. These seeds have been specifically prepared to be more resistant to the environment and have a higher yield, either through natural breeding practices or more commonly through genetic engineering. There is a need to devise such methods that are better for the environment and simultaneously lead to better and more sustainable crop yields. We cannot afford to compromise either of the two, as both are fundamental to our survival. To do this we will have to revamp the whole system of modern agriculture and renovate it so that, we can achieve these goals. Uncertainty will always exist when we talk about future climatic conditions and their effect on crop yield. Yet, we still should strive to make that uncertainty as small as possible and direct it in a more positive direction.

11.2 CLIMATE CHANGE INFLUENCE

Climate change is a multifaceted phenomenon that refers to local-, regional-, or global-scale alteration in Earth's climates and also the effects caused by these changes. These fluctuations in the Earth's climate, that may take millions of years, are primarily driven by human actions, predominantly the burning of fossil fuels and elimination of forests, consequently leading to global warming. Global warming, which is defined as rapid upsurge in atmospheric CO_2 concentration, is one of the largest anthropogenic disturbances ever placed on natural systems (Sala Osvaldo et al., 2000). Climate modification impacts include rising temperatures, atmospheric

CO_2 concentrations, and changes in rate of precipitation. The progression frequency of crops and weeds are directly affected by increased CO_2 concentrations. Also, the increased atmospheric CO_2 levels may alter warmth, precipitation, and daylight that eventually influence the growth rate of plants and animals. On the other hand, elevations in sea level might result in the loss of farmland by flooding and intensifying the salt concentrations of groundwater in littoral regions (MacCracken, 2008). Another practice playing a foremost role in determining the globe's climate is the greenhouse effect. It produces the warm environment close to the Earth's surface where life-forms became able to develop. However, because of anthropogenic activities, the level of greenhouse gases has increased, which ultimately has contributed to an increase in the Earth's temperature, leading to global warming (Mahato, 2014).

11.2.1 CLIMATE CHANGE INFLUENCE ON CROP PRODUCTION

The growth of plants is dependent on different biotic and abiotic factors. Climate comes under abiotic factors. Plants can only grow in conditions that are favorable for them. Any fluctuations can affect their development. Climate changes directly impact the yield and production area of crops. The alteration in seasons affects the phenology of cereal crops. The phenology of crops is set by environmental factors as well as agricultural practices like sowing time. It has been observed that due to the phonological shift, there has been a decrease in overall yield. This is mainly because the cropping time changes with climate change. A temperature rise shortens the vegetative and reproductive phases of plants, leading to decreased crop yield. So, both of them need to be aligned in a way that results in better crop production (Fatima et al., 2020). The yield also depends on the location the crop is being grown in. For instance, hot areas will pose stress for the plants if they have a rise in temperature as compared to sites on high latitudes where increased temperature will provide plants with warmer conditions for better development and higher crop harvest (Fei et al., 2017). These conditions will make early planting and harvesting of winter crops possible. However, for the three most grown crops—wheat, maize, and barely—the yield declined 2%–3% during 1981–2002 because of the temperature rise. This caused a loss of $5 billion/year in the USA (Lobell, 2007). The precipitation patterns also change with changing climatic conditions. This especially affects rain-fed crops. The rain either arrives earlier or later than usual. These changes can either lead to flooding or drought. To combat this, the sowing time of crops needs to be switched. In a research conducted in Henan, China, it was noticed that just a 1% increase in precipitation led to a decrease in non-rain-fed wheat crops (Lee, 2017). There was a decline seen in the rain-fed wheat in Victoria, Australia, when the region received little rainfall. This is because of reduced water availability for the crops to grow. Such dry conditions cause the stomata to close to avoid evaporation. This further leads to reduced carbon dioxide absorption. As CO_2 is essential for photosynthesis, decreased CO_2 intake will lead to reduced yield (Anwar et al., 2007). Climate warming has also led to drought conditions in warmer areas. Crops in such sites face delayed germination, stunted growth due to reduced cell elongation and mitosis, and shorter leaf area, which decreases the rate of photosynthesis. The enzymes leading to grain filling in cereal crops seem to have diminished activity. If millet is exposed to drought at the flowering stage, it will become infertile. All this leads to a low crop yield (Ansari, 2020).

11.2.2 CLIMATE CHANGE INFLUENCE ON SOIL PROPERTIES

Like all other aspects of crop production, climate change also affects the soil on which the crops are growing. Different facets of climate change exert varying influences on soil and its properties. Some are negative, while others may turn out to be positive. Soil erosion is one of the main problems that happen due to climate modification (I. FAO, 2015). Studies done on soil erosion predicted that 43 Pgyr^{-1} (billion tonnes per year) of soil is eroded (Borrelli, 2020). All this soil is wasted, leading to food insecurity. With no other source of food, we depend on land for more than 95% of our total food supply (FAO, 2015). Due to erosion of the topsoil, the most nutrient-rich part of the soil is destroyed. To compensate for the lost nutrients, farmers must add additional nutrients in the form of fertilizers, which comes at a hefty price and can become a burden on the already weak and struggling economies. Climate change, including rainfall, temperature, and CO_2 concentration, can also affect the actual physical and chemical properties of soil in different ways. CO_2 concentration is very important for plants as they depend on it for photosynthesis. Studies have shown that an increase in environmental CO_2 can lead to increased plant productivity due to a phenomenon called CO_2 fertilization. This happens because more CO_2 is present, and this positively influences the rate of photosynthesis, leading to increased productivity (Piao, 2006). This trend is predominant in northern areas. In tropical and subtropical regions, the climate becomes too hot because of the increased CO_2 concentration for the plants to reach their full potential. Elevated CO_2 also promotes mycorrhizal and other soil microbiome activity, which benefits the plant by increasing the total soil organic matter. Mycorrhiza also helps in the uptake of phosphate. Rainfall has a varying effect on the soil. Due to climate change, rainfall intensity has increased and is predicted to increase in the future (NIle, 2019). This will have diverse effects on different parts of the world. Arid climates may experience an increase in plant productivity, while humid and areas with heavy rainfall might experience flooding or the soil moisture content being too high, which also negatively affects plant productivity.

11.2.3 CLIMATE CHANGE IMPACT ON INSECT PESTS

Environmental change has a negative influence on organisms and overall biodiversity (FRANCO et al., 2006). Insects, due to their ectothermic lifestyle, are highly sensitive to increases in temperature (Deutsch et al., 2008). Among insects, plant pests are more adapted to altered ecological circumstances as compared to other species. This attribute makes them more capable of enduring extreme surroundings of agricultural ecosystem. They rapidly adapt to the territories in which they can attain great abundance. In nature, pests are influenced by many natural and anthropogenic aspects (Visser, 2008). The degree of climate change is directly proportional to different pest species that will be affected, and inversely linked with the width of ecological necessities required by each species. The occurrence of pests in an environment depends on the existence of their host species.

11.2.4 Climate Change Impact on Plant Pathogens

Plant diseases are the outcome of the interface between a susceptible host, pathogen, and the surroundings. The environment significantly influences plants, pathogens, and their antagonists. So, alterations in the environment are related to the degree of injuries caused by a disease. They are frequently responsible for the advent of new infections (Anderson et al., 1988). Changes linked with global warming may affect the severity of disease ultimately influencing the coevolution of plants and their pathogens (Eastburn et al., 2011). Studies focusing on the effects of modified climate on plant pathogens specified that it has influence on pathogen growth and endurance rates, thus changing the effects of diseases on crops (Harvell et al., 2002). Environmental factors including temperature and relative humidity can directly influence the biological aspects of pathogens such as propagation and growth rates of propagules (Colhoun, 1973). Pathogens at higher latitudes are able to tolerate wider temperature ranges than their physiological optima. Thus, global warming is positively influencing their abilities, which in turn will increase the threat of disease epidemics these pathogens are related with (Desprez-Loustau et al., 2007). A pathogen's biology can also be affected by the plant architecture, for example, increased leaf densities result in humid conditions that contribute to stimulating pathogen growth thus changing the microenvironment. Variations in temperature and humidity control the reproduction rate of pathogens and have strong influence on polycyclic pathogens (Caffarra et al., 2012).

11.2.5 Climate Change Impact on Yield and Food Security

Climate change has a direct influence on agriculture and global food production (Kotir, 2011). Among various effects of changing climate, the major threat is agronomy. It disturbs food production and the agronomic system in different ways (Godfray, 2011). Constantly increasing temperatures, CO_2 concentrations, precipitation regimes, growth periods, and the crop cycle are climatic and biological variables that are affecting crop production. The notable challenge in the 21st century is to adequately supply food for the rapidly growing world population while nourishing the already strained environment. According to the Food and Agriculture Organization (FAO), global food security is a condition that prevails when the entire population has reliable access to nutritious food having the potential to meet their nutritional requirements for a healthy life. Food security consists of four aspects: food stability, availability, utilization, and access (Kang et al., 2009) (Figure 11.3).

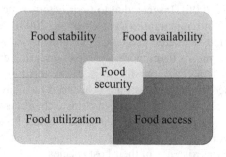

FIGURE 11.3 Aspects of food security.

FIGURE 11.4 Key challenges for future global food security.

The FAO (2008) stated that food security is influenced by the climate having effects on the components of national and international food systems. The Foresight Global Food and Farming Futures project analyzed the five challenges (Figure 11.4) of global food security.

11.3   GROWING CONDITIONS REQUIRED FOR DIFFERENT CEREAL CROPS

Crop production is affected by internal and external factors (Figure 11.5) as discussed in the following.

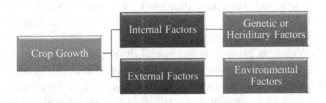

FIGURE 11.5 Factors affecting crop production.

11.3.1 INTERNAL FACTORS AFFECTING CROP PRODUCTION

The enhancement in crop yields and desirable attributes including high yield; grain and straw quality; and resistance to insects, pests, lodging, drought, and salinity are related to the genetic makeup of plants. These characteristics are not influenced by the ecological constraints.

11.3.2 EXTERNAL FACTORS AFFECTING CROP PRODUCTION

External factors include biotic, socioeconomic, climatic, soil, and physiographic factors affecting crop production. Furthermore, climatic factors affect about 50% of the yield. Atmospheric and edaphic variables that influence crop harvest include:

- Precipitation—Consists of total water falling from the atmosphere in the form of rainfall, dew, snow, and hail influencing vegetation.
- Temperature—Temperature measures the intensity of heat influencing the distribution of crop plants and vegetation.
- Atmospheric humidity—Humidity is the concentration of water vapors present in the air. It influences the water requirement of the crops.
- Solar radiation—Processes, including photosynthesis, photoperiodism, circadian rhythms, taking place within the plant require light. Solar radiation controls the distribution of temperature, hence controlling the distribution of crops in a region.
- Soil moisture—Moisture (water), a primary component of plant growth, is extracted from the ground soil. The soil water helps in mineralization and maintains the soil temperature.
- Soil air—Water absorption by the plant roots depends upon aeration of soil. Soil air is important for nutrient availability, decomposition of organic matter, and respiration of microbes present in soil.
- Soil temperature—It affects the amount of water absorption, germination of seeds, and the overall growth rate of plant.
- Soil mineral—The most important source of plant nutrients is soil. Minerals from soil are derived by the weathering of rocks (Tandzi & Mutengwa, 2020).

11.3.2.1 Paddy (Rice)

The first essential step in rice production, or any other crop, is selecting quality kernels of a variety that is feasible for the environment in which they will be cultivated. Kernels are the product that must be collected and administered to obtain the maximum yield of any rice variety. The best quality seeds are pure; viable; free of diseases, insects, and pathogens; uniform in size; and responsible for improving productivity by 5%–20%. The utilization of prime quality seed results in advance crop emergence, reduced replanting, lower seeding rates, and uniform and vigorous plant growth. Vigorous growth during initial developmental phases reduces weed issues and improve resistance to insect and infections (IRRI, 2015). Rice is grown in regions with normal temperatures between 20°C and 27°C in the course of the growing season. During early growth stages (the first four months) abundant sunlight is required. The temperature should not go below 15°C, as germination will not occur below that specific temperature. Paddy fields need more water as compared to other crops, so areas receiving a minimum 115 cm rainfall are suitable for cultivation. Clay loam soil, due to its higher water retention capacity, is the best for rice cultivation. During the germination phase, paddy requires an even ground level so that the fields are flooded equally. Hence the perfect habitats with no gradient are in the Menam Chao Phraya, Ganges, Irrawaddy, Siking, and Mekong. The three basic nutrients requirement for rice crops are nitrogen (N), potassium (K), and phosphorus (P). If these essential nutrients are present in a deficient amount, organic manure and artificial fertilizers need to be added (Orhan, 2013).

11.3.2.2 Pearl Millet

Pennisetum glaucum (pearl millet) is the most vital cereal after rice, wheat, maize, barley, and sorghum. It is considered the foremost drought-resistant crop in commercial production. It is cultivated in those regions that experience recurrent dry conditions during the vegetative or propagative period. It also has the capacity to tolerate sandy and acidic soils more than any other summer grain crop. The temperature required for pearl millet to germinate well is 75°F–90°F. Germination takes place within two to four days provided favorable conditions (Satyavathi et al., 2021).

11.3.2.3 Maize

Maize ranks as the third most-grown crop and is the highest-yielding cereal in the world. It is cultivated in deep soil in order to get high yields. Soil having a pH of 6.5 to 7.4, rich in organic matter, and water holding ability is preferred for obtaining great production. It requires substantial moisture and warmth from sprouting to flowering. The favorable temperature for maize plant germination and growth is 21°C and 32°C, respectively. After wheat and rice, *Zea mays* is the third most important cereal in Pakistan. In Northwest Pakistan (Khyber Pakhtunkhwa), maize ranked second after wheat in its importance (Amanullah et al., 2016).

11.3.2.4 Wheat

Wheat was cultivated 9,000 years ago in the Middle East. It is cultivated in warm temperatures (70–75°F) because it needs a lot of sunshine (Spilde, 1989). It can be cultivated in a wide variety of soils but grows best on well-drained loamy soil. The essential minerals required for flourishing its growth are N, P, sulfur (S), zinc (Zn), copper (Cu), boron (B), iron (K), and magnesium (Mg). Out of these nutrients, N is required in abundance as it is the major part of proteins in the plant (Abbate et al., 1995).

11.3.2.5 Sorghum

Sorghum is an essential food crop, importantly in the semiarid tropics throughout the globe. It is the main cereal crop in South Asia and sub-Saharan Africa having areas with limited food supply throughout the year. It is the world's fifth most valuable grain crop, after wheat, rice, maize, and barley. Sorghum grows on low-potential, shallow soil with increased clay concentration. However, it shows poor growth rate on sandy soil. Its tolerance level in basic salts is greater than other cereal crops and can be grown on soils within a pH range of 5.5–8.5. The percentage of clay in the soil should be optimally set at 10%–30% for the production of sorghum plants. The climatic requirements for germination and production of sorghum seeds are day length, water needs, and high temperature. The optimum minimum temperature for germination ranges from 7°C to 10°C. Almost 80 % of seeds can germinate within 10–12 days at 15°C. Plantation should be done in plenty of water with depth of 10 cm. Temperature plays the fundamental role in growth and development of plants after seed germination. The optimal photoperiod for flower formation induction ranges from 10 to 11 hours, as it is a short-day plant. Photoperiods longer than 11 to 12 hours induce vegetative growth. It can be grown in a varying range of soil as well as in irregular rainfall of approximately 400 mm (Editors of Encyclopedia Britannica, 2019).

11.3.2.6 Oat

In the global statistics of cereal production, oats ranked in the sixth position after wheat, maize, rice, barley, and sorghum. Oat has always been a valuable form of cattle feed, and offers a good resource of minerals, protein, and fibers. Oats are grown for grain consumption as well as straw for bedding, silage, haylage, chaff, and hay in different regions of the world. Livestock feed is however the primary usage of this crop (Welch, 1995). Oat, due to a well-developed root system, has a high assimilating ability. It develops to a depth of 120 cm and a width of up to 80 cm. It can extract nutrients from hardly soluble soil compounds. Oat prefers temperate climates. The seeds begin to germinate at a temperature of 2°C–3°C. Cold weather (15°C–18°C) is preferred during germination and tillering. Oats prefer wet soils as they are moisture-loving plants. The grains absorb 60% of water by their weight. The crop gives the best yields in wet years with precipitation in the first half of summer. The rapidly developing root system helps the crop to suffer less from spring droughts than spring wheat and barley. Oat can grow on sandy loam, loamy, clayey, and peat soils. It succeeds better than other grain crops in acidic (pH 5.0–6.0) soils and drained peatlands (Forsberg and Reeves, 1992).

11.4 CEREAL MODELING: POTENTIAL APPROACHES TO ENHANCE CEREAL CROPS PRODUCTION

Cereal is derived from the Latin term *cerealis*, which means "grain", and is botanically used to refer to a fruit or grain comprising of bran, germ, and endosperm. Cereal-producing plants are recognized as members of the monocotyledonous annual grass family known as Poaceae or Gramineae (Figure 11.6). These plants are characterized by their thin and long stalks as in wheat, rice, millet, barley, maize, oat, and sorghum. Worldwide, cereals are consumed in various forms including grains, flour, pasta, and bread, as staple components of a daily diet due to their high nutritional values (McKevith, 2004). It is estimated that almost all cereals are comprised of approximately 75% carbohydrates and about 15% protein, which provides over 50% of total calories consumed worldwide in a single day (Laskowski et al., 2019). Due to higher dependence upon them as staple nutritious food, cereals are grown on vast areas so that the yield achieved is sufficient to meet the global demands for cereals.

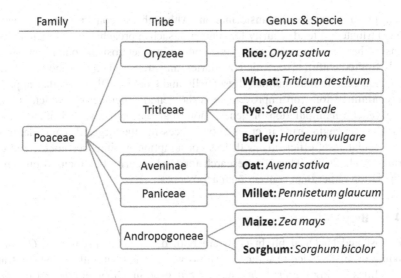

FIGURE 11.6 Taxonomic arrangement of some major cereals of the family Poaceae.

More than 2600 million tons of cereals were consumed in the year 2016–2017, and a mere 5 years later, global cereal consumption was expected to be over 2800 million tons in the year 2021–2022 (FAO, 2021). This abrupt increment in cereal consumption is directly proportional to the speedy growth in global population. Nearly 3 billion people lived on Earth in the 1960s, but there are now 7.7 billion people (Fróna et al., 2019). The world population is increasing continuously at a pace of 1% increment each year and if this ratio continues, it is expected that there will be 9.7 billion people at the mid of the current century (UNDESA, 2019). It is clearly aphoristic that increasing global population will unanimously increase food requirements; therefore, lots of scientific and technological efforts are dedicated to enhancing cereal production. However, attaining the mounting yield is mainly restricted and challenged by numerous factors collectively known as stress.

Plants are unable to relocate, therefore they, being sessile, are prone to a wide range of stresses. The term "stress" entails any abrupt change occurring in the external environment of plants that adversely affects development and yield (Verma et al., 2013). All the stresses affecting plant growth and productivity are divided into two key categories based on the source of stress known as biotic stress and abiotic stress. Biotic stresses are triggered by living organisms such as the diseases caused by pathogenic microorganisms or insects, while abiotic stresses are those stresses that occur due to nonliving environmental factors such as variation in the availability of nutrients, temperature, light, water, and salinity (Zhu, 2002). In order to fulfill dietary needs of the continuously growing population, it is necessary to protect plants from any type of stress to achieve higher yield. Several approaches are perceived and attempted with the aim to minimize the drastic impact of biotic as well as abiotic stresses upon plants and achieve maximum yield. One such approach is the conventional strategies that encompass the agricultural or horticultural practices adopted for yield maximization such as crossings among species based on phenotypic traits and soil transformation. Although the conventional approaches have contributed to food security for centuries, such approaches have practical limitations as they are expensive, more time consuming, heterosis is found in offspring, usually limited within taxonomic boundaries, and are based on morphological traits rather than genetic information (Breseghello and Coelho, 2013). Another approach widely admitted for crop improvement is the genetic engineering, which is based on the genetic manipulation of genes, molecular mechanisms, and other components. Genetic engineering in crops allows accessing the targeted manipulations at the molecular level efficiently with less consumption of time, resources, and cost (Parmar et al., 2017). Thus, the research for food security has focused on genetic manipulation rather than conventional approaches.

11.4.1 BREEDING

Rice blast and bacterial blight (BB) are both detrimental diseases of *O. sativa*. Breeding of an elite rice restorer line Hui_{773} was carried out with two stable transformed lines having the *cry1Ab* gene and *bar* gene in Zhongguo91, and the *Xa21* gene containing the Yujing6 line. These genes confer multiple actions such as insect resistance, herbicide tolerance, and resistance to bacterial blight disease, respectively. The results approved considerable heterosis by producing transformed hybrid of *O. sativa* cultivar via breeding (Wei et al., 2008). Marker-assisted backcross breeding was used to introduce two important BB-resistant genes (*Xa21* and *xa13*) as well as a significant gene (*Pi54*) for resistance in bacterial blast into the Indian *O. sativa* MTU1010 (Arunakumari et al., 2016). Broad range resistance against various rice diseases can be conferred by pyramiding R genes, thus polymorphic lines, namely, *Pigm/Pi54*, *Piz-t/Pi54*, *Pi2/Pi1*, and *Pi2/Pi54* are better than individual resistant-gene-containing lines (W. Li et al., 2020).

Fungal diseases of wheat result in yield losses of 15%–20% per year, which can be controlled by taking suitable measures (Figueroa et al., 2018). An elite *T. aestivum* cultivar was successfully integrated with a combination of R genes, i.e., *Pm2*, *Pm21*, and *Pm4a*, of powdery mildew utilizing molecular markers for double homozygote selection showing broad-spectrum effectiveness in near isogenic lines of Yang 158 cultivar (Liu et al., 2000). There are over 187 R genes conferring fungal resistance in wheat. Of them two pyramided single dominance executing R genes, i.e., *Yr15* and *Yr64*, of stripe rust showed broad-spectrum resistance against all tested strains of yellow rust (Qie et al., 2019). Pests, such as aphids, can cause significant yield losses in grains. As a result, breeding for aphid resistance is beneficial for improving yield. In four spring cultivars of *T. aestivum* (Astrid, Odeta, Alicia, and Libertina) and two emmer cultivars (Rudico and Tapiruz) measurement of the antibiosis level of *Metopolophium dirhodum* (Aphididae) was performed. The response of each cultivar and projected population growth to *M. dirhodum* was recorded after applying an age-stage, two-sex life table. Rudico was most vulnerable to *M. dirhodum* with one order of magnitude bigger than other wheat cultivated varieties (Platková et al., 2020). Maize inbred lines underwent a multivariate analysis in which protein of the glutathione S-transferase (GST) encoding gene was found to express pleiotropic function against three detrimental fungal diseases (Wisser et al., 2011).

Bemisia tabaci (white fly) mediated *Begomovirus* transmission in dicots can't be controlled by physical or chemical barriers. Tomato yellow leaf curl virus (TYLCV) resistant lines F_5 were obtained from a cross with *Lycopersicom peruvianum* cultivar (Lapidot & Friedmann, 2002). Breeding has been modernized to cisgenesis and intragenesis in which genetic transfer occurs in sexually compatible plants that can also breed naturally such that resistant genes are transferred to related species or genetic expression is enhanced (Mores et al., 2021). The cisgenic approach was developed for phytophthora disease in *Solanum tuberosum* (potato) in which multiple pyramided *R* genes were transferred to disease-susceptible cultivated varieties (Haverkort et al., 2009). Because it can speed up the abrupt mutation development and expand the pool of allelic variations accessible for genetic improvement, induced mutagenesis appears to be a potential option for generating resistant varieties of *O. sativa* (Viana et al., 2019).

11.4.2 RECOMBINANT TECHNOLOGY

Over the past decades, the advancement of cereals was accomplished mostly by conventional methods of breeding. Particularly, wide hybridization performed a key part in generating several cultivated varieties possessing better and novel agronomic traits (Jauhar & Chibbar, 1999; Jauhar & Peterson, 2001; Xiao et al., 1996). But, because of the continuous increase in human population, inadequate water availability, progressively depleted fossil energy sources, and the changed environmental scenario, advanced methods were immediately needed to deal with upcoming challenges (Shrawat & Lörz, 2006). Recombinant technology has gained much importance since the mid-1990s providing innovative prospects for a robust and targeted introduction of desirable agronomic characteristics in cereals (Jones, 2005; Kumlehn & Hensel, 2009). The application of genetic transformation has significantly developed an understanding of a plant's molecular processes vital for plant improvement, the orientation of spike, time period of flowering, and endurance to various biotic and abiotic stresses (Hensel, 2020). Recombinant DNA technology involves alteration in genetic material outside a living organism to acquire desired and improved characteristics in organisms or their products. This technique comprises the insertion of genome or DNA fragments from diverse sources, containing the desired gene sequence using an appropriate vector (Berk & Zipursky, 2001). Genetic transformation based on vector *Agrobacterium* and biolistic systems (i.e., direct gene delivery through microprojectile bombardment) has been successfully employed in cereals (Shrawat & Lörz, 2006). *Agrobacterium* is Gram-negative and soil-borne plant pathogenic bacteria that occur in several plant species. Its pathogenic strains induce crown galls in plants and this is principally associated with tumor-inducing (Ti) plasmid present in their genome (Kuzmanović et al., 2015). It comprises of a T-DNA region, which is enclosed by right and left repeats and assists in the relocation of DNA bounded by these two borders (Gelvin, 2003). This T-DNA region can be modified using any desirable single or many genes, and then can be transmitted to plant tissues or cells due to infection caused by *Agrobacterium*. Some species of this bacterium possess more than one T-DNA region on Ti plasmids leading to more than two borders of T-DNA from which T-DNA can be further processed (Gelvin, 2003).

The *Agrobacterium*-mediated genetic transformation facilitates the specific integration of genetic transcripts into the host genome and displays much better permanence for the transferred gene (Dai et al., 2001). Various steps involved in *Agrobacterium*-mediated transformation systems in cereals are shown in Figure 11.7. The unique capacity of *Agrobacterium* of delivering an isolated DNA fragment into the host plant genome is successfully manipulated in cereals (M. Cheng et al., 2004; Repellin et al., 2001).

Major cereals, namely, wheat, rice, barley, maize, and sorghum, were successfully transmitted by *Agrobacterium* (Hiei et al., 2014). The first transgenic approach in cereals was based on *Agrobacterium*-mediated direct transmission of gene in rice (Hiei et al., 1994) that generated several transgenic rice plants that successfully expressed transgenes. This study was then followed by a publication of a protocol for *Agrobacterium*-mediated genetically transformed maize production (Ishida et al., 1996). Various desirable genes have also been delivered to cereals via an *Agrobacterium*-facilitated transformation scheme which provided excellent resistance or forbearance. Biotic stress in crop plants is caused by viruses, bacteria, fungi, insects, nematodes, and wild plants. In comparison to abiotic stresses

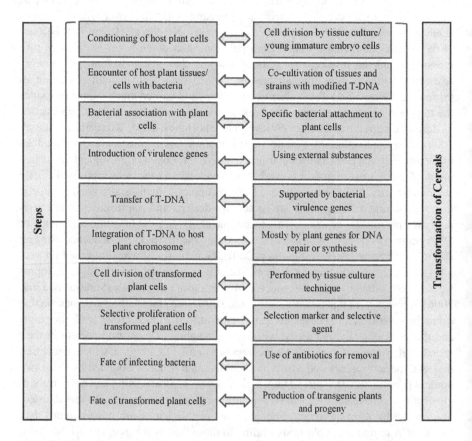

FIGURE 11.7   Steps in cereals transformation system mediated by *Agrobacterium*.

or environmental factors, the biotic stress agents are directly involved in affecting plant's growth by withdrawing their host of nourishment. This leads to diminished host immunity leading to death (Bakala et al., 2021). Plants do not have an established immune system in comparison to vertebrates. However, they have developed a succession of defense systems against biotic stresses. Genetic basis of such strategies is preserved in the genetic code of plants, and their genomes encode several resistance genes against biotic stress. Following the completion of genome sequences of various plants during the last years, an insight into biotic stress-resistance genes encoded within plant genomes of many important crops, including maize, rice, and sorghum, have been obtained (Singla et al., 2016). Like in rice (Y. Sun et al., 2016) and maize (Svitashev et al., 2016), the ALS gene conferring herbicide resistance has been well documented. Similarly, a wheat Roundup ready-resistant variety (Monsanto's MON71800) was released (Obert et al., 2004) and later in 2004 was approved for use in food by the U.S. Food and Drug Administration (FDA). However, little appreciation was given in Asia and Europe to such genetically modified plants resulting in substantial market losses and further market entry was stopped (Hensel, 2020). Also, the GS1 transferred gene showed tolerance to herbicide phosphinothricin in a wheat study (Huang et al., 2005).

A study reported an introduction of a stilbene synthase gene from *Vitis vinifera* to *O. sativa*. This gene was associated with the biological synthesis of a phytoalexin, i.e., an inducible antimicrobial compound, and provided a good resistance in the rice transformants against the pathogenic fungus *Pyricularia oryzae* (Stark-Lorenzen et al., 1997). Similarly, recombinant rice plants expressed along with a coat protein of rice dwarf virus have been documented (Zheng et al., 1997). The Rpg1 gene provided resistance against stem rust in transgenic barley plants (Horvath et al., 2000). Recently, a study discovered a novel approach presenting fungal resistance in barley. It has been shown that barley plants expressed inverted repeat structures whose sequence targets fungal genes. After transmission, they induced gene suppression (Nowara et al., 2010). If they are utilized against infection-causing genetic sequences of the fungus, the infection severity can be reduced or even can be prevented. *Agrobacterium*-mediated transformation in sorghum plants using *hph* gene expressed disease resistance against fungal disease (Indra Arulselvi et al., 2010).

A report indicated that leading Indica rice plants transformed with endotoxin encoding protein, namely, CryIAc, from bacterium *B. thuringiensis* showed effective resistance against the insect pest yellow stem borer (Nayak et al., 1997). Following this, a study reported high toxicity against the pests striped and yellow stem borer in transgenic *O. sativa* plants that expressed the synthetic genes cryIA(b) and cryIA(c) (X. Cheng et al., 1998). Later, *O. sativa* with the transmitted gene Ubi1:cry1Ab delivered from *B. thuringiensis* provided resistance to lepidopteran species of pests (Shu et al., 2000). Likewise, rice expressed with the Ubi1:cry1Ac gene was found resistant to yellow stem borer (Khanna & Raina, 2002). The RSs1:gna gene provided resistance against sap-sucking insects (Nagadhara et al., 2003), while the genes Ubi1:cryIA(b) and Ubi1:cryIA(c) (Nagadhara et al., 2004) and pCubi:mpi pCC1:mpi (Vila et al., 2005) gave resistance against stem borers (Nagadhara et al., 2004). Other studies reported the genes RSs1:gna (Nagadhara et al., 2004), Ubi1:cryIB, and Ubi1:cry1Aa (Breitler et al., 2004) offered effective resistance against different insect pests.

Abiotic stresses, predominantly water deficiency, high salt content, high temperature, and low temperature, can greatly affect the plant's growth and development stages. Experimentation has been conducted to recognize the mechanisms behind abiotic stress tolerance to enhance the corresponding plant's tolerance, specifically agriculture crop production (Jeyasri et al., 2021). Cereal plants have undergone evolution to survive in various environments where they often face several abiotic stress conditions (Giordano et al., 2021; Kumari et al., 2021). The potential of cereals to tolerate various dominant abiotic stresses (Figure 11.8) is extremely essential for yield resilience, and improvement in this aspect is a continuous target for breeders and researchers (Halford et al., 2015). Knowledge of different crops' responses and physiological processes at varying stages of growth are necessary for mitigation of abiotic stresses via management strategies or alteration of the cropping pattern in a region. Understanding such plant responses to diverse stresses at the molecular level is necessary for the improvement of crops possessing wider range of stress tolerance (Jeyasri et al., 2021). In crop breeding, the identification, characterization, and further transfer of the genes (via genetic engineering) responsible for expressing abiotic stress tolerance in crops has been a major challenge, and various studies have been conducted on diverse abiotic stresses in major cereals. Excessive soil salinity negatively affects the productivity of crops including rice. Usually, rice crops can tolerate a limited salt amount in water without affecting their growth and yield. But, the

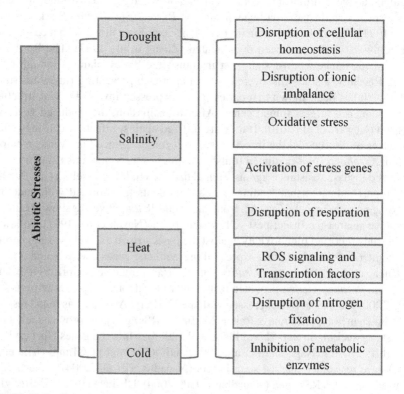

FIGURE 11.8 Cereal crop responses to different abiotic stresses.

tolerance is highly dependent on the rice type and species and also the plant growth stage (Hasanuzzaman et al., 2013). In rice, stress tolerance to drought, salinity, and cold have been achieved by expressing a stress-responsive MAPK gene (OsMAPK5) (Xiong & Yang, 2003). Earlier, in rice crops, the codA gene transferred from soil-borne bacterium *Arthrobacter globiformis* provided tolerance to salinity and low temperature (Sakamoto & Murata, 1998). Likewise, the CNAtr gene enhanced salt forbearance in transgenic *O. sativa* plants (Ma et al., 2005).

In a study, codA gene of *Arthrobacter globiformis* was transferred into *O. sativa* via *Agrobacterium* and this improved performance of the transgenics under drought stress (W. Sawahel, 2003). A similar enhanced tolerance to drought was achieved by expressing a pMnSOD gene from pea (F.-Z. Wang et al., 2005). An elevated expression of the *E. coli* nhaA gene in transgenic rice enhanced salinity and dehydration tolerance (Wu et al., 2005). Similarly, growth and yield of maize crops are severely influenced by various abiotic stresses. Among them, stress due to salinity results in various physicochemical alterations in plants, including disruption of ionic imbalance, cellular homeostasis, respiration, nitrogen fixation, and inhibition of different metabolic enzymes like photosynthetic enzymes (S. Iqbal et al., 2020). Various transgenic maize plants have been generated carrying foreign genes resistant to abiotic stresses, such as *E. coli* betA gene that encodes choline dehydrogenase improved cold as well as drought tolerance (Quan et al., 2004). To enhance the cold tolerance, a low temperature-tolerant PPDK cDNA was separated from *Flaveria brownii* and inserted into maize plants using *Agrobacterium* (Ohta et al., 2004).

Wheat plants are also affected by such stresses. Among these, heat and salinity stress severely affect their growth and yield. Increasing salt content in irrigation water pose an adverse effect on wheat yield. To enhance wheat performance and its yield under salinity stress, the most efficient strategy is to develop plant modifications under strain conditions (Shah et al., 2018). Transgenic wheat has been generated that expresses tolerance under salinity stress conditions, for example, in a study transgenic wheats were generated by delivering the P5CS gene (W. A. Sawahel & Hassan, 2002). Sorghum often displays tolerance to drought and waterlogging situations and is sown in diverse soil conditions (Calone et al., 2020). Severe drought conditions result in considerable sorghum yield loss by adversely affecting different plant physiological functions, development of inflorescence, and leaf growth (Abdel-Ghany et al., 2020; Djanaguiraman et al., 2020). Sorghum plants have also been transformed using *Agrobacterium* vector in several studies, for example, the use of TPS1 gene proved increased forbearance to abiotic stress conditions (Yellisetty et al., 2015). Likewise, the expression of transferred NPK1 gene via *Agrobacterium*-mediated transformation resulted in good sorghum growth and yield under water deficiency (Assem et al., 2017).

11.4.3 GENE EDITING

The biotic stress agents, including pests and pathogens, are major production and yield constraints causing 20% to 40% losses to world agricultural productivity, therefore causing major hurdles in food security and safety, which remains a profound agricultural challenge globally (Rodriguez-Moreno et al., 2017; Savary et al.,

2012). Also, the existing agriculture is mostly dependent on chemicals to avoid crop losses due to various plant pathogenic diseases. Their excessive treatment pose negative influence on human health and the surrounding ecosystem. To maintain a constant supply of these important staples in addition to improving in their nutritional contents and to address climate change pose a great opposition and demand the application of numerous novel crop breeding approaches. To overcome various biotic and abiotic challenges, various molecular techniques have been practiced for the improvement of cereal plants (Hillary & Ceasar, 2019). The enhancement of plant resistance against biotic stresses is an imperative task in sustaining production in order to meet an increase in the global population. Mutations, either natural or induced, may modify the interaction and restrict some steps in the infection process (Dracatos et al., 2018). Diversity in plant genetic resources (PGR) offers a possibility for plant breeders to generate novel and improved plant cultivars with desired functions, encompassing both breeder-preferred characteristics, such as tolerance to biotic stress and photosensitivity, as well as farmer-desired characteristics like yield potential and large-sized seeds (Govindaraj et al., 2015). Effective exploitation of such accessible genetic resources is essential for the genomic improvement of any crop plant (Ibrahim Bio Yerima & Achigan-Dako, 2021). The available genetic resources for various gene pools of crops, which include breeding research materials, crop landraces, and wild relatives, collectively form the base of modern plant breeding and are maintained in gene banks (Loskutov, 2021).

Gene editing can be defined as a process involving advanced techniques in molecular biology for site-specific, efficient, and precise modifications within a genome (Sedeek et al., 2019). The resultant plants can be precisely termed as genetically modified (GM) plants that occur through the transfer of a transgene (gene) of known function. Gene editing has extended tremendous potential and scope over conventional practices in improving a cereal's quality, protection, and traits. It provides remarkable opportunities in cereal crops by developing new GM plants or genotypes incorporated with the desired fragment of genome. So far several GM plants including cereals have been developed against various biotic and abiotic stresses (T. Wang et al., 2019). The process of genetic engineering or gene editing in plants starts with isolation of the desired gene from a living source, which is then incorporated within a suitable vector to make a recombinant DNA molecule, and finally this recombinant DNA molecule is inserted into the host's (plant) genome—thus integrating a new function within the GM plant (Parmar et al., 2017). One of the main components of gene editing tools used for production of GM crops that have the most significant impact upon the overall process of gene editing is selection of a suitable enzyme. Traditionally, enzymes like restriction endonucleases are used for cleaving DNA molecules, while ligases are employed to rejoin the fragments of DNA. Although the restriction endonucleases and ligases are helpful in manipulating small genomes such as those of viruses and bacteria, it is complicated to arbitrarily maneuver complex and larger genomic content of higher organisms, such as animals and plants, through these enzymes. Moreover, these enzymes can target only specific patterns or sequence within genome, therefore alternate enzymes are preferred that could provide more options in manipulation of genomes (Kamburova et al., 2017).

Gene editing through transcription activator-like effector nucleases (TALENs) (Christian et al., 2010) and zinc-finger nucleases (ZFNs) (Kim et al., 1996) have been known since the end of last century, however, they were focused on soon after the discovery of clustered regularly interspaced short palindromic repeats (CRISPR) (Jinek et al., 2012), which paved a simple and easy way for desired gene editing. These enzymes use the method of a typical sequence-specific nuclease (SSN) that can identify a specific DNA sequence to create double-stranded breaks (DSBs), and the plant's intrinsic repair mechanism joins the DSBs either through homologous recombination (HR), which causes insertions or replacements, or through non-homologous end joining (NHEJ), which causes gene knockouts by insertion or deletion of nucleotides. So far, these tools have proved effective and are widely desired in genetic manipulations of various plants for crop development (Zhang, Massel, et al., 2018). Keeping in view of both (1) the need of genetic modifications in cereal plants for stress resistance and yield improvement, as well as (2) the potentials offered by programmable TALENs and CRISPR, it seems plausible and effective to utilize TALENs and CRISPR for rapid and effective cereal improvement through genetic modeling or modifications.

11.4.4   Zinc-Finger Nucleases (ZFNs)

ZFNs are synthetically constructed restriction enzymes serving as an effective tool used for gene editing (GE). They can cut any long fragment of double-stranded DNA (dsDNA) sequences (Carroll, 2011; Osakabe et al., 2010). The introduction of DSBs by these site-specific endonucleases is then resulted by error-prone NHEJ repair, which creates minor insertions or deletions at the ZFN cleavage locus. ZFNs are chimeric proteins comprising two domains, i.e., a synthetic Cys2-His2 zinc-finger domain that joins to DNA and non-specific DNA cleavage domain of the *Flavobacterium okeanokoites* I (FokI) DNA restriction enzyme (Curtin et al., 2011; Fiaz et al., 2019). ZFNs have been applied for directed chromosomal deletions (H. J. Lee et al., 2010; Şöllü et al., 2010), removal of transgene (Petolino et al., 2010; Weinthal et al., 2013), and targeted incorporation of DNA (Ainley et al., 2013; Shukla et al., 2009). The studies that utilized ZFNs in plants are comparatively fewer, and the effectiveness of editing attained by these nucleases is usually low (Ahmar et al., 2020). Pertaining to the advancement of disease resistance in crops, ZFNs demonstrated a small influence by genetic modification in the development of diseases in host plants because they appear to be complex for engineering and a challenging to multiplex (de Galarreta & Lujambio, 2017; Jaganathan et al., 2018; Khandagale & Nadaf, 2016), have a lesser affinity for AT-rich sequence, and can attach to any region of DNA sequence other than the target site leading to off-target results (W. A. Ansari et al., 2020). In addition, vector construction for ZFNs is laborious and time consuming (Tang et al., 2017). However, synthetic zinc-finger proteins (AZPs) have been applied successfully and contributed toward an effective antiviral resistance in crop plants through blockage of DNA-binding sites of replication proteins of viruses (Sera, 2005; Takenaka et al., 2007).

GE using ZFNs was first documented in tobacco and *Arabidopsis*. Later, ZFNs were effectively used in cereals like wheat, rice, maize, and other crop plants (W. A.

Ansari et al., 2020). The initial effective application of ZFNs in cereals was reported for site-directed mutagenesis in maize. This study reported a targeted cleavage of a phytic acid biosynthesis gene called inositol-1,3,4,5,6-pentakisphosphate kinase 1 (IPK1). This protein encoding gene has the potential for both herbicide tolerance as well as required modification of the inositol phosphate profile in Z. *mays* seeds (Shukla et al., 2009). Later, a study (Petolino et al., 2010) also showed targeting of a herbicide-tolerance gene by ZFN. This resulted in regeneration of herbicide-resistant maize plants. Lately, ZFN-mediated genome editing has been effectively applied in allohexaploid wheat, and coding sequences of acetohydroxyacid synthase (AHAS) were directed to induce resistance against herbicides (imidazolinone) and this study resulted in a 2.9% recuperation in transformed plants (Ran et al., 2018). Likewise, using ZFN-modified AAD1 gene was directed in maize that exhibited potential for herbicide resistance (Ainley et al., 2013).

Herbicide tolerance has also been achieved in wheat (M. Zhang & Coaker, 2017) via base editing implemented to alter genes, namely, acetolactate synthase (ALS) and acetyl-coenzyme A carboxylase. Both genes provide good forbearance against various herbicides like imidazolinone, sulfonylurea, and aryloxyphenoxy propionate. The use of ZFN effectively generated mutation in SuR loci conferring herbicide resistance. Following this, ZFN-mediated, NHEJ-directed endogenous gene editing was reported in *T. aestivum* (Ran et al., 2018). In this study, a precise individual amino acid change was introduced into the coding gene fragment of acetohydroxyacid synthase (AHAS) and this resulted in modified wheat plants carrying resistance against imidazolinone herbicide (imazamox).

ZFNs appear to be highly programmable, and an adjustment can be done in the amino acids belonging to the zinc-finger domain for various genomic targets. Thus, their application is not dependent on the development of beneficiary plant lines in which preintroduction of target sites has been performed. Precise zinc-finger repeats have been created for many nucleotide triplets. Yet, the modular arrangement of these zinc-finger repeats for an efficient and sequence-specified ZFN demands screening and optimization, which is often difficult (Urnov et al., 2010).

11.4.5 TRANSCRIPTION ACTIVATOR-LIKE EFFECTOR NUCLEASES (TALENs)

TALENs quickly appeared as effective substitutes for ZFNs for incorporating targeted DSBs and desirable editing within genomes. TALENs are parallel in structure to ZFNs with an adjustable DNA-binding domain (DBD) attached with non-specific FokI nuclease domain. A complete TALEN consists of a DBD, nuclear localization signal (NLS), transcription activation domain (TAD) localized at the N terminus, and a C-terminal translocation signal. The central DBD consists of 33–35 amino acids repeats and each repeat can identify a single nucleotide in target sequence which helps in designing TALENs easier than ZFNs—a reason why TALENS are preferred over ZFNs. The highly conserved repeats of DBD in TALENs are derived from transcription activator-like effectors (TALEs), which are produced by *Xanthomonas* and are effective in altering genetic transcription within host plant cells (Boch & Bonas, 2010). These TALENs are inserted into plant cells through a type-III secretion system that allows binding to genomic DNA and alteration of transcription of targeted genes (Joung & Sander, 2013). So far, several plant genomes, including rice, maize, wheat, barley, potato, and *Arabidopsis*, have been edited using TALENs (Martínez-Fortún et al., 2017). Genome editing of cereals through TALENs for a particular attribute can be termed as cereal modeling. Various plants have been modeled including cereals that are developed through application of TALENs, which has effectively conferred resistance. TALENs have also been used for modification of other traits like fragrance and flowering (Shan et al., 2015; Z. Sun et al., 2013). Although several cereals are edited through TALENs for various trait improvements, only two stress-tolerant cereals are reported so far that are edited through TALEN specifically resistant against biotic stresses.

Rice genome was edited for the first time using TALENs and was made resistant to bacterial blight by disrupting the promoter region of the bacterial blight susceptibility gene *OsSWEET14*, also known as *Os11N3* (T. Li et al., 2012). The *OsSWEET14* gene encodes a type of sucrose-efflux transporter (SWEET) in rice, which is targeted by *Xanthomonas oryza* by using its own TALE PthXo3 or AvrXa7. Once the gene is activated by *X. oryza*, the sugar present in the plant's cell is utilized by *X. oryza* to fulfill its own nutritional requirements. The promoter region of *OsSWEET14* contains effector binding element (EBE) for AvrXa7, and its TATA box overlaps with EBE for PthXo3. Li et al. (2012) designed and applied two pairs of TALENS named pair 1 and pair 2 (each TALEN pair designed contained 24 repeat units for identification of 24 specific nucleotides at the target site) to induce mutations within overlapping EBEs present in the promoter region of *OsSWEET14*. This application of TALENs helped in editing the virulence factor of PthXo3 and AvrXa7 without compromising the developmental function of *OsSWEET14*. In a similar study, TALENs were used to knockout *TFIIAy5* gene in rice, which also made it resistant to *X. oryzae* or bacterial blight disease (Han et al., 2020). Wheat is the second cereal that has been modeled through TALEN-mediated gene editing specifically for biotic stress tolerance. The trait for resistance to powdery mildew disease is not found naturally in wheat. However, this polyploid cereal has been successfully edited through TALEN by inducing targeted mutations in three homeoalleles of *Mildew-Resistance Locus* encoding for MLO proteins. The heritable resistance was confirmed through stable response of *TaMLO* by causing mutations in the A1 allele (Y. Wang et al., 2014). Besides biotic stresses, several cereal crops have been edited through TALENs for various traits including reduced deposition of epicuticular wax in maize leaves (Char et al., 2015), fragrance in rice (Shan et al., 2015), and initiation of ploidy in maize (Kelliher et al., 2017). However, there is a big gap to be filled with application of TALENs to bring up GM cereals for abiotic stress resistance.

11.4.6 Clustered Regularly Interspaced Short Palindromic Repeats (CRISPR)

CRISPR is a most effective and recently invented biotechnological approach preferably employed for targeted editing within genomes. The CRISPR–Cas system and the repeats were initially characterized and experimented by Ishino and colleagues in the 1980s (Ishino et al., 1987). Later on, the term CRISPR was coined by Jansen and colleagues in 2002. The CRISPR–Cas system is an immune system adapted by archea and bacteria to get protection against invading phages. Based on their structure, Cas systems are classified into two classes, six types, and 19 subtypes (Shmakov et al., 2017). The classes are categorized on the basis of composition of effector nucleases; class 1 effector nucleases contain numerous proteins with diverse range of functions, while class 2 effector nucleases contain a single protein with many domains (Makarova et al., 2015). The most frequently and widely used variant is CRISPR/Cas9 (Type II-A) in which spCas9 effector, obtained from *Streptococcus pyogenesis*, is used because it efficiently generates DSBs. However, the application of spCas9 is limited due to the protospacer adjacent motif (PAM) of spCas9 is NGG, where N can be any nucleotide and G refers to guanine thus it restricts the use of spCas9 for AT-rich sequences and may result in off-target consequences. Such limitations are overcome by using Cas9 variants that arose through mutations, which reduces the interaction between the nuclease domain and non-specific DNA which ultimately minimizes the off-target consequences (Kleinstiver et al., 2016). The whole mechanism of action of the CRISPR/Cas system comprises of three phases: expression, interference, and adaptation. In the expression phase, the CRISPR array sequence homologous to the target sequence gets transcribed into pre-CRISPR RNA, known as pre-crRNA, which then binds to transactivating crRNA, known as tracrRNA. Once the pre-crRNA+tracrRNA complex is formed, the complex binds to Cas9 protein where RNase III cuts the long pre-crRNAs within the complex to form the crRNA+tracrRNA complex known as gRNA. In the next phase called interference, gRNA guides Cas9 complex to the target site and attaches to the target after PAM. Cas9 possesses both nuclease and helicase activities, therefore it unwinds the target sequence followed by cutting through NHN and the RuvC domain resulting in DSB within the target. In the final stage called adaptation, the DSB is repaired by HDR or NHEJ, thus the sequence is integrated in the genome and transcribed (Jackson Simon et al., 2017). CRISPR is preferred over TALENs and ZFNs because of its simplicity, lower cost, and easy handling attributes. Moreover, a multitarget approach can be adopted that enables manipulating multiple genes simultaneously. Due to this reason, CRISPR has been broadly applied to many plants including cereals for several trait improvements including yield, nutritional value, fruit quality, and forbearance against environmental stress conditions. Some of the well exemplified applications of CRISPR for stress tolerance in cereal crops are discussed as follows.

The ultimate aim of cereal modeling or genome editing in cereals is the acquisition of higher yield. CRISPR has remained successful in improving biotic stress resistance in cereals. Using CRISPR/Cas9, knockout wheat mutants for the *Taedr1* gene were generated through alteration of three *EDR1* homeologs, which resulted in wheat tolerant to powdery mildew and could not induce cell death from mildew (Y. Zhang et al., 2017). Likewise, *O. sativa* plants resistant to bacterial blight were developed by mutating *OsSWEET13* and *OsERF922* through CRISPR/Cas9 (F. Wang et al., 2016; Zhou et al., 2015). CRISPR/Cas9 has remained efficient in enhancing abiotic stress forbearance as well. Insertions, deletions, and/or replacements in genes through CRISPR/Cas9 has aided in developing drought-resistant maize. For this purpose, the *ARGOS8* gene was overexpressed by inserting GOS_2 promoter or by replacing endogenous promoter of *ARGOS8* with GOS_2 at 5′ UTR. Higher levels of *ARGOS8* transcripts were observed, which conferred enhanced water deficit tolerance to *Z. mays* (Shi et al., 2017). In addition to producing biotic and abiotic stress-tolerant cereals, CRISPR/Cas9 has also been used to genetically modify some cereals for herbicide tolerance including maize (Svitashev et al., 2015), soybean (Z. Li et al., 2015), rice (Butt et al., 2017), and some other characteristics such as generating haploid rice (Yao et al., 2018), enhancing amylase content in rice (Y. Sun et al., 2017), and increasing the grain quality and morphology in rice (M. Li et al., 2016) and wheat (Zhang, Li, et al., 2018). It is obvious that although the genome editing tools like TALENs and CRISPR have proved highly effective in targeted genome editing, there is still a lot to accomplish particularly with focus on biotic and abiotic stresses. Also, cereals other than rice, wheat, and maize shall be studied to fill the gap of relevant studies of them.

11.5 CONCLUSION AND FUTURE DIRECTIONS

Cereals are considered staple meals around the globe. In order to feed the world's estimated population of 9.8 billion people by 2050, cereal production must expand by 70%–100%. Since the beginning of the Industrial Revolution, we have observed expanded pollution that has shown its influence on all living beings, causing climate change, which has further boosted surface temperature and loss of soil moisture. These variations have affected the agricultural industry by influencing crop production thus posing a threat to world food security. Genome editing is a revolutionizing biotechnological approach with vast uses in various fields of plant sciences including crop improvement in agriculture. It has recently modernized improvement in various cereal crops in combination with other breeding. It has been efficiently used to achieve novel and desired agronomic and quality characteristics in cereals. These encompass various crop adaptable characters to mitigate the outcomes of climate change, tolerance to several biotic and abiotic stresses, higher crop yields, ideal plant orientation, improvement in grain quality, better dietary content, and much safer agriculture products. Employment of GE techniques in cereals will thus allow the opportunity to advance novel cereal varieties comprising improved production and superiority. However, this technique still needs to meet many technical as well as regulatory challenges in order to display its full potential for novel traits achievement. Several GE techniques have been applied in cereals such as meganucleases, ZFNs, CRISPR–Cas system, and TALENs. These are the site-specific endonuclease-based systems that permit site-directed mutagenesis in the genome via producing DSBs in desirable genes associated with a minute level of side effects. Such editing techniques have been applied so far for engineering resistance against diseases in crops and provided an innovative defense against several plant viruses.

REFERENCES

Abbate, P. E., Andrade, F. H., & Culot, J. P. (1995). The effects of radiation and nitrogen on number of grains in wheat. *The Journal of Agricultural Science, 124*(3), 351–360.
Abdel-Ghany, S. E., Ullah, F., Ben-Hur, A., & Reddy, A. S. (2020). Transcriptome analysis of drought-resistant and drought-sensitive sorghum (*Sorghum bicolor*) genotypes in response to PEG-induced drought stress. *International Journal of Molecular Sciences, 21*(3), 772. doi:10.3390/ijms21030772

Ahmar, S., Saeed, S., Khan, M. H. U., Ullah Khan, S., Mora-Poblete, F., Kamran, M., Faheem, A., Maqsood, A., Rauf, M., & Saleem, S. (2020). A revolution toward gene-editing technology and its application to crop improvement. *International Journal of Molecular Sciences, 21*(16), 5665. doi:10.3390/ijms21165665

Ahsan, F., Chandio, A. A., & Fang, W. (2020). Climate change impacts on cereal crops production in Pakistan: Evidence from cointegration analysis. *International Journal of Climate Change Strategies and Management, 12*, 257–269. doi:10.1108/IJCCSM-04-2019-0020

Ainley, W. M., Sastry-Dent, L., Welter, M. E., Murray, M. G., Zeitler, B., Amora, R., Corbin, D. R., Miles, R. R., Arnold, N. L., & Strange, T. L. (2013). Trait stacking via targeted genome editing. *Plant Biotechnology Journal, 11*(9), 1126–1134. doi:10.1111/pbi.12107

Amanullah, I., Irfanullah, A., & Hidayat, Z. (2016). Potassium management for improving growth and grain yield of maize (*Zea mays* L.) under moisture stress condition. *Scientific Reports, 6*(1), 1–12

Anderson, A. J., Habibzadegah-Tari, P., & Tepper, C. S. (1988). Molecular studies on the role of a root surface agglutinin in adherence and colonization by *Pseudomonas putida*. *Applied and Environmental Microbiology, 54*(2), 375–380.

Ansari, M. S. I. A. K. S. I. (2020). *Effect of drought stress on crop production*. Singapore: Springer. doi:10.1007/978-981-15-1322-0_3

Ansari, W. A., Chandanshive, S. U., Bhatt, V., Nadaf, A. B., Vats, S., Katara, J. L., Sonah, H., & Deshmukh, R. (2020). Genome editing in cereals: Approaches, applications and challenges. *International Journal of Molecular Sciences, 21*(11), 4040. doi:10.3390/ijms21114040

Anwar, M. R., O'Leary, G., McNeil, D., Hossain, H., & Nelson, R. (2007). Climate change impact on rainfed wheat in south-eastern Australia. *Field Crops Research, 104*(1), 139–147. doi:10.1016/j.fcr.2007.03.020

Arunakumari, K., Durgarani, C., Satturu, V., Sarikonda, K., Chittoor, P., Vutukuri, B., Laha, G., Nelli, A., Gattu, S., & Jamal, M. (2016). Marker-assisted pyramiding of genes conferring resistance against bacterial blight and blast diseases into Indian rice variety MTU1010. *Rice Science, 23*(6), 306–316. doi:10.1016/j.rsci.2016.04.005

Assem, S. K., Zamzam, M. M., Saad, M. E., Hussein, B. A., & Hussein, E. H. (2017). The impact of over-expression of NPK1 gene on growth and yield of sorghum under drought stress. *African Journal of Biotechnology, 16*(49), 2267–2277. doi:10.5897/AJB2017.16202

Bakala, H. S., Mandahal, K. S., Sarao, L. K., & Srivastava, P. (2021). Breeding wheat for biotic stress resistance: Achievements, challenges and prospects. In *Current Trends in Wheat Research*. IntechOpen, London, http://dx.doi.org/10.5772/intechopen.97359.

Berk, A., & Zipursky, S. L. (2001). Molecular cell biology. *Biochemistry and Molecular Biology Education, 29*, 126–133.

Boch, J., & Bonas, U. (2010). Xanthomonas AvrBs3 family-type III effectors: Discovery and function. *Annual Review of Phytopathology, 48*(1), 419–436. doi:10.1146/annurev-phyto-080508-081936

Borrelli, P. e. a. (2020). Land use and climate change impacts on global soil erosion by water (2015–2070). *PNAS, , 17*(36), 21994–22001.

Breitler, J. C., Vassal, J. M., del Mar Catala, M., Meynard, D., Marfà, V., Melé, E., Royer, M., Murillo, I., San Segundo, B., & Guiderdoni, E. (2004). Bt rice harbouring cry genes controlled by a constitutive or wound-inducible promoter: Protection and transgene expression under Mediterranean field conditions. *Plant Biotechnology Journal, 2*(5), 417–430. doi:10.1111/j.1467-7652.2004.00086.x

Breseghello, F., & Coelho, A. S. G. (2013). Traditional and modern plant breeding methods with examples in rice (*Oryza sativa* L.). *Journal of Agricultural and Food Chemistry, 61*(35), 8277–8286. doi:10.1021/jf305531j

Butt, H., Eid, A., Ali, Z., Atia, M. A., Mokhtar, M. M., Hassan, N., Lee, C. M., Bao, G., & Mahfouz, M. M. (2017). Efficient CRISPR/Cas9-mediated genome editing using a chimeric single-guide RNA molecule. *Frontiers in Plant Science, 8*, 1441.

Caffarra, A., Rinaldi, M., Eccel, E., Rossi, V., & Pertot, I. (2012). Modelling the impact of climate change on the interaction between grapevine and its pests and pathogens: European grapevine moth and powdery mildew. *Agriculture, Ecosystems and Environment. 148*, 89–101.

Calone, R., Sanoubar, R., Lambertini, C., Speranza, M., Vittori Antisari, L., Vianello, G., & Barbanti, L. (2020). Salt tolerance and Na allocation in *Sorghum bicolor* under variable soil and water salinity. *Plants, 9*(5), 561. doi:10.3390/plants9050561

Carroll, D. (2011). Genome engineering with zinc-finger nucleases. *Genetics, 188*(4), 773–782. doi:10.1534/genetics.111.131433

Challinor, A., Watson, J., et al. (2014). A meta-analysis of crop yield under climate change and adaptation. *Nature Climate Change, 4*(4), 287–291

Char, S. N., Unger-Wallace, E., Frame, B., Briggs, S. A., Main, M., Spalding, M. H., Vollbrecht, E., Wang, K., & Yang, B. (2015). Heritable site-specific mutagenesis using TALENs in maize. *Plant Biotechnology Journal, 13*(7), 1002–1010.

Cheng, M., Lowe, B. A., Spencer, T. M., Ye, X., & Armstrong, C. L. (2004). Factors influencing Agrobacterium-mediated transformation of monocotyledonous species. *In Vitro Cellular & Developmental Biology-Plant, 40*(1), 31–45. doi:10.1079/IVP2003501

Cheng, X., Sardana, R., Kaplan, H., & Altosaar, I. (1998). Agrobacterium-transformed rice plants expressing synthetic cryIA (b) and cryIA (c) genes are highly toxic to striped stem borer and yellow stem borer. *Proceedings of the National Academy of Sciences, 95*(6), 2767–2772. doi:10.1073/pnas.95.6.2767

Christian, M., Cermak, T., Doyle, E. L., Schmidt, C., Zhang, F., Hummel, A., Bogdanove, A. J., & Voytas, D. F. (2010). Targeting DNA double-strand breaks with TAL effector nucleases. *Genetics, 186*(2), 757–761.

Colhoun, J. (1973). Effects of environmental factors on plant disease. *Annual Review of Phytopathology*, 343–364.

Curtin, S. J., Zhang, F., Sander, J. D., Haun, W. J., Starker, C., Baltes, N. J., Reyon, D., Dahlborg, E. J., Goodwin, M. J., & Coffman, A. P. (2011). Targeted mutagenesis of duplicated genes in soybean with zinc-finger nucleases. *Plant Physiology, 156*(2), 466–473. doi:10.1104/pp.111.172981

Dai, S., Zheng, P., Marmey, P., Zhang, S., Tian, W., Chen, S., Beachy, R. N., & Fauquet, C. (2001). Comparative analysis of transgenic rice plants obtained by Agrobacterium-mediated transformation and particle bombardment. *Molecular Breeding, 7*(1), 25–33. doi:10.1023/A:1009687511633

de Galarreta, M. R., & Lujambio, A. (2017). Therapeutic editing of hepatocyte genome in vivo. *Journal of hepatology, 67*(4), 818–828. doi:10.1016/j.jhep.2017.05.012

Desprez-Loustau, M.-L., Robin, C., Reynaud, G., Déqué, M., Badeau, V., Piou, D., ... Marçais, B. (2007). Simulating the effects of a climate-change scenario on the geographical range and activity of forest-pathogenic fungi. *Canadian Journal of Plant Pathology, 29*(2), 101–120.

Deutsch, C. A., Tewksbury, J. J., Huey, R. B., Sheldon, K. S., Ghalambor, C. K., Haak, D. C., & Martin, P. R. (2008). Impacts of climate warming on terrestrial ectotherms across latitude. *Proceedings of the National Academy of Sciences. 105*(18), 6668–6672.

Djanaguiraman, M., Prasad, P. V., Ciampitti, I., & Talwar, H. S. (2020). Impacts of abiotic stresses on sorghum physiology. In *Sorghum in the 21st century: Food–fodder–feed–fuel for a rapidly changing world* (pp. 157–188). Springer Nature, Singapore

Dracatos, P. M., Haghdoust, R., Singh, D., & Park, R. F. (2018). Exploring and exploiting the boundaries of host specificity using the cereal rust and mildew models. *New Phytologist, 218*(2), 453–462. doi:10.1111/nph.15044

Dubey, M., Mishra, A., Singh, R. (2021). Climate change impact analysis using bias-corrected multiple global climate models on rice and wheat yield. *Journal of Water & Climate Change, 12*(4), 1282–1296.

Eastburn, D. M., McElrone, A. J., & Bilgin, D. D. (2011). Influence of atmospheric and climatic change on plant–pathogen interactions. *Plant Pathology, 60*(1), 54–69.

Editors of Encyclopedia Britannica. (2019). Sorghum. *In Encyclopedia Britannica.*

FAO. (2015). *Healthy soils are the basis for healthy food production.* Food and Agriculture Organization of the United Nations.

FAO (2008) *Food and Agriculture Organization of the United Nations.* FAO statistical yearbook 2007–2008. FAO, Rome

FAO. (2021). *Crop prospects and food situation - Quarterly global report no. 4, December 2021.* Food and Agriculture Organization of the United Nations. doi:10.4060/cb7877en

FAO, I. (2015). *The status of the world's soil resources (main report).*

FAOSTAT. (2021). *Suite of food security indicators.*

Farooq, M., Wahid, A., Kobayashi, N., Fujita, D., & Basra, S. M. A.. (2009). Plant drought stress: Effects, mechanisms and management. *Agronomy for Sustainable Development, 29,* 185–212.

Fatima, Z., Ahmed, M., Hussain, M., Abbas, G., Ul-Allah, S., Ahmad, S., Ahmed, N., Ali, M. A., Sarwar, G., Haque, E, Iqbal, P., & Hussain, S. (2020). The fingerprints of climate warming on cereal crops phenology and adaptation options. *Scientific Reports, 10*(1), 18013. doi:10.1038/s41598-020-74740-3

Fei, C. J., McCarl, B. A., & Thayer, A. W. (2017). Estimating the impacts of climate change and potential adaptation strategies on cereal grains in the United States. *Frontiers in Ecology and Evolution, 5*(62). doi:10.3389/fevo.2017.00062

Fiaz, S., Ahmad, S., Noor, M. A., Wang, X., Younas, A., Riaz, A., Riaz, A., & Ali, F. (2019). Applications of the CRISPR/Cas9 system for rice grain quality improvement: Perspectives and opportunities. *International Journal of Molecular Sciences, 20*(4), 888. doi:10.3390/ijms20040888

Figueroa, M., Hammond-Kosack, K. E., & Solomon, P. S. (2018). A review of wheat diseases—a field perspective. *Molecular plant pathology, 19*(6), 1523–1536. doi:10.1111/mpp.12618

Forsberg, R. A., & Reeves, D. L. (1992). Breeding oat cultivars for improved grain quality. *Oat Science and Technology, 33,* 751–775.

Franco, A. M. A., Hill, J. K., Kitschke, C., Collingham, Y. C., Roy, D. B., Fox, R., … Thomas, C. D. (2006). Impacts of climate warming and habitat loss on extinctions at species' low-latitude range boundaries. *Global Change Biology, 12*(8), 1545–1553.

Fróna, D., Szenderák, J., & Harangi-Rákos, M. (2019). The challenge of feeding the world. *Sustainability, 11*(20), 5816.

Gelvin, S. B. (2003). Agrobacterium-mediated plant transformation: The biology behind the "gene-jockeying" tool. *Microbiology and Molecular Biology Reviews, 67*(1), 16–37. doi:10.1128/MMBR.67.1.16-37.2003

Giordano, M., Petropoulos, S. A., & Rouphael, Y. (2021). Response and defence mechanisms of vegetable crops against drought, heat and salinity stress. *Agriculture, 11*(5), 463. doi:10.3390/agriculture11050463

Godfray, H. C. (2011). Ecology. Food and biodiversity. *Science, 333*(6047),1231–1232.

Govindaraj, M., Vetriventhan, M., & Srinivasan, M. (2015). Importance of genetic diversity assessment in crop plants and its recent advances: An overview of its analytical perspectives. *Genetics Research International, 2015.* doi:10.1155/2015/431487

Halford, N. G., Curtis, T. Y., Chen, Z., & Huang, J. (2015). Effects of abiotic stress and crop management on cereal grain composition: Implications for food quality and safety. *Journal of Experimental Botany, 66*(5), 1145–1156. doi:10.1093/jxb/eru473

Han, J., Xia, Z., Liu, P., Li, C., Wang, Y., Guo, L., Jiang, G., & Zhai, W. (2020). TALEN-based editing of TFIIAy5 changes rice response to *Xanthomonas oryzae* pv. *oryzae*. *Scientific Reports, 10*(1), 2036. doi:10.1038/s41598-020-59052-w

Harvell, C. D., Mitchell, C. E., Ward, J. R., Altizer, S., Dobson, A. P., Ostfeld, R. S., & Samuel, M. D. (2002). Climate warming and disease risks for terrestrial and marine biota. *Science, 296*(5576), 2158–2162.

Hasanuzzaman, M., Nahar, K., Fujita, M., Ahmad, P., Chandna, R., Prasad, M., & Ozturk, M. (2013). Enhancing plant productivity under salt stress: Relevance of poly-omics. *Salt Stress in Plants*, 113–156. doi:10.1007/978-1-4614-6108-1_6

Haverkort, A. J., Struik, P., Visser, R., & Jacobsen, E. (2009). Applied biotechnology to combat late blight in potato caused by *Phytophthora infestans*. *Potato Research, 52*(3), 249–264.

Hensel, G. (2020). Genetic transformation of Triticeae cereals–Summary of almost three-decade's development. *Biotechnology Advances, 40*, 107484. doi:10.1016/j.biotechadv. 2019.107484

Hiei, Y., Ishida, Y., & Komari, T. (2014). Progress of cereal transformation technology mediated by *Agrobacterium tumefaciens*. *Frontiers in Plant Science, 5*, 628. doi:10.3389/fpls.2014.00628

Hiei, Y., Ohta, S., Komari, T., & Kumashiro, T. (1994). Efficient transformation of rice (*Oryza sativa* L.) mediated by Agrobacterium and sequence analysis of the boundaries of the T-DNA. *The Plant Journal, 6*(2), 271–282. doi:10.1046/j.1365-313x.1994.6020271.x

Hillary, V. E., & Ceasar, S. A. (2019). Application of CRISPR/Cas9 genome editing system in cereal crops. *The Open Biotechnology Journal, 13*(1). doi:10.2174/1874070701913010173

Horvath, H., Huang, J., Wong, O., Kohl, E., Okita, T., Kannangara, C. G., & von Wettstein, D. (2000). The production of recombinant proteins in transgenic barley grains. *Proceedings of the National Academy of Sciences, 97*(4), 1914–1919. doi:10.1073/pnas.030527497

Huang, Q.-M., Liu, W.-H., Sun, H., Deng, X., & Su, J. (2005). *Agrobacterium tumefaciens*-mediated transgenic wheat plants with glutamine synthetases confer tolerance to herbicide. *Chinese Journal of Plant Ecology, 29*(2), 338. doi:10.17521/cjpe.2005.0044

Ibrahim Bio Yerima, A. R., & Achigan-Dako, E. G. (2021). A review of the orphan small grain cereals improvement with a comprehensive plan for genomics-assisted breeding of fonio millet in West Africa. *Plant Breeding, 140*(4), 561–574. doi:10.1111/pbr.12930

Indra Arulselvi, P., Michael, P., Umamaheswari, S., & Krishnaveni, S. (2010). Agrobacterium mediated transformation of *Sorghum bicolor* for disease resistance. *International Journal of Pharma and Bio Sciences, 1*, 4. doi:10.1.1.185.1061

Iqbal, M., Raja, N. I., Mashwani, Z.-U.-R., Hussain, M., Ejaz, M., & Yasmeen, F. (2019). Effect of silver nanoparticles on growth of wheat under heat stress. *Iranian Journal of Science and Technology, Transactions A: Science, 43*(2), 387–395. doi:10.1007/s40995-017-0417-4

Iqbal, S., Hussain, S., Qayyaum, M. A., & Ashraf, M. (2020). The response of maize physiology under salinity stress and its coping strategies. In *Plant stress physiology*. IntechOpen , London, 1–25.

IRRI. (2015). *Rice production manual*. Retrieved from Los Baños (Philippines).

Ishida, Y., Saito, H., Ohta, S., Hiei, Y., Komari, T., & Kumashiro, T. (1996). High efficiency transformation of maize (*Zea mays* L.) mediated by *Agrobacterium tumefaciens*. *Nature Biotechnology, 14*(6), 745–750. doi:10.1038/nbt0696-745

Ishino, Y., Shinagawa, H., Makino, K., Amemura, M., & Nakata, A. (1987). Nucleotide sequence of the iap gene, responsible for alkaline phosphatase isozyme conversion in *Escherichia coli*, and identification of the gene product. *Journal of Bacteriology, 169*(12), 5429–5433. doi:10.1128/jb.169.12.5429-5433.1987

Jackson Simon, A., McKenzie Rebecca, E., Fagerlund Robert, D., Kieper Sebastian, N., Fineran Peter, C., & Brouns Stan, J. J. (2017). CRISPR-Cas: Adapting to change. *Science, 356*(6333), eaal5056. doi:10.1126/science.aal5056

Jaganathan, D., Ramasamy, K., Sellamuthu, G., Jayabalan, S., & Venkataraman, G. (2018). CRISPR for crop improvement: An update review. *Frontiers in Plant Science, 9*, 985. doi:10.3389/fpls.2018.00985

Jansen, R., Embden, J. D. A.v. , Gaastra, W., & Schouls, L. M. (2002). Identification of genes that are associated with DNA repeats in prokaryotes. *Molecular Microbiology, 43*(6), 1565–1575. doi:10.1046/j.1365-2958.2002.02839.x

Jauhar, P. P., & Chibbar, R. N. (1999). Chromosome-mediated and direct gene transfers in wheat. *Genome, 42*(4), 570–583. doi:10.1139/g99-045

Jauhar, P. P., & Peterson, T. S. (2001). Hybrids between durum wheat and *Thinopyrum junceiforme*: Prospects for breeding for scab resistance. *Euphytica, 118*(2), 127–136. doi:10.1023/A:1004070006544

Jeyasri, R., Muthuramalingam, P., Satish, L., Pandian, S. K., Chen, J.-T., Ahmar, S., Wang, X., Mora-Poblete, F., & Ramesh, M. (2021). An overview of abiotic stress in cereal crops: Negative impacts, regulation, biotechnology and integrated omics. *Plants, 10*(7), 1472. doi:10.3390/plants10071472

Jinek, M., Chylinski, K., Fonfara, I., Hauer, M., Doudna, J. A., & Charpentier, E. (2012). A programmable dual-RNA–guided DNA endonuclease in adaptive bacterial immunity. *Science, 337*(6096), 816–821.

Jones, H. D. (2005). Wheat transformation: Current technology and applications to grain development and composition. *Journal of Cereal Science, 41*(2), 137–147. doi:10.1016/j.jcs.2004.08.009

Joung, J. K., & Sander, J. D. (2013). TALENs: A widely applicable technology for targeted genome editing. *Nature Reviews. Molecular Cell Biology, 14*(1), 49–55. doi:10.1038/nrm3486

Kamburova, V. S., Nikitina, E. V., Shermatov, S. E., Buriev, Z. T., Kumpatla, S. P., Emani, C., & Abdurakhmonov, I. Y. (2017). Genome editing in plants: An overview of tools and applications. *International Journal of Agronomy, 2017*, 7315351. doi:10.1155/2017/7315351

Kang, Y., Khan, S., & Ma, X. (2009). Climate change impacts on crop yield, crop water productivity and food security – A review. *Progress in Natural Science, 19*(12), 1665–1674.

Kelliher, T., Starr, D., Richbourg, L., Chintamanani, S., Delzer, B., Nuccio, M. L., Green, J., Chen, Z., McCuiston, J., & Wang, W. (2017). MATRILINEAL, a sperm-specific phospholipase, triggers maize haploid induction. *Nature, 542*(7639), 105–109.

Khandagale, K., & Nadaf, A. (2016). Genome editing for targeted improvement of plants. *Plant Biotechnology Reports, 10*(6), 327–343. doi:10.1007/s11816-016-0417-4

Khanna, H., & Raina, S. (2002). Elite indica transgenic rice plants expressing modified Cry1Ac endotoxin of *Bacillus thuringiensis* show enhanced resistance to yellow stem borer (*Scirpophaga incertulas*). *Transgenic Research, 11*(4), 411–423. doi:10.1023/a:1016378606189

Kim, Y.-G., Cha, J., & Chandrasegaran, S. (1996). Hybrid restriction enzymes: Zinc finger fusions to Fok I cleavage domain. *Proceedings of the National Academy of Sciences, 93*(3), 1156–1160.

Kleinstiver, B. P., Pattanayak, V., Prew, M. S., Tsai, S. Q., Nguyen, N. T., Zheng, Z., & Joung, J. K. (2016). High-fidelity CRISPR–Cas9 nucleases with no detectable genome-wide off-target effects. *Nature, 529*(7587), 490–495. doi:10.1038/nature16526

Kotir, J. (2011). Climate change and variability in Sub-Saharan Africa: A review of current and future trends and impacts on agriculture and food security. *Environment, Development and Sustainability: A Multidisciplinary Approach to the Theory and Practice of Sustainable Development, 13*(3), 587–605.

Kumari, V. V., Roy, A., Vijayan, R., Banerjee, P., Verma, V. C., Nalia, A., Pramanik, M., Mukherjee, B., Ghosh, A., & Reja, M. (2021). Drought and heat stress in cool-season food legumes in sub-tropical regions: Consequences, adaptation, and mitigation strategies. *Plants, 10*(6), 1038. doi:10.3390/plants10061038

Kumlehn, J., & Hensel, G. (2009). Genetic transformation technology in the Triticeae. *Breeding Science, 59*(5), 553–560. doi:10.1270/jsbbs.59.553

Kuzmanović, N., Puławska, J., Prokić, A., Ivanović, M., Zlatković, N., Jones, J. B., & Obradović, A. (2015). *Agrobacterium arsenijevicii* sp. nov., isolated from crown gall tumors on raspberry and cherry plum. *Systematic and Applied Microbiology, 38*(6), 373–378. doi:10.1016/j.syapm.2015.06.001

Lapidot, M., & Friedmann, M. (2002). Breeding for resistance to whitefly-transmitted geminiviruses. *Annals of Applied Biology, 140*(2), 109–127.

Laskowski, W., Górska-Warsewicz, H., Rejman, K., Czeczotko, M., & Zwolińska, J. (2019). How important are cereals and cereal products in the average polish diet? *Nutrients, 11*(3), 679. doi:10.3390/nu11030679

Lee, H. J., Kim, E., & Kim, J.-S. (2010). Targeted chromosomal deletions in human cells using zinc finger nucleases. *Genome Research, 20*(1), 81–89. doi:10.1101/gr.099747.109.

Lee, S. Z. G. S. Y. Q. X. Y. J. (2017). Modeling the impacts of climate change and technical progress on the wheat yield in inland China: An autoregressive distributed lag approach. *PLOS ONE. 12*(9), e0184474, doi:10.1371/journal.pone.0184474

Li, M., Li, X., Zhou, Z., Wu, P., Fang, M., Pan, X., Lin, Q., Luo, W., Wu, G., & Li, H. (2016). Reassessment of the four yield-related genes Gn1a, DEP1, GS3, and IPA1 in rice using a CRISPR/Cas9 system. *Frontiers in Plant Science, 7*, 377.

Li, T., Liu, B., Spalding, M. H., Weeks, D. P., & Yang, B. (2012). High-efficiency TALEN-based gene editing produces disease-resistant rice. *Nature Biotechnology, 30*(5), 390–392.

Li, W., Deng, Y., Ning, Y., He, Z., & Wang, G.-L. (2020). Exploiting broad-spectrum disease resistance in crops: From molecular dissection to breeding. *Annual Review of Plant Biology, 71*, 575–603.

Li, Z., Liu, Z.-B., Xing, A., Moon, B. P., Koellhoffer, J. P., Huang, L., Ward, R. T., Clifton, E., Falco, S. C., & Cigan, A. M. (2015). Cas9-guide RNA directed genome editing in soybean. *Plant Physiology, 169*(2), 960–970.

Liu, J., Liu, D., Tao, W., Li, W., Wang, S., Chen, P., Cheng, S., & Gao, D. (2000). Molecular marker-facilitated pyramiding of different genes for powdery mildew resistance in wheat. *Plant Breeding, 119*(1), 21–24. doi:10.1046/j.1439-0523.2000.00431.x

Lobell, D. (2007). Field Ch. B.: Global scale climate–crop yield relationship and the impacts of recent warming. *Environmental Research Letters, 2*(1), 014002.

Loskutov, I. G. (2021). *Advances in cereal crops breeding Plants, 10*(8), 1705.

Ma, X., Qian, Q., & Zhu, D. (2005). Expression of a calcineurin gene improves salt stress tolerance in transgenic rice. *Plant Molecular Biology, 58*(4), 483–495. doi:10.1007/s11103-005-6162-7

MacCracken, M. C. (2008). Prospects for future climate chnage and the reasons for early action. *Journal of the Air and Waste Management Association, 58*(6), 735–786.

Mahato, A. (2014). Climate change and its impact on agriculture. *International Journal of Scientific and Research Publications (IJRSP).*

Makarova, K. S., Wolf, Y. I., Alkhnbashi, O. S., Costa, F., Shah, S. A., Saunders, S. J., Barrangou, R., Brouns, S. J. J., Charpentier, E., Haft, D. H., Horvath, P., Moineau, S., Mojica, F. J. M., Terns, R. M., Terns, M. P., White, M. F., Yakunin, A. F., Garrett, R. A., van der Oost, J., Backofen, R., & Koonin, E. V. (2015). An updated evolutionary classification of CRISPR–Cas systems. *Nature Reviews Microbiology, 13*(11), 722–736. doi:10.1038/nrmicro3569

Martínez-Fortún, J., Phillips, D. W., & Jones, H. D. (2017). Potential impact of genome editing in world agriculture. *Emerging Topics in Life Sciences, 1*(2), 117–133.

McKevith, B. (2004). Nutritional aspects of cereals. *Nutrition Bulletin, 29*(2), 111–142.

Mores, A., Borrelli, G. M., Laidò, G., Petruzzino, G., Pecchioni, N., Amoroso, L. G. M., Desiderio, F., Mazzucotelli, E., Mastrangelo, A. M., & Marone, D. (2021). Genomic approaches to identify molecular bases of crop resistance to diseases and to develop future breeding strategies. *International Journal of Molecular Sciences, 22*(11), 5423. doi:10.3390/ijms22115423

Nagadhara, D., Ramesh, S., Pasalu, I., Rao, Y. K., Krishnaiah, N., Sarma, N., Bown, D., Gatehouse, J., Reddy, V., & Rao, K. (2003). Transgenic indica rice resistant to sap-sucking insects. *Plant Biotechnology Journal*, *1*(3), 231–240. doi:10.1046/j.1467-7652.2003.00022.x

Nagadhara, D., Ramesh, S., Pasalu, I., Rao, Y. K., Sarma, N., Reddy, V., & Rao, K. (2004). Transgenic rice plants expressing the snowdrop lectin gene (gna) exhibit high-level resistance to the whitebacked planthopper (*Sogatella furcifera*). *Theoretical and Applied Genetics*, *109*(7), 1399–1405. doi:10.1007/s00122-004-1750-5

Nayak, P., Basu, D., Das, S., Basu, A., Ghosh, D., Ramakrishnan, N. A., Ghosh, M., & Sen, S. K. (1997). Transgenic elite indica rice plants expressing CryIAc∂-endotoxin of *Bacillus thuringiensis* are resistant against yellow stem borer (*Scirpophaga incertulas*). *Proceedings of the National Academy of Sciences*, *94*(6), 2111–2116. doi:10.1073/pnas.94.6.2111

Nile, B. K., Hassan, W. H., & Alshama, G. A. (2019). Analysis of the effect of climate change on rainfall intensity and expected flooding by using ANN and SWMM programs. *ARPN Journal of Engineering and Applied Sciences*, *14*(5), 974–984.

Nowara, D., Gay, A., Lacomme, C., Shaw, J., Ridout, C., Douchkov, D., Hensel, G., Kumlehn, J., & Schweizer, P. (2010). HIGS: Host-induced gene silencing in the obligate biotrophic fungal pathogen *Blumeria graminis*. *The Plant Cell*, *22*(9), 3130–3141. doi:10.1105/tpc.110.077040

Obert, J. C., Ridley, W. P., Schneider, R. W., Riordan, S. G., Nemeth, M. A., Trujillo, W. A., Breeze, M. L., Sorbet, R., & Astwood, J. D. (2004). The composition of grain and forage from glyphosate tolerant wheat MON 71800 is equivalent to that of conventional wheat (*Triticum aestivum* L.). *Journal of Agricultural and Food Chemistry*, *52*(5), 1375–1384. doi:10.1021/jf035218u

Ohta, S., Ishida, Y., & Usami, S. (2004). Expression of cold-tolerant pyruvate, orthophosphate dikinase cDNA, and heterotetramer formation in transgenic maize plants. *Transgenic Research*, *13*(5), 475–485. doi:10.1007/s11248-004-1452-4

Orhan, D. (2013). Land suitability assessment for rice cultivation based on GIS modeling. *Turkish Journal of Agriculture and Forestry*, *37*(3), 326–334.

Osakabe, K., Osakabe, Y., & Toki, S. (2010). Site-directed mutagenesis in Arabidopsis using custom-designed zinc finger nucleases. *Proceedings of the National Academy of Sciences*, *107*(26), 12034–12039. doi:10.1073/pnas.1000234107

Parmar, N., Singh, K. H., Sharma, D., Singh, L., Kumar, P., Nanjundan, J., Khan, Y. J., Chauhan, D. K., & Thakur, A. K. (2017). Genetic engineering strategies for biotic and abiotic stress tolerance and quality enhancement in horticultural crops: A comprehensive review. *3 Biotech*, *7*(4), 239–239. doi:10.1007/s13205-017-0870-y

Petolino, J. F., Worden, A., Curlee, K., Connell, J., Moynahan, T. L. S., Larsen, C., & Russell, S. (2010). Zinc finger nuclease-mediated transgene deletion. *Plant Molecular Biology*, *73*(6), 617–628. doi:10.1007/s11103-010-9641-4.

Piao, S. (2006). Effect of climate and CO2 changes on the greening of the Northern Hemisphere over the past two decades. *Geophysical Research Letters*. *33*(23), L23402.

Platková, H., Skuhrovec, J., & Saska, P. (2020). Antibiosis to *Metopolophium dirhodum* (Homoptera: Aphididae) in spring wheat and emmer cultivars. *Journal of Economic Entomology*, *113*(6), 2979–2985. doi:10.1093/jee/toaa234

Qie, Y., Liu, Y., Wang, M., Li, X., See, D. R., An, D., & Chen, X. (2019). Development, validation, and re-selection of wheat lines with pyramided genes Yr64 and Yr15 linked on the short arm of chromosome 1B for resistance to stripe rust. *Plant Disease*, *103*(1), 51–58.

Quan, R., Shang, M., Zhang, H., Zhao, Y., & Zhang, J. (2004). Engineering of enhanced glycine betaine synthesis improves drought tolerance in maize. *Plant Biotechnology Journal*, *2*(6), 477–486. doi:10.1111/j.1467-7652.2004.00093.x

Ran, Y., Patron, N., Kay, P., Wong, D., Buchanan, M., Cao, Y. Y., Sawbridge, T., Davies, J. P., Mason, J., & Webb, S. R. (2018). Zinc finger nuclease-mediated precision genome editing of an endogenous gene in hexaploid bread wheat (*Triticum aestivum*) using a DNA repair template. *Plant Biotechnology Journal, 16*(12), 2088–2101. doi:10.1111/pbi.12941.

Repellin, A., Båga, M., Jauhar, P. P., & Chibbar, R. N. (2001). Genetic enrichment of cereal crops via alien gene transfer: New challenges. *Plant Cell, Tissue and Organ Culture, 64*(2), 159–183. doi:10.1023/A:1010633510352

Rodriguez-Moreno, L., Song, Y., & Thomma, B. P. (2017). Transfer and engineering of immune receptors to improve recognition capacities in crops. *Current Opinion in Plant Biology, 38*, 42–49. doi:10.1016/j.pbi.2017.04.010

Roser, H. R. a. M. (2020). CO_2 and greenhouse gas emissions. OurWorldInData.org. Retrieved from https://ourworldindata.org/co2-and-other-greenhouse-gas-emissions.

Sakamoto, A., & Murata, A. N. (1998). Metabolic engineering of rice leading to biosynthesis of glycinebetaine and tolerance to salt and cold. *Plant Molecular Biology, 38*(6), 1011–1019. doi:10.1023/a:1006095015717

Sala Osvaldo, E., Stuart Chapin, F., III., Armesto, J. J., Berlow, E., Bloomfield, J., Wall, D. H. (2000). Global biodiversity scenarios for the year 2100. *Science, 287*, 1770–1774.

Satyavathi, C. T., Ambawat, S., Khandelwal, V., & Srivastava, R. K. (2021). Pearl Millet: A climate-resilient nutricereal for mitigating hidden hunger and provide nutritional security. *Frontiers in Plant Science, 12*, 659938.

Savary, S., Ficke, A., Aubertot, J.-N., & Hollier, C. (2012). Crop losses due to diseases and their implications for global food production losses and food security. *Food Security, 4*, 519–537.

Sawahel, W. (2003). Improved performance of transgenic glycinebetaine-accumulating rice plants under drought stress. *Biologia Plantarum, 47*(1), 39–44. doi:10.1023/A:1027372629612

Sawahel, W. A., & Hassan, A. H. (2002). Generation of transgenic wheat plants producing high levels of the osmoprotectant proline. *Biotechnology Letters, 24*(9), 721–725. doi:10.1023/A:1015294319114

Sedeek, K. E., Mahas, A., & Mahfouz, M. (2019). Plant genome engineering for targeted improvement of crop traits. *Frontiers in Plant Science, 10*, 114. doi:10.3389/fpls.2019.00114

Sera, T. (2005). Inhibition of virus DNA replication by artificial zinc finger proteins. *Journal of Virology, 79*(4), 2614–2619. doi:10.1128/JVI.79.4.2614-2619.2005

Shah, T., Xu, J., Zou, X., Cheng, Y., Nasir, M., & Zhang, X. (2018). Omics approaches for engineering wheat production under abiotic stresses. *International Journal of Molecular Sciences, 19*(8), 2390. doi:10.3390/ijms19082390

Shan, Q., Zhang, Y., Chen, K., Zhang, K., & Gao, C. (2015). Creation of fragrant rice by targeted knockout of the OsBADH2 gene using TALEN technology. *Plant Biotechnology Journal, 13*(6), 791–800. doi:10.1111/pbi.12312

Shi, J., Gao, H., Wang, H., Lafitte, H. R., Archibald, R. L., Yang, M., Hakimi, S. M., Mo, H., & Habben, J. E. (2017). ARGOS 8 variants generated by CRISPR-Cas9 improve maize grain yield under field drought stress conditions. *Plant Biotechnology Journal, 15*(2), 207–216.

Shmakov, S., Smargon, A., Scott, D., Cox, D., Pyzocha, N., Yan, W., Abudayyeh, O. O., Gootenberg, J. S., Makarova, K. S., Wolf, Y. I., Severinov, K., Zhang, F., & Koonin, E. V. (2017). Diversity and evolution of class 2 CRISPR–Cas systems. *Nature Reviews Microbiology, 15*(3), 169–182. doi:10.1038/nrmicro.2016.184

Shrawat, A. K., & Lörz, H. (2006). Agrobacterium-mediated transformation of cereals: A promising approach crossing barriers. *Plant Biotechnology Journal, 4*(6), 575–603. doi:10.1111/j.1467-7652.2006.00209.x

Shu, Q., Ye, G., Cui, H., Cheng, X., Xiang, Y., Wu, D., Gao, M., Xia, Y., Hu, C., & Sardana, R. (2000). Transgenic rice plants with a synthetic cry1Ab gene from *Bacillus thuringiensis* were highly resistant to eight lepidopteran rice pest species. *Molecular Breeding, 6*(4), 433–439. doi:10.1023/A:1009658024114

Shukla, V. K., Doyon, Y., Miller, J. C., DeKelver, R. C., Moehle, E. A., Worden, S. E., Mitchell, J. C., Arnold, N. L., Gopalan, S., & Meng, X. (2009). Precise genome modification in the crop species Zea mays using zinc-finger nucleases. *Nature, 459*(7245), 437–441. doi:10.1038/nature07992

Singla, J., Krattinger, S., Wrigley, C., Faubion, J., Corke, H., & Seetharaman, K. (2016). *Biotic stress resistance genes in wheat.* doi:10.5772/intechopen.97359

Şöllü, C., Pars, K., Cornu, T. I., Thibodeau-Beganny, S., Maeder, M. L., Joung, J. K., Heilbronn, R., & Cathomen, T. (2010). Autonomous zinc-finger nuclease pairs for targeted chromosomal deletion. *Nucleic Acids Research, 38*(22), 8269–8276. doi:10.1093/nar/gkq720.

Spilde, L. A. (1989). Influence of seed size and test weight on several agronomic traits of barley and hard red spring wheat. *Journal of Production Agriculture, 2*(2), 169–172.

Stark-Lorenzen, P., Nelke, B., Hänßler, G., Mühlbach, H., & Thomzik, J. (1997). Transfer of a grapevine stilbene synthase gene to rice (*Oryza sativa* L.). *Plant Cell Reports, 16*(10), 668–673. doi:10.1007/s002990050299

Sun, Y., Jiao, G., Liu, Z., Zhang, X., Li, J., Guo, X., Du, W., Du, J., Francis, F., & Zhao, Y. (2017). Generation of high-amylose rice through CRISPR/Cas9-mediated targeted mutagenesis of starch branching enzymes. *Frontiers in Plant Science, 8*, 298.

Sun, Y., Zhang, X., Wu, C., He, Y., Ma, Y., Hou, H., Guo, X., Du, W., Zhao, Y., & Xia, L. (2016). Engineering herbicide-resistant rice plants through CRISPR/Cas9-mediated homologous recombination of acetolactate synthase. *Molecular Plant, 9*(4), 628–631. doi:10.1016/j.molp.2016.01.001

Sun, Z., Li, N., Huang, G., Xu, J., Pan, Y., Wang, Z., Tang, Q., Song, M., & Wang, X. (2013). Site-specific gene targeting using transcription activator-like effector (TALE)-based nuclease in *Brassica oleracea. Journal of Integrative Plant Biology, 55*(11), 1092–1103. doi:10.1111/jipb.12091

Svitashev, S., Schwartz, C., Lenderts, B., Young, J. K., & Cigan, A. M. (2016). Genome editing in maize directed by CRISPR–Cas9 ribonucleoprotein complexes. *Nature Communications, 7*(1), 1–7. doi:10.1038/ncomms13274

Svitashev, S., Young, J. K., Schwartz, C., Gao, H., Falco, S. C., & Cigan, A. M. (2015). Targeted mutagenesis, precise gene editing, and site-specific gene insertion in maize using Cas9 and guide RNA. *Plant Physiology, 169*(2), 931–945.

Takenaka, K., Koshino-Kimura, Y., Aoyama, Y., & Sera, T. (2007). *Inhibition of tomato yellow leaf curl virus replication by artificial zinc-finger proteins.* Paper presented at the Nucleic Acids Symposium Series.

Tandzi, L. N., & Mutengwa, C. S. (2020). Factors affecting yield of crops. In *Agronomy -climate change & food security.* 9, Intechopen, London.

Tang, X., Lowder, L. G., Zhang, T., Malzahn, A. A., Zheng, X., Voytas, D. F., Zhong, Z., Chen, Y., Ren, Q., & Li, Q. (2017). A CRISPR–Cpf1 system for efficient genome editing and transcriptional repression in plants. *Nature Plants, 3*(3), 1–5. doi:doi.org/10.1038/nplants.2017.18

UNDESA. (2019). *World population prospects 2019: Highlights.*

Urnov, F. D., Rebar, E. J., Holmes, M. C., Zhang, H. S., & Gregory, P. D. (2010). Genome editing with engineered zinc finger nucleases. *Nature Reviews Genetics, 11*(9), 636–646. doi:10.1038/nrg2842.

Verma, S., Nizam, S., & Verma, P. K. (2013). Biotic and abiotic stress signaling in plants. In *Stress signaling in plants: Genomics and proteomics perspective, Volume 1* (pp. 25–49). Springer, New York.

Viana, V. E., Pegoraro, C., Busanello, C., & Costa de Oliveira, A. (2019). Mutagenesis in rice: The basis for breeding a new super plant. *Frontiers in Plant Science, 10,* 1326. doi:10.3389/fpls.2019.01326

Vila, L., Quilis, J., Meynard, D., Breitler, J. C., Marfà, V., Murillo, I., Vassal, J. M., Messeguer, J., Guiderdoni, E., & San Segundo, B. (2005). Expression of the maize proteinase inhibitor (mpi) gene in rice plants enhances resistance against the striped stem borer (Chilo suppressalis): Effects on larval growth and insect gut proteinases. *Plant Biotechnology Journal, 3*(2), 187–202. doi:10.1111/j.1467-7652.2004.00117.x

Visser, M. E. (2008). *Keeping up with a warming world; assessing the rate of adaptation to climate change.*

Wang, F., Wang, C., Liu, P., Lei, C., Hao, W., Gao, Y., Liu, Y.-G., & Zhao, K. (2016). Enhanced rice blast resistance by CRISPR/Cas9-targeted mutagenesis of the ERF transcription factor gene OsERF922. *PLOS ONE, 11*(4), e0154027.

Wang, F.-Z., Wang, Q.-B., Kwon, S.-Y., Kwak, S.-S., & Su, W.-A. (2005). Enhanced drought tolerance of transgenic rice plants expressing a pea manganese superoxide dismutase. *Journal of Plant Physiology, 162*(4), 465–472. doi:10.1016/j.jplph.2004.09.009

Wang, J., Vanga, S. K., Saxena, R., Orsat, V., & Raghavan, V. (2018). Effect of climate change on the yield of cereal crops: A review. *Climate, 6*(2), 41. doi:10.3390/cli6020041

Wang, T., Zhang, H., & Zhu, H. (2019). CRISPR technology is revolutionizing the improvement of tomato and other fruit crops. *Horticulture Research, 6*(1), 77. doi:10.1038/s41438-019-0159-x

Wang, Y., Cheng, X., Shan, Q., Zhang, Y., Liu, J., Gao, C., & Qiu, J.-L. (2014). Simultaneous editing of three homoeoalleles in hexaploid bread wheat confers heritable resistance to powdery mildew. *Nature Biotechnology, 32*(9), 947–951. doi:10.1038/nbt.2969

Wei, Y., Yao, F., Zhu, C., Jiang, M., Li, G., Song, Y., & Wen, F. (2008). Breeding of transgenic rice restorer line for multiple resistance against bacterial blight, striped stem borer and herbicide. *Euphytica, 163*(2), 177–184. doi:10.1007/s10681-007-9614-0

Weinthal, D. M., Taylor, R. A., & Tzfira, T. (2013). Nonhomologous end joining-mediated gene replacement in plant cells. *Plant Physiology, 162*(1), 390–400. doi:10.1104/pp.112.212910

Welch, R. W. (1995). The chemical composition of oats. In *The oat crop: Production and utilization.* Springer, Dordrecht, https://doi.org/10.1007/978-94-011-0015-1_10

Wisser, R. J., Kolkman, J. M., Patzoldt, M. E., Holland, J. B., Yu, J., Krakowsky, M., Nelson, R. J., & Balint-Kurti, P. J. (2011). Multivariate analysis of maize disease resistances suggests a pleiotropic genetic basis and implicates a GST gene. *Proceedings of the National Academy of Sciences, 108*(18), 7339–7344. doi:10.1073/pnas.1011739108

Wu, L., Fan, Z., Guo, L., Li, Y., Chen, Z.-L., & Qu, L.-J. (2005). Over-expression of the bacterial nhaA gene in rice enhances salt and drought tolerance. *Plant Science, 168*(2), 297–302. doi:10.1016/j.plantsci.2004.05.033

Xiao, J., Grandillo, S., Ahn, S. N., McCouch, S. R., Tanksley, S. D., Li, J., & Yuan, L. (1996). Genes from wild rice improve yield. *Nature (London), 384*(6606), 223–224. doi:10.1038/384223a0

Xiong, L., & Yang, Y. (2003). Disease resistance and abiotic stress tolerance in rice are inversely modulated by an abscisic acid–inducible mitogen-activated protein kinase. *The Plant Cell, 15*(3), 745–759. doi:10.1105/tpc.008714

Yao, L., Zhang, Y., Liu, C., Liu, Y., Wang, Y., Liang, D., Liu, J., Sahoo, G., & Kelliher, T. (2018). OsMATL mutation induces haploid seed formation in indica rice. *Nature Plants, 4*(8), 530–533.

Yellisetty, V., Reddy, L., & Mandapaka, M. (2015). In planta transformation of sorghum (*Sorghum bicolor* (L.) Moench) using TPS1 gene for enhancing tolerance to abiotic stresses. *Journal of Genetics, 94*(3), 425–434. doi:10.1007/s12041-015-0540-y

Zaidi, S. S.-e.-A., Tashkandi, M., Mansoor, S., & Mahfouz, M. M. (2016). Engineering plant immunity: Using CRISPR/Cas9 to generate virus resistance. *Frontiers in Plant Science, 7*, 1673. doi:10.3389/fpls.2016.01673

Zhang, M., & Coaker, G. (2017). Harnessing effector-triggered immunity for durable disease resistance. *Phytopathology, 107*(8), 912–919. doi:10.1094/PHYTO-03-17-0086-RVW.

Zhang, Y., Bai, Y., Wu, G., Zou, S., Chen, Y., Gao, C., & Tang, D. (2017). Simultaneous modification of three homoeologs of Ta EDR 1 by genome editing enhances powdery mildew resistance in wheat. *The Plant Journal, 91*(4), 714–724.

Zhang, Y., Li, D., Zhang, D., Zhao, X., Cao, X., Dong, L., Liu, J., Chen, K., Zhang, H., & Gao, C. (2018). Analysis of the functions of Ta GW 2 homoeologs in wheat grain weight and protein content traits. *The Plant Journal, 94*(5), 857–866. doi:10.1111/tpj.13903

Zhang, Y., Massel, K., Godwin, I. D., & Gao, C. (2018). Applications and potential of genome editing in crop improvement. *Genome Biology, 19*(1), 210. doi:10.1186/s13059-018-1586-y

Zheng, H., Li, Y., Yu, Z., Li, W., Chen, M., Ming, X., Casper, R., & Chen, Z. (1997). Recovery of transgenic rice plants expressing the rice dwarf virus outer coat protein gene (S8). *Theoretical and Applied Genetics, 94*(3–4), 522–527. doi:10.1007/s001220050446

Zhou, J., Peng, Z., Long, J., Sosso, D., Liu, B., Eom, J. S., Huang, S., Liu, S., Vera Cruz, C., & Frommer, W. B. (2015). Gene targeting by the TAL effector PthXo2 reveals cryptic resistance gene for bacterial blight of rice. *The Plant Journal, 82*(4), 632–643.

Zhu, J.-K. (2002). Salt and drought stress signal transduction in plants. *Annual Review of Plant Biology, 53*(1), 247–273.

12 Genetic Transformation Methods in Cereal Crops

Noor-ul-Ain Malik, Faiza Munir,
Saba Azeem, Rabia Amir, Maria Gillani,
Alvina Gul, Aneela Mustafa, and Nosheen Fatima

CONTENTS

DOI: 10.1201/9781003250845-12

12.1 INTRODUCTION

Cereals are monocot grasses, such as wheat, barley, maize, rice, sorghum, millet, oat, rye, triticale (hybrid of wheat and rye), and sugarcane, belonging to the family Gramineae/Poaceae. Cereal grains have been consumed by humans since the Neolithic period, around 10,000 years ago, and currently account for two-thirds of the global food supply (Repellin et al., 2001). Cereals are an essential source of nutrients and contribute 50% of global caloric intake. Approximately half of the world's agricultural land is presently used to produce 2.5 billion metric tons of cereal grains every year. However, the world's population is expected to be doubled by 2050, globally increasing the demand for cereal crop production (Singer et al., 2019).

Several factors, notably climate change and population growth, threaten the production of cereal crops and shifted the demand–supply balance resulting in global food insecurity. If the same situation persists, in the next 40 years approximately 30% more food will be required; this needs to be addressed by coping with an already stressed environment for crops. Cereal crop production must expand in a sustainable manner due to their crucial role in human and animal feed. This can be addressed by using standard soil and crop management practices to lessen the environmental effect, or by genetically modifying crop cultivars (Razzaq et al., 2019).

In terms of plant improvement, cereal breeders have made it a priority to develop more hardy cultivars capable of producing better yields with fewer inputs (water, fertilizers, etc.). For the success of future breeding projects, a deeper understanding of the genetic foundation of crop performance, particularly under constrained or unfavorable conditions, is required. Clearly, the use of sophisticated biotechnology techniques would undoubtedly be required to produce such next-generation varieties. Understanding and utilizing the genetic diversity contained in vast collections of germplasm for different species of cereal grains will be extremely beneficial for gene discovery and genome-assisted crop development. Targeted gene transfers into plants using biotechnological methods are critical for effective breeding in the 21st century and consequently for survival of humanity (Giraldo et al., 2019). Genetic transformation of crop plants was not common a few decades ago, and it was nearly impossible with essential cereals like barley and wheat. However, several advancements have made cereal crops more accessible to gene transfer techniques, with better understanding of their cell biology and tissue culture offering a link between molecular biology and plant breeding (Cocking & Davey, 1987).

Plant transformation results in the random integration of novel sequences into plant genomes. Over the last few years, remarkable improvements have given plant cells more control over integration and allowed for precise, targeted alterations to DNA sequences (genome editing). In genome editing, sequence-specific nucleases (SSNs) that induce a DNA double-strand break (DSB) at a specific genomic target are used. Depending on how the cell repairs the break, these sites allow for targeted mutagenesis (Altpeter et al., 2016).

In angiosperms, non-homologous end joining (NHEJ) is the most used cellular mechanism for repairing the break. NHEJ frequently causes minor alterations at the repaired location and can be utilized to carry out targeted mutagenesis to modify gene expression or function. SSNs are either transiently delivered to protoplasts or stably inserted into the genome as a transgene to induce targeted mutagenesis (Puchta, 2005; Voytas & Gao, 2014). In case of stable incorporation, the SSN occasionally mutates the lineages eventually integrated into reproductive cells, allowing mutations to be passed down to progeny during the development of transgenic plants. In successive generations, the nuclease transgene can be segregated away, to obtain a non-transgenic plant with mutations in the target locus of interest (Altpeter et al., 2016).

Cells can fix DSB sites by homologous recombination (HR) by utilizing a homologous strand or maybe a user-supplied sequence. A user-supplied repair template is provided by exogenously means along with the SSN and may contain specific modifications in genomes ranging from single base changes that alter protein to several transgenes incorporated at the site of break (Altpeter et al., 2016). The simultaneous delivery of both the SSN and the repair template is a challenge for HR-mediated gene editing. Moreover, to boost the efficiency of HR, virus base vectors are used to enhance SSN titer and repair templates that are delivered to cells (Baltes & Voytas, 2015; Voytas & Gao, 2014).

The introduction of CRISPR/Cas and similar reagents, which employ guide RNAs to detect target DNA sequences via Watson–Crick base pairing, substantially improved reagent design (Cong et al., 2013; Jinek et al., 2012; Mali et al., 2013). Therefore, the use of CRISPR/Cas has made genome modification more approachable, resulting in widespread implementation and advanced technologies.

12.2 GENETIC IMPROVEMENT OF CEREALS THROUGH BREEDING AND CLASSICAL CYTOGENETICS

12.2.1 HYBRIDIZATION WITHIN THE PRIMARY GENE POOL

Plant breeding was practiced long before the invention of genetics and cytogenetics. The genetic and cytogenetic techniques were formulated around the start of the 20th century and hugely accelerated plant breeding initiatives (Repellin et al., 2001). Cereal crop landraces include a diverse set of genes in their seeds for agronomically significant characteristics. These landraces are in the same gene pool as their respective crop species and may be easily crossed. Alien gene transfer into crop species, such as pearl millet, is facilitated by complete or high matching between chromosomes of the landrace and those of the crop plant. The genetic richness in wild forms has resulted from free gene flow between wild and farmed varieties of pearl millet (Jauhar & Hanna, 1998). A wide range of genes for agronomically relevant traits can be found in cereal landraces. There is a lot of variety in seed and other properties between and within landraces. Independent domestications and migration events that result in geographical isolation are responsible for such variation (Repellin et al., 2001).

12.2.2 DISTANT HYBRIDIZATION AND CHROMOSOMAL MANIPULATION

Distant hybridization refers to the crossing between two distinct species or genera to produce new crop species. It is a unique approach for introducing favorable agronomical features. Wide relatives of cereal crops depict superior trait genes, and some of these genes have been transferred through intergeneric and interspecific hybridization (Jauhar & Hanna, 1998). Chromosomal pairing among wild relatives and cereal crops provides the basis for gene introgression. The process is comparatively simple in diploid cereals such as barley, maize, and pearl millet, where only one genome needs to be reconstructed. The degree of chromosomal homology that occurs between these species determines the effectiveness of gene transfer from the donor to the recipient. For example, for a wheat–alien exchange to be usable it must have the following characteristics: the exchanged segment should be small, better integration of the alien genetic material for the wheat genetic material that it replaces, and normal Mendelian transmission of the introgressed genetic material to the progeny (Repellin et al., 2001).

12.2.3 Methods of Direct Gene Transfer in Cereals

12.2.3.1 Biolistic Transformation

Direct gene transfer for plant genetic transformation is often accomplished by using a device known as a gene gun, and the technique is referred to as particle bombardment or biolistic transformation (Baltes et al., 2017). Genetic transformation via particle bombardment allows larger-sized genes to be transferred into the nucleus of a cell. The target genes are coated on inert metal particles (platinum, gold, or tungsten), which are then driven by an electric discharge or a stream of compressed helium and then directed to plant tissues. The metal particles enter the plant cell and transfer target genes (Danilova, 2007). Particle bombardment-mediated genetic transformation allows large sized DNA to be transferred into the nucleus of a cell. This approach can also be used to effectively target various explants or tissues, but may lead to messy integration (Ramasamy et al., 2018).

Preparation of explants is an essential prerequisite for effective gene transfer via biolistic transformation. To begin, explants are grown in a medium containing osmotic chemicals such as mannitol, sorbitol, NaCl, or any other osmotic compound. Explants have shown a maximum efficiency rate of biolistic transformation when exposed to osmotic media both before and after the particle bombardment, as the osmotic media exposure leads to reduced oxidative stress (Kemper et al., 1996).

Researchers found that the way to make the process of micro-bombardment less laborious and to enhance its effectiveness is by adding certain improvements in biolistic transformation devices. Biolistic transformation technology is frequently employed in the production of successful cereal crop transformations. For example, biolistic-based cane transformation and micro-propagation resulted in successful production of an abiotic stress-tolerant variety along with high-value by-products (Ramasamy et al., 2018). Other crops, e.g., wheat, maize, rice, rye, oat, and barley, have all been transformed using this method. However, the drawback of this method is that the quantity of inserted copies cannot be controlled. If a large quantity of target genes is inserted in a nucleus of a single cell it leads to gene silencing. Furthermore, biolistic technology leads to messy integration and may allow restricted size of inserts. Despite the drawbacks, the biolistic transformation technique is presently the most widely utilized method for direct DNA transfer into cereal crops (Danilova, 2007).

12.2.3.2 Protoplast Transformation

The use of protoplasts was one of the first technologies for direct gene transfer into plant cells. Single plant cells without a cell wall are known as protoplasts. They can easily be manipulated for genetic transformation. The technique has been used to modify a vast number of plant species; the only need is that the procedures for cultivating protoplasts and understanding of their subsequent regeneration into plants. The protoplast solution is supplemented with foreign DNA. To enhance DNA penetration across the plasma membrane, a polyethylene glycol (PEG) solution with an alkaline pH greater than 2+ is used in this procedure (Aulinger et al., 2003). PEG solution helps in the deformation of plant cell membranes for easy penetration of the gene of interest.

Direct treatment of protoplasts with target DNA has various drawbacks, the most notable of which is the difficulty in obtaining protoplasts and regenerating them. Furthermore, there is a significant risk of soma clonal variations during in vitro culture of protoplasts. However, this approach is frequently employed for temporary expression, which refers to the production of RNA or the activity of a protein in a host cell without the insertion of the gene into the plant genome. The transient expression technique is frequently used to evaluate the potential of various genetic constructs, such as plant gene promoter regulatory elements (Danilova, 2007).

The protoplast transformation of cereals is not common, but transformations in a few crops showed positive results such as in rice, maize, and wheat. Rice is the model system for molecular genetic study among the major grains, due to its tiny diploid genome and relative ease in cell and tissue culture, including protoplast techniques. Furthermore, due to its economic importance as a major food supply, rice has received a lot of research attention. In the last few years, significant progress has been made in barley, maize, and wheat protoplast research (Krautwig & Lörz, 1995).

12.2.3.3 Liposome Fusion Method

In recent research, liposomes are used as vectors to achieve plant transformations. The liposome fusion method ensures targeted delivery of genes via protoplasts. This method ensures safe delivery of encapsulated DNA to specific cells. For this purpose, a variety of liposomes are used for gene delivery in multiple plant types, but cationic or neutral liposomes are mostly preferred among polyanionic types of DNA. On the other hand, anionic liposomes are restricted for plant transformation experiments (Shi et al., 2017).

Liposome-mediated plant cell transformations exhibit a variety of unique characteristics such as they offer protection to nucleic acids against different types of nucleases present in them. Optimum conditions for efficient fusion of protoplasts and liposomes enhance targeted delivery of genes and their expression. For example, negatively charged liposomes were used for encapsulated covering of DNA and were introduced into the protoplasts of tomato and tobacco. Expression of the CAT gene was examined in tobacco protoplasts for one week after the transformation experiment (Low et al., 2018). Moreover, plasmid DNA sequences were monitored for ten days that were unintegrated in genomes of plant cells. CAT gene activity was observed in callus tissues of tobacco after the inoculation stage and cloned in other callus tissues. In addition to these, degradation products were analyzed after transformation experiments. All these studies suggest that liposome-mediated gene transformation is used for viral replication and transient gene expression in a variety of plant cells (Adams et al., 2019). The process is indicated in Figure 12.1.

12.2.3.4 Genetic Manipulation via Electroporation

The term *electroporation* was coined in the 1980s. It is a transformation technique that entails high-intensity electrical pulses to generate transient pores in cell membranes to enable access of foreign gene constructs (Keshavareddy et al., 2018). The transient pores created on the lipid bilayer through the applied electrical field can channel the movement of molecules in or out of cells. The phenomenon of electroporation can be reversible and irreversible depending upon the cell response. If the strength of the electrical field is high, it may cause cell death, an irreversible change. Contrary to this, if the cell survives the electrical pulse strength, the membrane pores will be temporary, and the phenomenon is named reversible electroporation. Normally, 50%–70% of cells exposed to the electrical pulse field are killed. The fatal concerns vary from cell to cell, and do not depend on density or energy input. Instead, cell death is dependent on field intensity and duration of treatment (Ozyigit, 2020).

Electroporation is accomplished with the help of an electroporator device, which is made up of three parts: (1) a power supply, (2) electroporation cuvettes, and (3) electrodes. The power supply contains all control units. First, the cell suspension is pipetted into electroporation glass cuvettes. Second, the electric pulse field is applied to the cells through electrical conductors. These processes facilitate the direct contact between cell suspension and electrodes. Every single cell has a different field strength based on applied parameters (e.g., capacitance, voltage, and resistance). The process of electroporation depends on cell size, cell-mediated media, type, electrical conductivity, etc. However, ideal electroporation is hard to obtain, because tissues have different cell types, which also vary in spatial organization due to the presence of gap junctions (Ozyigit, 2020). Despite the homogeneous application of the electrical field, it leads to more electroporation of some cells than others due to the non-homogeneous environment inside cells.

For efficient transformation, the value of field strength should be 1–20 kV/cm and pulse duration should lie between 1 and 30 ms. The level of efficiency tends to decrease as the thickness of membrane covering the recipient DNA increases, for example, Gram-positive bacteria (thick membrane) have lower transformation efficiency as compared to Gram-negative bacteria (thin membrane). Likewise, the electroporation efficiency is higher for circular double-strand DNA than circular single-strand DNA. Moreover, pretreatment with different chemicals can increase the electroporation efficiency (Kumar et al., 2019).

Transgenic rice plants have been reconstructed through somatic embryogenesis. The procedure involves cell suspension-derived protoplast electroporation with plasmid containing nptII under promoter 35s CMV (cauliflower mosaic virus). Bittersweet nightshade is an alkaloid producing woody plants found in the different temperate regions of Europe, America, and Asia. Protoplasts of this plant have been transformed through electroporation, and electroporated protoplast tissues stimulate

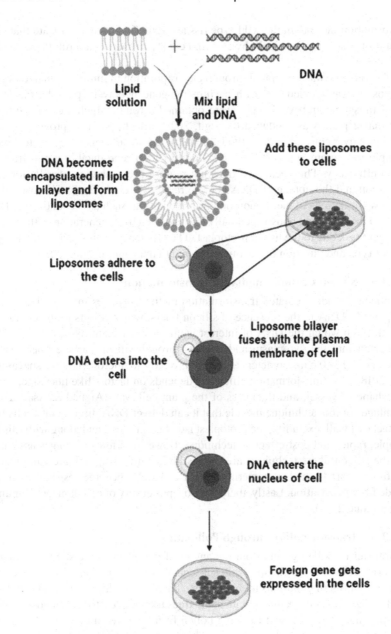

FIGURE 12.1 Flow diagram illustrating liposome fusion for plant transformation. DNA fragments of our desired gene are placed in a lipid solution. Lipid bilayer encapsulates the DNA fragments of interest and form liposome. Liposomes get attached to the cells of the cell line when added to them. Then the DNA gets incorporated into the nucleus of the cells and the expression of foreign gene is visible in the cells in which liposomes have entered.

root formation more than the wild-type tissues. Experiments demonstrate that electroporation also stimulates cell division and colony formation in protoplasts (Zhang et al., 1988).

DNA delivery was facilitated through electroporation in immature maize zygotic embryos. It was carried out with a chimeric gene neomycin phosphotransferase (neo). Transgenic embryonic cells were obtained through which a complete transgenic maize plant was produced (Ozyigit, 2020). Electroporation provides several advantages over conventional methods. Some of the advantages being the technique is simple to operate and cost effective, and has high reproducibility and transformation efficiency. This technique does not require preincubation of cells for electroporation and the uptake of DNA is quite rapid (Shi et al., 2017). However, there are a few limitations of electroporation transformation, such as the amount of DNA delivered into the plant body is generally low and even the regeneration of the plant is a stringent task due to the use of protoplast. The efficiency of this technique depends on plant type and duration of treatment applied (Thangadurai et al., 2020).

12.2.3.5   Silicon Carbide-mediated Transformation

The silicon carbide-mediated transformation method involves mixing plant tissues with plasmid DNA in the presence of silicon fibers, which results in the creation of pores through which the DNA of interest is taken up by plant tissues. The DNA is transported into the cytoplasm and nucleus through silicon carbide fibers having diameters of 0.3–0.6 micrometers and lengths of 10–80 micrometers (Keshavareddy et al., 2018). The transformation efficiency depends on factors like fiber size, vortex time, shape of vessel, and thickness of the plant cell wall (Arshad & Asad, 2019). Advantages of this technique include that it can deliver DNA directly into cells with an intact cell wall, excluding the protoplast isolation process, and along with this it is a simple, rapid, and cost-effective technique. However, a few challenges associated with the silicon fibers technique are the health hazards, i.e., they are carcinogenic in nature. Apart from that, sometimes plant cell walls become resistant to silicon carbide fiber penetration. Lastly, there is also a possibility of cell damage (Begum & Jayawardana, 2021).

12.2.3.6   Transformation through Pollination

The transfer of pollen grains from the anther of one plant to the stigma of another plant often results in fertilization and subsequent seed formation. Nectar serves the function of attracting animals to the flowers. There are multiple reasons for pollination (fertilization). For example, with some trees like maples, there is a possibility of cross-pollination if an insect carrying pollen from one tree flies to another nearby tree of a similar species in that season. This process increases genetic diversity by mixing genes from two different parents and allows formation of more seeds to increase crop yield.

Two methods have been utilized for gene transfer by pollen transformation in plants: gene transfer through soaking pollen in DNA and gene delivery through growing pollen tubes. Gene transfer is accomplished by soaking pollen grains in DNA solution. This method is simple, but there has not been any practical evidence reported for gene transfer done via soaking pollen grains (Jing-Xue et al., 2001).

In the second method, after pollination, the stigmas were cut to expose the pollen tubes, and DNA was placed into the incision site, which likely dispersed into the ovule through the developing pollen tube. This procedure is easy and promising. Through this method, both stable and transient expression can be accomplished (Eapen, 2011).

12.2.4 Indirect Method of Gene Transfer

12.2.4.1 *Agrobacterium* Gene Transformation in Cereals

Agrobacterium-mediated transformation in wheat was reported by Cheng and his co-workers in 2021. Using this method, they stabilized the integration of transgene via T-DNA transfer and permanent expression in progeny was confirmed. This study confirmed that monocots can be transformed via *Agrobacterium* just as dicots. This can be done by editing various factors such as explant, culture conditions, inoculation technique, modification of strain, and plasmid of *Agrobacterium tumefaciens* (Cheng et al., 2021).

Recent research reveals that the core polarity of the *Agrobacterium* T-DNA complex may be important for the subsequent integration of the T-DNA strand into the plant genome (Prasanna & Panda, 1997). Genetic studies have shown that this integration is a polar process (Keshavareddy et al., 2018). This data contributes to elucidating the differentiated behavior of the pathogen–host relationship between mono- and dicotyledonous plants (Adams et al., 2019). Another factor involved in the capacity for infection is the number of phenolic compounds released by the host plants to interact with the bacteria (Adams et al., 2019). Upon the injury of cells, phenolic compounds like acetosyringone activate *vir* genes that are responsible for the transfer of T-DNA from *Agrobacterium* into the host cells. Monocot tissues probably do not produce these compounds or do so at insufficient levels (Kumar et al., 2019).

However, researchers believe that the inoculation of *A. tumefaciens* in monocots that are pretreated with phenolic compounds brings a significant increase in transformation (Wang et al., 2017). When evaluating the induction of *vir* gene expression in wheat and oats, it was found that these cereals contain a gene-inducing substance different from the usual phenolic compound (acetosyringone) (Ozyigit, 2020). In monocots, most studies are in the agroinfection phase, and more efforts have been directed toward identifying bacterial strains that infect different genotypes and plant species, as well as plasmids that are more effective in incorporating genes of agronomic interest in the host genome (Van Eck, 2018).

Certain strains of *A. tumefaciens* and plasmid types, i.e., GV2260 (strain) and A281 and A348 (plasmids), are reported to infect some wheat genotypes. The successful transformation of monocots via *A. tumefaciens* depends on the use of the appropriate strain, as there is a significant interaction between the plant genotype and the bacterial strain, both for monocotyledons and for dicots. A few decades ago, it was believed that the use of *A. tumefaciens* to obtain transgenic cereals was not possible. However, research conducted in 1994 obtained transgenic rice plants using the LBA4404 bacterium with the plasmid pTok233 (Bhatt et al., 2021).

Some studies indicated that the association of different transformation techniques can be used to increase the frequency of transformation in plants, e.g., the association of biolistic with *A. tumefaciens* (Danilova et al., 2006). For obtaining transformed cereal lines via *A. tumefaciens* there should be clear identification of objectives involved in genetic manipulation, whether transformation is being done for physiological or biochemical studies, or to generate variability not obtained by conventional methods in breeding programs. Evaluate the possible consequences that these modified plants can bring to the environment before proceeding with genetic transformation. Identify the type of explant that is most suitable for the infection and regeneration of plants. Assess the resistance or susceptibility of plant tissue to selective factors such as temperature and the presence or absence of light during transformation. For example, less contamination is observed when explants are exposed to bacteria that are subcultured under continuous light as compared to explants that are exposed to bacterial subculture carried out in the dark. Confirmation of plant transformation must be done with a combination of different activities, such as the GUS (beta-glucuronidase) activity that is used to detect the presence of the gene of interest, and expression tests of the transgene to ensure the correct selection of transgenic lines. These transgene expression tests should be applied to the progeny of these transformed plants to confirm the stable transformation (Hensel, 2020). The process is illustrated in Figure 12.2.

The commonly used strategies to promote bacterial penetration in monocotyledonous plants require mechanical wounding. Such methods make it easier for *Agrobacterium* genes to pass through plant cell walls. However, these mechanical treatments may occasionally harm or worsen the explant's physiological state to the point where growth retardation and impaired regeneration potential occur. To avoid these issues, alternative approaches such as an agrobacterial monolayer could be used. To make a monolayer, *Agrobacterium* is cultivated overnight, then 1 ml of the suspension is placed in petri dishes with agar-solidified nutrition media and dispersed equally. After that, the petri dishes are placed in a laminar flow for 10–15 minutes to dry slightly. Plant tissues are then infected and co-cultured over the bacterial monolayer (Danilova et al., 2006). For maize transformation, this technique worked well. It was also used in the *Agrobacterium*-mediated transformation of *Dierama erectum*, a monocotyledonous ornamental geophyte, and results showed that it is a better gene delivery mechanism compared to traditional methods (Danilova et al., 2009). This approach is particularly useful for maize lines and hybrids that are more sensitive to *Agrobacterium* toxicity and transformation-related operations.

12.3 COMBINATIONS OF GENETIC TRANSFORMATION TECHNOLOGIES

Combinations of different genetic engineering techniques in plants have paved new avenues for humanity. The plant genome can be edited by inserting new genes from a foreign source like fungi, other plants, bacteria, and even animal and human genes. Such an insertion can provide resistance to extreme environmental conditions and to any microbial or viral disease (Punja, 2001). The plants created by using genetic engineering are transgenic plants. A new avenue is the utilization of such transgenic plants as bioreactors to create proteins that are costly to produce in the industries (Stöger et al., 2000; Torres et al., 1999). Genetic transformations can be carried out

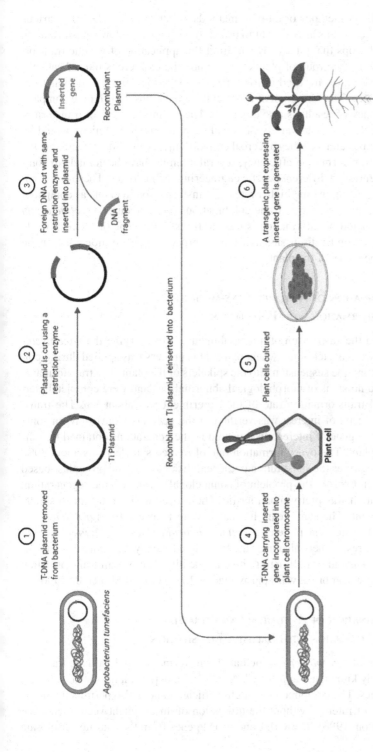

FIGURE 12.2 Mechanism of *Agrobacterium*-mediated transformation. The T-DNA plasmid is obtained from *Agrobacterium tumefaciens* bacteria. A single restriction enzyme is used to cut the Ti plasmid as well as the DNA of interest. A recombinant plasmid is formed that is then again inserted into the bacterium and then that bacterium infects the plant of interest. Upon infection the Ti plasmid having the gene of interest incorporates itself into the chromosome of the plant. Then the plant cells are cultured to visualize the gene of interest being expressed. It is further carried to create transgenic plants.

either through direct methods or indirect methods of gene transfer. Despite various strategies, the cereal crops are difficult to transform as they have low transformation efficiency cereal crops that ultimately can limit the application of genetic transformation. Triticeae is the group of plants that include the cash crops like wheat, rye, and barley. These crops are used all over the world, particularly for animal feed, food, and beverages. The formation of transgenic cereal crops is still quite complex, and the competency of results is also quite low. The technique used for transformation depends on the type and variety of cereal crop; no single technique would be efficient for all types and varieties of cereal crops (Punja, 2001). So, to overcome the problem of low transformation efficiency, several attempts have been made to combine the aforementioned different genetic engineering technologies. These combinations resulted in accomplishing higher rates of transformation in plants as compared to a single technique. In nearly all the combinations of techniques, *Agrobacterium tumefaciens*-mediated transformation is used to transfer the gene of interest, while all other techniques under discussion assist in the process of penetration either in the cell wall or in the ovary of the plant.

12.3.1 COMBINATION OF *A. TUMEFACIENS*-MEDIATED TRANSFORMATION WITH POLLINATION

A study reported the suspension of a recombinant vector carrying the *kanamycin*-resistance gene from bacteria into wheat plant. This process was applied during pollination by pipetting the suspension into the spikelets of the plant. The transformants obtained showed no signs of morphological abnormalities and were completely fertile. The overall transformation rate of the experiment was about 1%. The transformation of the gene of interest was verified in the genome. Conversely, in some circumstances, the gene of interest disappeared in the generations obtained by self-crossing of the plant. The ovary formation rate of treated spikelets was about 80%. This technique signifies the substitution of the leaf disk method and protoplast-based transformation as it solved the problem of soma clonal variations and regeneration. The next generation was pretty much fertile. The procedure is pretty much simple and highly efficient. The same team has carried out a successful *Agrobacterium*-mediated gene transfer via pollen in *Petunia hybrida* (Hess & Dressier, 1989). Another group of researchers also used the technique to apply *Agrobacterium* to the pistils in wheat plants after pollinating them artificially. The overall transformation rate of the spring wheat bread was approximately 2.7% (Pukhal'Skii et al., 1996).

12.3.2 COMBINATION OF *A. TUMEFACIENS*-MEDIATED TRANSFORMATION WITH BALLISTIC TRANSFECTION

The combination of *Agrobacterium*-mediated transformation and ballistic transfection is collectively known as "agrolistic". As the name depicts, it has the advantages of both techniques. The combination is highly efficient, and it also permits the insertion of the gene of interest without the utilization of an unsought vector sequence (Hansen & Chilton, 1996). The virD1 and virD2 genes from the cauliflower mosaic virus are inserted in a plasmid under the control of a 35S constitutive promoter along with the target genes. Then these plasmids are used to coat the particles to carry out the bombardment in the plant tissues. These virulence genes that were artificially created aided the transfer of T-DNA into the nucleus of the cell, and the gene of interest became incorporated into the genome of the plant (Filichkin & Gelvin, 1993). The efficiency is increased in this technique as there are a more predictable number of copies of a gene that are being transferred into the plant cells. Agrolistically transformed tobacco was achieved and a selectable marker flanked by border sequences that were added along with virD1 and virD2 genes. The efficiency of agrolistically inserted genes was 20% as compared to biolistic inserts (Hansen & Chilton, 1996).

12.3.3 COMBINATION OF *A. TUMEFACIENS*-MEDIATED TRANSFORMATION BY USING SILICON CARBIDE FIBERS

In 1999 Singh and Chawla experimented with *Agrobacterium*-mediated transformation by using silicon carbide fibers. The silicon carbide fibers were used to create wounds in immature wheat embryos by adding the immature wheat embryos in 5% by weight of fibers and *Agrobacterium* LBA4404 strain in the suspension. The efficiency of transformation increased from 2.4% to 33.3% with the use of silicon carbide fibers. The *Agrobacterium* strain possessed the binary vectors pBI121 and PTOK233 having GUS as a reporter gene and some other selectable marker genes. It was verified that using silicon carbide fibers enhanced the frequency of transformation in plants (Danilova & Dolgikh, 2005). Another team of researchers from Japan exposed immature maize embryos with the LBA4404 strain of *Agrobacterium*. This strain contained a plasmid with a GUS gene that was controlled by a 35S promoter sequence. The activation of the virulence genes from *Agrobacterium* was performed by the external application of acetosyringone. Then the maize embryos were kept initially in an antibiotic free medium and later were transferred into an antibiotic media. This transfer induced the formation of calli in the embryos. GUS gene expression confirmed the successful transformation of maize plants. This combination gave rise to reasonably stable integration and expression. More than 70% of the transformants generated were fertile and morphologically normal. The inheritance of the gene of interest in the next generation was confirmed by various genetic and molecular analysis (Ishida et al., 1996). A study reported that genetically transformed maize through embryonic calli can be infected by *Agrobacterium*. The embryonic calli were used as explants and were co-cultivated with the *A. tumefaciens* LBA4404 strain that carried a plasmid containing nptII and uidA genes. Use of the vacuum infiltration strategy and activation of *Agrobacterium* with acetosyringone application enhanced the kanamycin resistance in plants (Danilova & Dolgikh, 2005).

12.4 GENE TRANSFER USING CRISPR–CAS9

To improve the crops in the presence of different stresses, e.g., climate change, and biotic and abiotic stresses, it is required to develop more efficient strategies that improve crops with an enhanced productivity rate and limit global food insecurity.

For this purpose, primarily classic breeding strategies have been used to accumulate all the genes that are linked with agronomic traits in one plant (Abdelrahman et al., 2015). But some major drawbacks of these classical strategies are that they are very challenging and time-consuming processes. Alternatively, modern tools available for genetic engineering can integrate any foreign gene of interest very precisely into the genome of any organism. These genetic engineering techniques can be accurate to such an extent that they can substitute an existing allele to its alternative form. These editing strategies have remarkable efficiency and are able to manipulate the genome at targeted sites of such plants that have complex genomes (Feng et al., 2013).

Modern genetic engineering techniques include TALENs, ZFNs, MNs, and CRISPR–Cas9. TALENs, ZFNs, and MNs are considered first-generation editing tools. They comprise of lengthy protocols to accomplish specificity of the target site and are laborious. On the contrary CRISPR–Cas9 is a second-generation editing tool. CRISPR is robust, easy to design, and is cost effective as compared to first-generation sequencing strategies. CRISPR–Cas9 utilizes engineered nucleases, and this strategy gained popularity by the end of 2011 (Kim & Kim, 2014). CRISPR–Cas9 is principally the adaptive immune system of type II prokaryotes and it protects them against invading foreign organisms like viruses. It functions in a three-step process: spacer acquisition, biogenesis, and target degradation of the invading organism (Jinek et al., 2012). This editing technology is modified from the natural system of bacteria and archaea. It includes a toolbox of nucleases that are engineered (Cong et al., 2013). The Cas9 system comprises of two parts: a single guide RNA (sgRNA) that recognizes the target DNA sequence and the Cas9 endonuclease that produces a double-stranded break (DSB) at the target site. In the CRISPR–Cas9 system the sgRNA can be designed according to the target site by changing the nucleotide sequence at the start, which makes this technique a lot simpler than the first-generation TALENs and ZFNs (Kim & Kim, 2014; Matres et al., 2021; Yu et al., 2021). Figure 12.3 illustrates a basic mechanism the of the CRISPR–Cas9 technique applied for genome editing.

12.5   TRAITS INTRODUCED IN CEREALS

Some of the main traits that are targeted by the CRISPR genome editing technique in nearly all cereals include tolerance and resistance against abiotic and biotic stresses, improvement in male sterility, and improving the yield and the quality of the grain (Borisjuk et al., 2019). The editing of genomes can be done by insertion, deletion, or gene replacement mutagenesis. Wheat lines were developed by the *Agrobacterium*-mediated CRISPR–Cas9 system in which the function of homeoalleles of Qsd1 was disrupted. This mechanism was previously identified in barley to control seed dormancy. The transformants exhibited a significantly greater period of seed dormancy than wild wheat. This technique resultantly decreases the sprouting of the grains in preharvest conditions. Such editing methods serve to improve cereals, especially genetically recessive traits (Y. Zhang et al., 2016). In cereals many genes have been reportedly edited to enhance the quality and yield of the plant. In 2015 Lawrenson and his team designed a Cas9 nuclease guided by RNA to get a dwarf phenotype of barley. The nuclease targeted the HvPM19 gene. This example includes

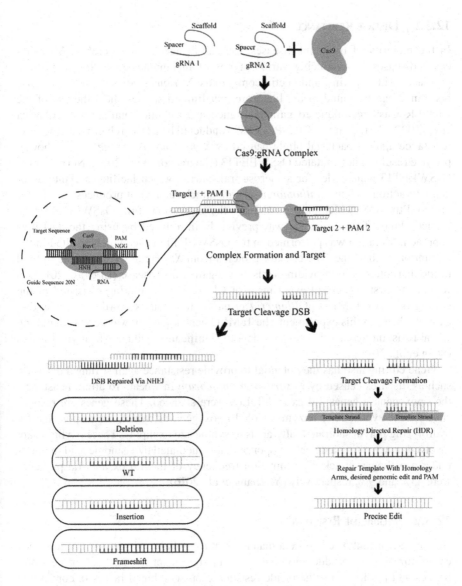

FIGURE 12.3 Basic mechanism of CRISPR–Cas9 technique. To enhance the specificity of the CRISPR technique two gRNA are designed so that they can target the opposite strands of the DNA within the target DNA. The DSBs are repaired through two different mechanisms: the error prone but efficient non-homologous end joining (NHEJ) mechanism, and the more accurate and less efficient homology directed repair (HDR) mechanism. The mutations under NHEJ include indel mutations and frame shift mutations.

gene knockout mechanism for trait improvement (Lawrenson et al., 2015). In rice, leaf morphology was targeted, which is considered an important agronomic trait in developing drought tolerance. CRISPR–Cas9 was used to develop the mutants with SRL1 and SRL2 genes. The resultant mutants displayed elevated yield, panicle, and grain per plant (Liao et al., 2019).

12.5.1 DISEASE RESISTANCE

In the majority of the plant diseases, the causal agents are microbes. Microbes cause diseases by producing various types of metabolites including hormones, polysaccharides, toxins, and pectin compounds. Various studies related to disease resistance can be found in the literature, but limited studies show the use of the CRISPR–Cas9 technique to gain resistance against microbial diseases (Ansari et al., 2020). A study using CRISPR–Cas9 conducted by a research group to increase resistance against bacterial blight caused by *Xanthomonas oryzae* in rice showed positive results. They mutated OsSWEET13 (Lawrenson et al., 2015). Normally the OsSWEET13 gene codes for a sucrose transporter, which facilitates plant pathogen interaction. When *Xanthomonas oryzae* attacks a plant, it produces an effector protein PthXo2X, which in turn enhances the expression of the OsSWEET13 gene in rice plants. This resistance was previously introduced by using the TALENs approach. Mutation was performed in the OsSWEET14 gene, which resulted in the disruption of the gene. The effector proteins from *X. oryzae* were not able to bind to the disrupted gene, providing resistance against the disease (Li et al., 2012). In wheat, the first successful experiment of CRISPR–Cas9 editing consisted of the editing of locus TaMLO of *Blumeria graminis* that causes powdery mildew disease in wheat. In this experiment, the TaMLO gene knockout was achieved making wheat resistant against powdery mildew to significantly reduce wheat yield losses (Shan et al., 2013).

Gene editing also has the potential to provide resistance against fungal diseases, such as diseases caused by *Fusarium graminearum* in wheat. To attain resistance, the lipoxygenase genes TaLpx and TaLox were silenced. These genes are responsible for the synthesis of enzymes called lipoxygenases, which are responsible for hydrolyzing polyunsaturated fatty acids to synthesize oxylipins. These oxylipins are involved in activating a defense response that is mediated by jasmonic acid in plants under a pathogen attack. The mutation frequency of the TaLox and TaLpx genes were 45% and 9%, respectively (Y. Zhang et al., 2016).

12.5.2 HERBICIDE RESISTANCE

Herbicide-resistant weeds pose a major threat to global food security because they can diminish crop productivity and result in significant losses. Lack of new herbicides and production of herbicide-resistant crops combined favors in controlling weeds and because of reduced crop phytotoxicity. The developed CRISPR/Cas-mediated genome editing techniques allow for effective targeted modification and have a lot of potential for developing herbicide-resistant plants (Han & Kim, 2019). Zong and his team produced wheat that was herbicide resistant by fusing Cas9 nickase with human cytidine deaminase. In this experiment, the TaALS gene was edited. Editing is in the form of amino acid substitution, which results in the resistance to sulfonylurea herbicides. The ALS gene normally codes for an enzyme known as acetolactate synthase. Acetolactate synthase is the enzyme required for the biosynthesis of branched-chain amino acids. The frequency of mutated wheat was quite high, i.e., 22.5% (Zong et al., 2018).

12.5.3 Pest Resistance

Fields are continuously exposed to insects, which increases the use of insecticides. Such a large scale and repeated application of pesticides impose a threat to the environment as well as human health. To overcome this problem, agriculturists are moving toward biological methods to control pests, and the cultivation of pest-resistant crop cultivars is also gaining importance (Han & Kim, 2019). One of the most important pathogens of wheat is cereal cyst nematode (CCN) and it is the reason for a huge loss of yield. AeVTDC1 and AeVTDC2 genes, which are involved in synthesis of tryptophan decarboxylase, were induced in cereal crops. The resistance was achieved by regulating the secondary metabolites after attack (Huang et al., 2018).

12.5.4 Improvement of Cereal Grain Quality

Editing of the genes through the CRISPR–Cas9 system has also helped in improving yield and grain size. Some of the genes reported to be associated with the grain length in wheat include TaGW, TaDEP, and TaGASR. The highly effective method for TaGASR editing was the RNP-mediated CRISPR–Cas9 method (Liang et al., 2017). Another study revealed that the knockout of the TaGW gene in wheat is responsible in modifying the biosynthesis pathway of gibberellin hormone. This gene knockout increased the grain, area, weight, and width in both rice and wheat plants (W. Wang et al., 2018). Another team reported that a gene was edited at locus ARGOS8 using CRISPR–Cas9 in maize plant. The ARGOS8 gene is related to the yield and quality of the grain. The mutations enhanced the transcription of the gene resulting in enhanced grain yield, which was proved by field experiments (Adams et al., 2019). To develop high-yielding plants, genes like TaGASR7 were targeted through CRISPR in wheat. This gene is linked to grain length and panicle erectness. The mutations demonstrated enhanced kernel weight in wheat (Y. Zhang & Gao, 2017). Enhancing yield and size of rice was achieved by editing the gene IsPPKL1. The heterosis of wheat and rice can be improved by targeting genes like Gn1a and Gs3 through modern gene editing techniques (Ansari et al., 2020). Bio-fortification refers to enhancing micronutrients levels in crops and alleviating the deficiencies in individuals consuming them. Genetic engineering of the TaVIT2 gene in wheat can enhance the levels of iron in wheat plants (Connorton et al., 2017). Targeting the SBEllb gene in rice results in elevation of amylase content in rice (Sun et al., 2016).

12.6 CONCLUSION

In the last few decades, remarkable progress has been made in the transformation of cereal crops. *A. tumefaciens*-mediated transformation is a widely used method for genetic modifications and it is successful in a variety of grain and grass species, but its efficiency is low. Some other genetic modification techniques like ZFNs and TALENs have initially revolutionized genetic engineering to quite an extent and recently the CRISPR gene editing technique has gained a lot of popularity as it is a cost-effective and precise technique as compared to traditional techniques. There is a hope that CRISPR–Cas9 has the potential to transform crop production for the increasing population of the world. Apart from increasing crop production, genetic engineering can be used to enhance the nutritional value of crops and also to enhance the resistance against abiotic and biotic stresses. But it is still challenging to apply these techniques in cereal crops. One of the major problems is the optimization of transformation needs to be conducted separately for every cultivating variety of cereal crops. Another issue is the identification of particular photospacer adjacent motif (PAM) sequences. Global regulations are also limiting the application of this technique in cereal crops. CRISPR–Cas9, being the latest and most efficient technique, presents some shortcomings like off-target editing and inefficient delivery systems, as the procedures are limited to specific tissue and certain varieties. The solution of off-target mutations can be solved to some extent by using long PAM sequences, but still this requires constant effort to enhance the adaptability of the experiment. By overcoming these problems related to genetic engineering, the goal of zero hunger can be easily achieved. For future perspectives, genome editing can be used to domesticate the wild varieties resulting in a new cereal variety that can bear the huge environmental stresses. This domestication is now reported in rice. Modern gene editing is improved by introducing SNPs that present as alleles and can be replaced. This is advantageous, as it decreases the chances of off-target mutations.

REFERENCES

Abdelrahman, M., Sawada, Y., Nakabayashi, R., Sato, S., Hirakawa, H., El-Sayed, M., Hirai, M. Y., Saito, K., Yamauchi, N., & Shigyo, M. (2015). Integrating transcriptome and target metabolome variability in doubled haploids of *Allium cepa* for abiotic stress protection. *Molecular Breeding, 35*(10), 1–11.

Adams, S., Pathak, P., Shao, H., Lok, J. B., & Pires-daSilva, A. (2019). Liposome-based transfection enhances RNAi and CRISPR-mediated mutagenesis in non-model nematode systems. *Scientific Reports, 9*(1), 1–12.

Altpeter, F., Springer, N. M., Bartley, L. E., Blechl, A. E., Brutnell, T. P., Citovsky, V., Conrad, L. J., Gelvin, S. B., Jackson, D. P., & Kausch, A. P. (2016). Advancing crop transformation in the era of genome editing. *The Plant Cell, 28*(7), 1510–1520.

Ansari, W. A., Chandanshive, S. U., Bhatt, V., Nadaf, A. B., Vats, S., Katara, J. L., Sonah, H., & Deshmukh, R. (2020). Genome editing in cereals: Approaches, applications and challenges. *International Journal of Molecular Sciences, 21*(11), 4040.

Arshad, M., & Asad, S. (2019). Embryogenic calli explants and silicon carbide whisker-mediated transformation of cotton (*Gossypium hirsutum* L.). In *Transgenic Cotton* (pp. 75–91). Springer.

Aulinger, I. E., Peter, S. O., Schmid, J. E., & Stamp, P. (2003). Gametic embryos of maize as a target for biolistic transformation: Comparison to immature zygotic embryos. *Plant Cell Reports, 21*(6), 585–591.

Baltes, N. J., Gil-Humanes, J., & Voytas, D. F. (2017). Genome engineering and agriculture: Opportunities and challenges. *Progress in Molecular Biology and Translational Science, 149*, 1–26.

Baltes, N. J., & Voytas, D. F. (2015). Enabling plant synthetic biology through genome engineering. *Trends in Biotechnology, 33*(2), 120–131.

Begum, R., & Jayawardana, N. U. (2021). *A review of nanotechnology as a novel method of gene transfer in plants. Journal of Agricultural Sciences – Sri Lanka, 16*(2), 300–316.

Bhatt, R., Asopa, P. P., Jain, R., Kothari-Chajer, A., Kothari, S. L., & Kachhwaha, S. (2021). Optimization of Agrobacterium mediated genetic transformation in *Paspalum scrobiculatum* L.(Kodo Millet). *Agronomy, 11*(6), 1104.

Borisjuk, N., Kishchenko, O., Eliby, S., Schramm, C., Anderson, P., Jatayev, S., Kurishbayev, A., & Shavrukov, Y. (2019). Genetic modification for wheat improvement: From transgenesis to genome editing. *BioMed Research International*, https://doi.org/10.1155/2019/6216304.

Cheng, Y., Wang, X., Cao, L., Ji, J., Liu, T., & Duan, K. (2021). Highly efficient *Agrobacterium rhizogenes*-mediated hairy root transformation for gene functional and gene editing analysis in soybean. *Plant Methods*, *17*(1), 1–12.

Cocking, E. C., & Davey, M. R. (1987). Gene transfer in cereals. *Science*, *236*(4806), 1259–1262.

Cong, L., Ran, F. A., Cox, D., Lin, S., Barretto, R., Habib, N., Hsu, P. D., Wu, X., Jiang, W., & Marraffini, L. A. (2013). Multiplex genome engineering using CRISPR/Cas systems. *Science*, *339*(6121), 819–823.

Connorton, J. M., Jones, E. R., Rodríguez-Ramiro, I., Fairweather-Tait, S., Uauy, C., & Balk, J. (2017). Wheat vacuolar iron transporter TaVIT2 transports Fe and Mn and is effective for biofortification. *Plant Physiology*, *174*(4), 2434–2444.

Danilova, S. A. (2007). The technologies for genetic transformation of cereals. *Russian Journal of Plant Physiology*, *54*(5), 569–581. https://doi.org/10.1134/S1021443707050019

Danilova, S. A., da Silva, J. A. T., & Kusnetsov, V. V. (2006). Novel approaches for Agrobacterium-mediated transformation of maize and ornamental grasses. *Floriculture, Ornamental and Plant Biotechnology*, *2*, 66–69.

Danilova, S. A., & Dolgikh, Y. I. (2005). Optimization of agrobacterial (*Agrobacterium tumefaciens*) transformation of maize embryogenic callus. *Russian Journal of Plant Physiology*, *52*(4), 535–541.

Danilova, S. A., Kusnetsov, V. V, & Dolgikh, Y. I. (2009). A novel efficient method for maize genetic transformation: Usage of agrobacterial monolayer. *Russian Journal of Plant Physiology*, *56*(2), 258–263.

Eapen, S. (2011). Pollen grains as a target for introduction of foreign genes into plants: An assessment. *Physiology and Molecular Biology of Plants*, *17*(1), 1–8.

Feng, Z., Zhang, B., Ding, W., Liu, X., Yang, D.-L., Wei, P., Cao, F., Zhu, S., Zhang, F., & Mao, Y. (2013). Efficient genome editing in plants using a CRISPR/Cas system. *Cell Research*, *23*(10), 1229–1232.

Filichkin, S. A., & Gelvin, S. B. (1993). Formation of a putative relaxation intermediate during T-DNA processing directed by the Agrobacterium tumefaciens VirD1, D2 endonuclease. *Molecular Microbiology*, *8*(5), 915–926.

Giraldo, P., Benavente, E., Manzano-Agugliaro, F., & Gimenez, E. (2019). Worldwide research trends on wheat and barley: A bibliometric comparative analysis. In *Agronomy* (Vol. 9, Issue 7). https://doi.org/10.3390/agronomy9070352

Han, Y.-J., & Kim, J.-I. (2019). Application of CRISPR/Cas9-mediated gene editing for the development of herbicide-resistant plants. *Plant Biotechnology Reports*, *13*(5), 447–457.

Hansen, G., & Chilton, M.-D. (1996). "Agrolistic" transformation of plant cells: Integration of T-strands generated *in planta*. *Proceedings of the National Academy of Sciences*, *93*(25), 14978–14983. https://doi.org/10.1073/pnas.93.25.14978

Hensel, G. (2020). Genetic transformation of Triticeae cereals–Summary of almost three-decade's development. *Biotechnology Advances*, *40*, 107484.

Hess, D., & Dressier, D. K. (1989). Tumor transformation of *Petunia hybrida* via pollen cocultured with *Agrobacterium tumefaciens*. *Botanica Acta*, *102*(3), 202–207.

Huang, Q., Li, L., Zheng, M., Chen, F., Long, H., Deng, G., Pan, Z., Liang, J., Li, Q., & Yu, M. (2018). The tryptophan decarboxylase 1 gene from *Aegilops variabilis* No. 1 regulate the resistance against cereal cyst nematode by altering the downstream secondary metabolite contents rather than auxin synthesis. *Frontiers in Plant Science*, *9*, 1297.

Ishida, Y., Saito, H., Ohta, S., Hiei, Y., Komari, T., & Kumashiro, T. (1996). High efficiency transformation of maize (*Zea mays* L.) mediated by *Agrobacterium tumefaciens*. *Nature Biotechnology*, *14*(6), 745–750. https://doi.org/10.1038/nbt0696-745

Jauhar, P. P., & Hanna, W. W. (1998). Cytogenetics and genetics of pearl millet. *Advances in Agronomy, 64*(1), 2–21.

Jinek, M., Chylinski, K., Fonfara, I., Hauer, M., Doudna, J. A., & Charpentier, E. (2012). A programmable dual-RNA–guided DNA endonuclease in adaptive bacterial immunity. *Science, 337*(6096), 816–821.

Jing-Xue, W., Yi, S., Gui-Mei, C., & Jing-Jing, H. (2001). Transgenic maize plants obtained by pollen-mediated transformation. *Journal of Integrative Plant Biology, 43*(3), 275.

Kemper, E. L., Silva, M. J., & Arruda, P. (1996). Effect of microprojectile bombardment parameters and osmotic treatment on particle penetration and tissue damage in transiently transformed cultured immature maize (*Zea mays* L.) embryos. *Plant Science, 121*(1), 85–93.

Keshavareddy, G., Kumar, A. R. V, & Ramu, V. S. (2018). Methods of plant transformation—a review. *International Journal of Current Microbiology and Applied Sciences, 7*(7), 2656–2668.

Kim, H., & Kim, J.-S. (2014). A guide to genome engineering with programmable nucleases. *Nature Reviews Genetics, 15*(5), 321–334.

Krautwig, B., & Lörz, H. (1995). Cereal protoplasts. *Plant Science, 111*(1), 1–10.

Kumar, P., Nagarajan, A., & Uchil, P. D. (2019). Electroporation. *Cold Spring Harbor Protocols, 2019*(7), pdb-top096271.

Lawrenson, T., Shorinola, O., Stacey, N., Li, C., Østergaard, L., Patron, N., Uauy, C., & Harwood, W. (2015). Induction of targeted, heritable mutations in barley and *Brassica oleracea* using RNA-guided Cas9 nuclease. *Genome Biology, 16*(1), 258. https://doi.org/10.1186/s13059-015-0826-7

Li, T., Liu, B., Spalding, M. H., Weeks, D. P., & Yang, B. (2012). High-efficiency TALEN-based gene editing produces disease-resistant rice. *Nature Biotechnology, 30*(5), 390–392.

Liang, Z., Chen, K., Li, T., Zhang, Y., Wang, Y., Zhao, Q., Liu, J., Zhang, H., Liu, C., & Ran, Y. (2017). Efficient DNA-free genome editing of bread wheat using CRISPR/Cas9 ribonucleoprotein complexes. *Nature Communications, 8*(1), 1–5.

Liao, S., Qin, X., Luo, L., Han, Y., Wang, X., Usman, B., Nawaz, G., Zhao, N., Liu, Y., & Li, R. (2019). CRISPR/Cas9-induced mutagenesis of semi-rolled Leaf1, 2 confers curled leaf phenotype and drought tolerance by influencing protein expression patterns and ROS scavenging in rice (*Oryza sativa* L.). *Agronomy, 9*(11), 728.

Low, L.-Y., Yang, S.-K., Andrew Kok, D. X., Ong-Abdullah, J., Tan, N.-P., & Lai, K.-S. (2018). Transgenic plants: Gene constructs, vector and transformation method. In *New Visions in Plant Science* (pp. 41–61). Intech Open.

Mali, P., Yang, L., Esvelt, K. M., Aach, J., Guell, M., DiCarlo, J. E., Norville, J. E., & Church, G. M. (2013). RNA-guided human genome engineering via Cas9. *Science, 339*(6121), 823–826.

Matres, J. M., Hilscher, J., Datta, A., Armario-Nájera, V., Baysal, C., He, W., Huang, X., Zhu, C., Valizadeh-Kamran, R., Trijatmiko, K. R., Capell, T., Christou, P., Stoger, E., & Slamet-Loedin, I. H. (2021). Genome editing in cereal crops: An overview. *Transgenic Research, 30*(4), 461–498. https://doi.org/10.1007/s11248-021-00259-6

Ozyigit, I. I. (2020). Gene transfer to plants by electroporation: Methods and applications. *Molecular Biology Reports, 47*(4), 3195–3210.

Prasanna, G. L., & Panda, T. (1997). Electroporation: Basic principles, practical considerations and applications in molecular biology. *Bioprocess Engineering, 16*(5), 261–264.

Puchta, H. (2005). The repair of double-strand breaks in plants: Mechanisms and consequences for genome evolution. *Journal of Experimental Botany, 56*(409), 1–14.

Pukhal'Skii, V. A., Smirnov, S. P., Korostyleva, T. V, Bilinskaya, E. N., & Va, A. A. (1996). Genetic transformation of wheat (*Triticum aestivum* L.) by agrobacterium tumefaciem. *Генетика, 32*(11), 1596–1600.

Punja, Z. K. (2001). Genetic engineering of plants to enhance resistance to fungal pathogens—a review of progress and future prospects. *Canadian Journal of Plant Pathology, 23*(3), 216–235.

Ramasamy, M., Mora, V., Damaj, M. B., Padilla, C. S., Ramos, N., Rossi, D., Solís-Gracia, N., Vargas-Bautista, C., Irigoyen, S., & DaSilva, J. A. (2018). A biolistic-based genetic transformation system applicable to a broad-range of sugarcane and energycane varieties. *GM Crops & Food, 9*(4), 211–227.

Razzaq, A., Saleem, F., Kanwal, M., Mustafa, G., Yousaf, S., Imran Arshad, H. M., Hameed, M. K., Khan, M. S., & Joyia, F. A. (2019). Modern trends in plant genome editing: An inclusive review of the CRISPR/Cas9 toolbox. *International Journal of Molecular Sciences, 20*(16), 4045.

Repellin, A., Båga, M., Jauhar, P. P., & Chibbar, R. N. (2001). Genetic enrichment of cereal crops via alien gene transfer: New challenges. *Plant Cell, Tissue and Organ Culture, 64*(2), 159–183.

Shan, Q., Wang, Y., Li, J., Zhang, Y., Chen, K., Liang, Z., Zhang, K., Liu, J., Xi, J. J., & Qiu, J.-L. (2013). Targeted genome modification of crop plants using a CRISPR-Cas system. *Nature Biotechnology, 31*(8), 686–688.

Shi, L., Chen, D., Xu, C., Ren, A., Yu, H., & Zhao, M. (2017). Highly-efficient liposome-mediated transformation system for the basidiomycetous fungus Flammulina velutipes. *The Journal of General and Applied Microbiology, 63*, 2010–2016.

Singer, S. D., Foroud, N. A., & Laurie, J. D. (2019). Molecular Improvement of Grain: Target Traits for a Changing World. *Encyclopedia of Food Security and Sustainability, 2*, 545–555.

Stöger, E., Vaquero, C., Torres, E., Sack, M., Nicholson, L., Drossard, J., Williams, S., Keen, D., Perrin, Y., & Christou, P. (2000). Cereal crops as viable production and storage systems for pharmaceutical scFv antibodies. *Plant Molecular Biology, 42*(4), 583–590.

Sun, Y., Zhang, X., Wu, C., He, Y., Ma, Y., Hou, H., Guo, X., Du, W., Zhao, Y., & Xia, L. (2016). Engineering herbicide-resistant rice plants through CRISPR/Cas9-mediated homologous recombination of acetolactate synthase. *Molecular Plant, 9*(4), 628–631.

Thangadurai, D., Shettar, A. K., Sangeetha, J., Raju, C. S., Islam, S., Al-Tawaha, A. R. M. S., Habeeb, J., Wani, S. A., & Baqual, M. F. (2020). Transformation techniques and molecular analysis of transgenic rice. In *Rice Research for Quality Improvement: Genomics and Genetic Engineering* (pp. 221–245). Springer.

Torres, E., Vaquero, C., Nicholson, L., Sack, M., Stöger, E., Drossard, J., Christou, P., Fischer, R., & Perrin, Y. (1999). Rice cell culture as an alternative production system for functional diagnostic and therapeutic antibodies. *Transgenic Research, 8*(6), 441–449.

Van Eck, J. (2018). The status of Setaria viridis transformation: Agrobacterium-mediated to floral dip. *Frontiers in Plant Science, 9*, 652.

Voytas, D. F., & Gao, C. (2014). Precision genome engineering and agriculture: Opportunities and regulatory challenges. *PLoS Biology, 12*(6), e1001877.

Wang, K., Liu, H., Du, L., & Ye, X. (2017). Generation of marker-free transgenic hexaploid wheat via an Agrobacterium-mediated co-transformation strategy in commercial Chinese wheat varieties. *Plant Biotechnology Journal, 15*(5), 614–623.

Wang, W., Pan, Q., He, F., Akhunova, A., Chao, S., Trick, H., & Akhunov, E. (2018). Transgenerational CRISPR-Cas9 activity facilitates multiplex gene editing in allopolyploid wheat. *The CRISPR Journal, 1*(1), 65–74.

Yu, H., Lin, T., Meng, X., Du, H., Zhang, J., Liu, G., Chen, M., Jing, Y., Kou, L., & Li, X. (2021). A route to de novo domestication of wild allotetraploid rice. *Cell, 184*(5), 1156–1170.

Zhang, H. M., Yang, H., Rech, E. L., Golds, T. J., Davis, A. S., Mulligan, B. J., Cocking, E. C., & Davey, M. R. (1988). Transgenic rice plants produced by electroporation-mediated plasmid uptake into protoplasts. *Plant Cell Reports, 7*(6), 379–384. https://doi.org/10.1007/BF00269517

Zhang, Y., & Gao, C. (2017). Recent advances in DNA-free editing and precise base editing in plants. *Emerging Topics in Life Sciences*, *1*(2), 161–168.

Zhang, Y., Liang, Z., Zong, Y., Wang, Y., Liu, J., Chen, K., Qiu, J.-L., & Gao, C. (2016). Efficient and transgene-free genome editing in wheat through transient expression of CRISPR/Cas9 DNA or RNA. *Nature Communications*, *7*(1), 1–8.

Zong, Y., Song, Q., Li, C., Jin, S., Zhang, D., Wang, Y., Qiu, J.-L., & Gao, C. (2018). Efficient C-to-T base editing in plants using a fusion of nCas9 and human APOBEC3A. *Nature Biotechnology*, *36*(10), 950–953.

13 Genome-Edited Cereal Characterization Using Metabolomics

Sania Zaib, Shomaila Mehmood, Misbah Naz,
Muhammad Tariq, and Sadia Banaras

CONTENTS

13.1 INTRODUCTION

The term *cereal* is derived from the Latin word *cerealis*, which means "grain". Grain contains three parts: endosperm, germ and bran. Cereals are annual grass plants belonging to the monocotyledonous Poaceae family. These plants, such as rice, wheat, maize, barley, rye, millet and sorghum, usually have thin long stalks and their starchy grains are the source of food. The term *cereal* is also used for foodstuff like breads, flours and pasta prepared from these grains (Sarwar et al. 2013). Grains of cereal are a staple food as they are grown in bulk and provide one-half of the calories daily consumed in the world. Complex carbohydrates are abundantly found in cereals, providing ample amount of energy. As a whole grain, they are a rich source of carbohydrates, protein, fats, vitamins, minerals and abundant fibers. However, refined grains, i.e., endosperm after the removal of germ and bran, mostly contains carbohydrates and lacks other important nutrients. Cereals are also used as animal feed for poultry and livestock rendering as poultry, meat and dairy products being used by humans. They also serve as raw material for the production of various substances such as glucose, oils, alcohols and adhesives (Sarwar et al. 2013). Hence, for global food security cereals are very essential.

DOI: 10.1201/9781003250845-13

According to the Food and Agriculture Organization (FAO), the world population is expected to reach about nine billion by the year 2050, and in order to feed such a large population there is the need to produce 50% more food than produced today (FAO 2009). However, increasing the food production faces several challenges such as desertification, shortage of land for agriculture purposes as the agriculture lands are being converted for non-agricultural purposes as well as global climate change (Godfray et al. 2010; Beddington et al. 2011; Zaib, Ahmad, and Shakeel 2020). Increasing uncertainties of climate change has resulted in the degradation of our agriculture system (Foley et al. 2011). For meeting the challenges faced by the agriculture system, there is need to produce crops with improved quality and higher yields with minimum input requirements (Tilman et al. 2011). For the improvement of crops, traditional breeding is widely used nowadays, however, it is costly, labor intensive and its progress from screening to commercial varieties takes a long time. Likewise, genetically modified (GM) crops generated by gene transfer carry beneficial traits for crop improvement. In spite of the fact that GM crops will play a very important role in controlling global food security, their use is largely affected by unsubstantiated safety concerns to health and the environment. Significant cost barriers to the widespread adoption of new genetically modified traits have been imposed by government regularity frameworks aiming to protect humans as well as the environment (Prado et al. 2014). Consequently, benefits of genetically modified traits have been limited to fewer cultivated crops (Yi Zhang, Massel, et al. 2018). Regardless of their economic importance, commercialization and authorization, genetically modified organisms have always been contentious among the scientific community and the public sector. Various factors such as risk assessment, labeling, marketing and traceability are strictly regulated in various countries. As GM crops make the larger portion of the modern agriculture sector worldwide, in accordance with the substantial equivalence principle, a genetically modified crop would be supposed to be safe if it is equivalently safe to its conventional species. Genome editing, on the other hand, facilitates efficient, precise and targeted modifications of genome loci (K. Chen and Gao 2014; C. Gao 2015).

13.2 GENOME EDITING

Genome editing is a disruptive technique that plays an important role in the agriculture sector for crop improvement by precise mutagenesis and gene targeting (Yi Zhang, Li, et al. 2018; Najera et al. 2019; Yangfan Zhang et al. 2019). For two decades, zinc-finger nucleases (ZFNs) and transcription activator-like effector nucleases (TALENs) have been used for genome editing (Y.-G. Kim, Cha, and Chandrasegaran 1996; Christian et al. 2010). However, clustered regularly interspaced short palindromic repeats (CRISPR)/Cas systems are widely being used providing simple and easily targeted gene editing (Jinek et al. 2012).

13.2.1 CRISPR–Cas9 System

CRISPR–Cas9 is a versatile, efficient and low-cost technique. There are multiple uses of CRISPR–Cas9 in the field of plant genome editing. It is economical, highly accurate, effective and easy to use, and even effective for editing multiple genes (M. Wang, Wang, et al. 2018). Multiplex genome editing allows manipulation of several genes at multiple gene loci. These features make CRISPR–Cas9 effective for rapid exploitation in plants and may help in encountering various problems of plant breeding (J.-F. Li et al. 2013; C. Gao 2018). In these technologies sequence-specific nucleases are being used that, after recognizing specific sequences of DNA, generate double-strand breaks (DSBs). The host's repair machinery then fixes these DSBs either by homologous recombination (HR) or by non-homologous end joining (NHEJ). NHEJ causes gene knockouts by deletion of nucleotides, while HR causes gene replacement or insertions, so it results in modifications of targeted sites (Zetsche et al. 2015) (Figure 13.1 and Figure 13.2a). In NHEJ, the host performs the repair without having a donor template, whereas in HDR a donor template is needed for repairing double-strand breaks. In plants, the use of HDR is challenging due to lower frequency of HDR than NHEJ (Puchta 2005).

13.2.2 CRISPR–Cpf1/Cas12a

Cpf1 recognizes T-rich protospacer adjacent motifs (PAMs) (Zetsche et al. 2015; Alok et al. 2020). Unlike Cas9, double-standard breaks created by Cas12 are staggered cuts in nature (Figure 13.2b), enabling the genome editing of AT-rich regions like promoter and untranslated regions. However, this system also has a drawback in that it cannot recognize the sites lacking TTTN motifs, hence restricting its usage in plants. However, this restraint has been overcome by the development of modified Cas12a that can recognize other PAMs too (L. Gao et al. 2017).

13.2.3 Base Editing

The CRISPR–Cas9-mediated base editing technique is a new efficient approach for the precise conversion of one DNA base into another without the need for a donor

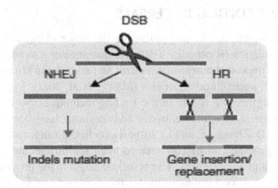

FIGURE 13.1 Genome editing techniques produce DSBs, which are fixed by a plant's endogenous repairing systems using NHEJ or HR. Small indels introduced by NHEJ in DSBs result in premature stop codons or frameshift mutations. Whereas the HR in the presence of a homologous donor DNA template causes insertions or gene replacement (Yi Zhang, Massel, et al. 2018).

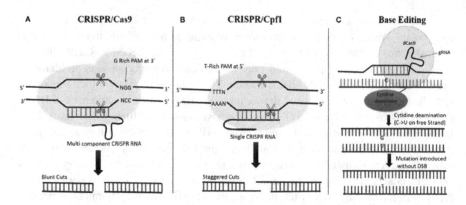

FIGURE 13.2   (a) In CRISPR–Cas9, Cas9a multicomponent protein after recognizing a G-rich protospacer adjacent motif (PAM) at the 3′ end of the target sequence generates a DSB; as a result blunt ends are produced. (b) In the CRISPR–Cpf1 system, a T-rich PAM region is being recognized by Cpf1 (a single component protein) at the 5′ end of the target site, which creates a DSB resulting in staggered cuts (Zafar et al. 2020).

repair template or DSBs, so as a result, desirable mutations occur at the specifically targeted sites (Hess et al. 2017; B. Yang et al. 2017). These technologies make use of Cas9 nickase (nCas9) or dead Cas9 (dCas9) fused with base editors enzymes. For instance, the fusion of Cas9 nickase and cytidine deaminase (CD) enzyme converts cytosine (C) to uracil (U). In the subsequent DNA repair process, U is treated as T (thymine) so allowing C → T substitution. Similarly, adenine deaminase (AD) converts adenine (A) to inosine (I), which is then treated as guanine (G) by polymerase resulting in A → G substitution (Gaudelli et al. 2017; H. Kim et al. 2017) (Figure 13.3). So, overall various genome editing methods and tools are used to investigate the phenotypes of genetically edited crops (Figure 13.4).

13.3   GENOME EDITING IN CEREALS

Among the cereals, wheat, rice and maize are the major (42%) sources of calories for the world population (Ricepedia 2020). Maintenance of a steady supply of these staple foods with improved nutritional content and addressing climate change need the use of modern agricultural strategies (Matres et al. 2021). In modern biotechnology, genetic engineering is one of the leading techniques. So far several cereal crops, including wheat, rice, maize, barley and sorghum, have been edited by using these techniques (D. Zhang, Li, and Li 2016; Ricroch, Clairand, and Harwood 2017). In cereal crops, genome editing has been adopted for various nutritional yields as well as for biotic and abiotic traits. The principal target is the provision of nutrition as well as to enhance food productivity for the growing world population (Ansari et al. 2020). The advancement of the editing approach CRISPR–Cas system combined with open source data of genes and single nucleotide polymorphism (SNPs) involved in vital traits in cereals resulted in a surge of publications of gene editing for crop improvement (Matres et al. 2021). CRISPR has been successfully used for maize (Shi et al. 2017), wheat (Yi Zhang et al. 2016), rice (J. Li et al. 2016) and

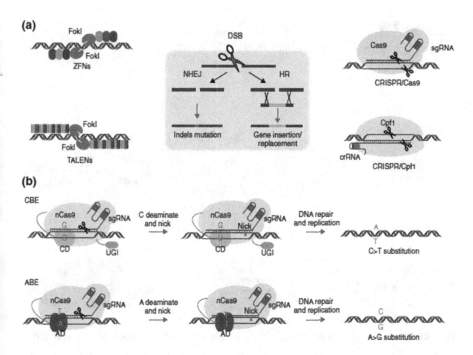

FIGURE 13.3 Illustration of CRISPR/Cas9-mediated base editing. In the CBE (cytidine deaminase-mediated base editing), cytidine deaminase fused with nCas9 is used for targeting the location, and without any DSB, cytosine is converted into uracil and then in the DNA repair or replication processes uracil is changed to thymine. In the ABE (adenine deaminase-mediated base editing) system, adenine deaminase fused with nCas9 converts adenine in the targeted region into inosine, which is then changed to guanine by polymerase enzyme (Yi Zhang, Massel, et al. 2018).

barley (Lawrenson et al. 2015). In cereal crops, CRISPR–Cas9 genome editing was mainly applied for improving resistance against bacteria, fungi, viruses and insects, as well as for the improvement of nutritional quality and tolerance against herbicides (Ansari et al. 2020).

Worldwide every year severe losses in agriculture production occur due to various diseases. The use of chemicals in agriculture is harmful for human health as well as for the environment. So in recent times, the development of resistant cultivars through genome editing has been a noteworthy goal (Ansari et al. 2020). With the help of the CRISPER–Cas9 technique, the *OsSWEET13* gene was mutated in rice to achieve resistance against bacterial blight disease (Zhou et al. 2015). In wheat, simultaneous targeting of three genes—*TaMLO-A1*, *TaMLO-B1* and *TaMLO-D1*—using TALEN and CRISPR helped in developing resistance against powdery mildew. In rice plants, mutagenesis of *eIF4G* was found to be effective for developing resistance against rice tungro spherical virus (Macovei et al. 2018). In the same way in wheat, base editing of *ALS* and acetyl-coenzyme A carboxylase genes conferred tolerance against herbicides (Ansari et al. 2020). Likewise in cereals, CRISPR–Cas9 has been employed for editing several genes related to quality and yield enhancement (Ansari et al. 2020).

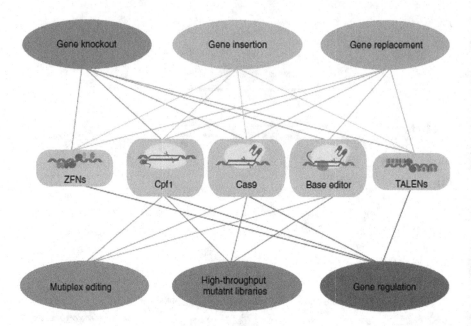

FIGURE 13.4   The network of genome-editing methods and tools (Yi Zhang, Massel, et al. 2018).

Precise modification of *ARGOS8* using CRISPR in maize helped in enhancing the grain yield of maize (Shi et al. 2017). In wheat, iron content was enhanced by engineering of the TaVT2 gene (Connorton et al. 2017). Through *SBEIIb* gene editing, amylase content was increased in rice (Sun et al. 2016). Knockout mutants of wheat genes created by multiplexed genome editing showed increased kernel weight and seed size (W. Wang, Pan, et al. 2018). It is expected that in cereal plants, use of CRISPR–Cas9 to target several genes simultaneously will increase rapidly over the coming years (Ansari et al. 2020). A general strategy for plant genome editing is shown in Figure 13.5.

13.3.1 METABOLOMICS

As a result of plant metabolism, large numbers of metabolites accumulate in fruits and cereals. There are almost 250,000 metabolites in plants (H.K. Kim, Choi, and Verpoorte 2010). These metabolites, such as secondary metabolites, organic acids, steroids, peptides, hormones, aldehydes, ketones, vitamins, lipids and amino acids, play a role in providing protection against herbivores and pathogens (Pott et al. 2019). Sugars, amino acids and organic acids are the primary metabolites produced as a result of primary metabolism. They act as building components as well as serve as precursors for secondary metabolism. Secondary metabolism is extremely complicated and produces a large number of compounds, such as terpenoids, polyphenols and sulfur/nitrogen containing metabolites (Aharoni and Galili 2011; Pott, Osorio, and Vallarino 2019). Several of these metabolites play a role in health-promoting traits (Vickers 2017; Pott, Osorio, and Vallarino 2019). Secondary metabolites being

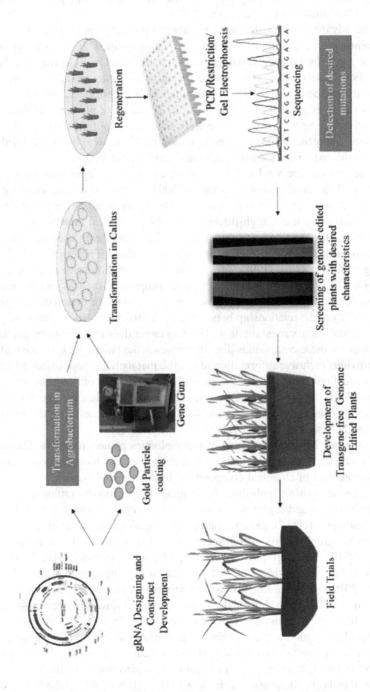

FIGURE 13.5 General strategy illustrating successive steps in plant genome editing (Zafar et al. 2020).

small molecules in nature are usually considered as nonessential for the general survival of plants; however, they play very important roles in adapting plants to abiotic and biotic stress situations (L. Yang et al. 2018).

With the advancement of the latest metabolite profiling techniques, thousands of metabolites can be simultaneously detected and quantified. Metabolite profiling reduces the gap between genotype and phenotype and helps in dissecting metabolic pathways. For the identification of genetic variants underlying metabolic content, it is important that these techniques be combined with genomic and transcriptomic techniques (K. Li et al. 2019; Labadie et al. 2020). Bases of plant metabolism can be dissected and complex traits could be associated with genotype through integrating metabolic and genetic information. In addition, this strategy is also helpful for breeding nutritionally rich and high-yielding crops (Wen et al. 2018).

Omics techniques are vital tools for understanding how organisms respond to genetic as well as environmental changes (Valdés et al. 2013). One of the most recent omics techniques—metabolomics—was first introduced at the beginning of the third millennium. It is a high-throughput assessment of all metabolites present in an organism. Targeted and non-targeted techniques are being used by scientists for the analysis of endogenous and exogenous metabolites (Frederich et al. 2016). According to Oliver et al. (1998), metabolome is the comprehensive quantitative evaluation of all of the low molecular weight compounds such as sugars, amino acids, organic acids and nucleotides present in a cell, tissue or organism at a given moment to analyze the relationship between genetic structure, gene expression, protein function and environmental effect. As being either the substrates or by-products of cell processes, these compounds directly influence the phenotypes. Hence metabolomics provides inclusive information about the metabolite composition, which is helpful for functional screening of the state of a cell, tissue or of their environment. Being the end products of either transcription or translation, changes in profusion of metabolites are considered as the key attribute of plant interaction with the environment (Tohge, Scossa, and Fernie 2015).

During the last 20 years the study of metabolomics gained attention. The combination of advanced analytical methods and bioinformatics tools in metabolomics provides ample data of chemical composition. In plant metabolomics, a metabolite profile is generated using high-throughput analytical chromatography techniques coupled with precise and highly sensitive mass spectrometric tools such as nuclear magnetic resonance (NMR), gas chromatography–mass spectrometry (GC-MS) and liquid chromatography–mass spectrometry (LC-MS), direct injection mass spectrometry (DIMS), and high-performance liquid chromatography (HPLC) (Obata and Fernie 2012). Among these, LC-MS provides wide-ranging outcomes (Alseekh and Fernie 2018). High-resolution mass spectrometry helps in the identification of the exact chemical composition of the analyte by providing sufficient mass accuracy, while the multiple rounds of mass spectrometry help in structural assignment through the identification and assembly of the consequential metabolite fragments. Purification of metabolites can assist in the identification of the absolute structure of metabolites through NMR (Liu et al. 2018). This also results in the enrichment of various metabolite databases such as KEGG, NIEST and GOLM. For data processing there are some free web tools such as offline software METLIN and

Metaboanalyst, and for plant based metabolites the important data processing tools include Met-Align, MET-XAlign, ChromaTOF and MET-COFEA (Kumar et al. 2017). In addition to various statistical tools like statistical analysis tools, Cytoscape and MetaboAnalyst help in simple statistical analysis like principle component analysis (PCA), heatmap, partial least squares (PLS), boxplots, K-means clustering and reconstructing metabolic pathways (Xie et al. 2015) (Figure 13.6). The availability of these tools helps in the analysis of various metabolites (primary and secondary) under various growth and stress conditions. In systems biology, metabolite profiling aims to provide a comprehensive understanding of plant responses to genetic and environmental perturbations (Saito and Matsuda 2010).

Various research areas, such as genomics, functional genomics, molecular plant breeding and biotechnology, have been supported through the application of metabolomics (Guijas et al. 2018). Understanding the genes involved in metabolism and dissecting metabolic pathways are important for improving plant adaptation to biotic and abiotic stress, for making food quality better, and for increasing crop yield, all being essential factors for securing global human nutrition (Gong et al. 2013). Since metabolomics directly measures the biochemical activities by monitoring substrate and their transformed products, metabolomics studies easily relate to phenotypes as compared to genomics, transcriptomics and proteomics studies. The reason is that the functions of genomic regions, expressed sequences and proteins can go through epigenetic regulation, post-translational modification (Gong et al. 2013) and activity impairment. Phenotypes of different metabolites in crops can be holistically explored (Fernie and Schauer 2009). Metabolomic quantitative trait loci (mQTLs) mapping and metabolic genome-wide association studies (mGWAS) are vital strategies for identifying genetic variants of traits linked with metabolism (Gong et al. 2013). Application of GWAS has made it easy to understand the genetic basis of complex traits like grain quality (Fiaz et al. 2019).

Molecular markers and mQTL mapping studies help in identifying candidate genes and associated genomic regions. Metabolic markers are very important for uncovering and investigating complex biological pathways liable for distinctive phenotypes (Fernandez et al. 2016). mQTLs on the other hand assist in connecting genomes and metabolomes, and in providing greater insight into genetic function. It also helps in investigating variation in phenotypes by metabolite profiling and through comprehensive analysis of gene expression (Wen et al. 2016). For candidate gene discovery, advances in genomic techniques through high-density maps have enabled mQTL detections (Scossa et al. 2016). Various candidate genes regulating biosynthesis of metabolites have been identified by multi-omics approaches via forward and reverse genetics methods (Beleggia et al. 2016).

On the other hand, metabolomic analysis also faces some challenges, as in plants more diversity of metabolites is found as compared to other organisms, so it makes it difficult to find the actual size of plant metabolomes (Dixon et al. 2006). Chemical diversity of metabolites, the relationship of each metabolite to the matrix and the broad dynamic range of the abundance of a given metabolite are the problems associated with metabolomics. Currently, no single technique is available for covering the entire metabolome. So, in order to achieve adequate metabolite coverage, it is necessary that various extraction techniques as well as analytical methods should be combined (Johnson, Ivanisevic, and Siuzdak 2016).

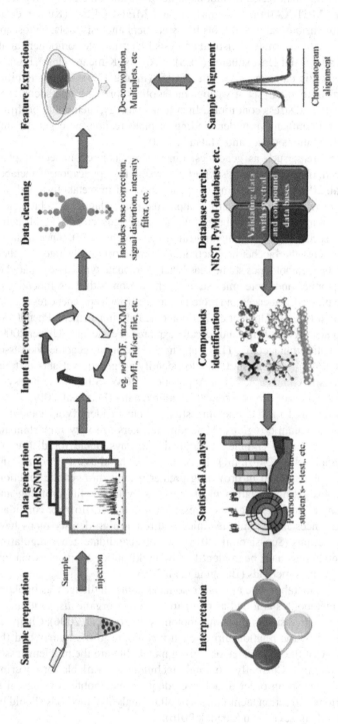

FIGURE 13.6 Schematic representation of the process of high-throughput data analysis. After converting into desired formats, raw data files are read. In order to remove false-positive results, background correction and noise reduction is done to clean up the input files. For differentiating individual peaks from the overlapped ones, the use of feature extraction is done. By the analysis of spectra and chemical structure available in databases or the metabolomics library, compound identification is done (Kumar et al. 2017).

13.3.1.1 Role of Metabolomics in Improving Cereal Crops Traits

For quantifying variation in metabolites, cereals have been broadly studied (W. Chen et al. 2016). In order to overcome the problems of insects, pests, disease and abiotic stress (e.g., drought, salt, temperature) responsible for crop yield reduction, significant progress has been made in biotechnology. In the study of rice, metabolomics proved to be very useful for determining undesirable and unexpected accumulation of compounds in genetically modified rice (Oikawa et al. 2008). Different research groups make use of metabolomics for exploring metabolic diversity between various varieties of rice and their natural variants (Hu et al. 2016; Okazaki and Saito 2016). Likewise, in maize, studies of metabolome helped researchers in differentiating and consequently selecting the superior genotypes containing high nutrition content (Wen et al. 2014; Venkatesh et al. 2016). A metabolomic approach was also being used for surveying chemical diversity between various rice and maize varieties and their natural variants (W. Chen et al. 2016). It has been reported that in maize, amino acid metabolism plays a role in regulating drought stress (Obata et al. 2015). Under drought stress, photorespiration is being tightly regulated due to the upregulation of serine and glycine amino acids. Moreover, it is reported that the buildup of myo-inositol and glycine under drought stress has some relation with grain size of maize, so for the identification of drought-tolerant maize, these metabolites could be considered as potential markers (Obata et al. 2015). Similarly in rice-tolerant plants, some compounds such as glucose, galactaric acid, gluconic acid, allantoin, salicylic acid and glucopyranoside were drastically induced, which could serve as metabolite markers to deal with drought stress in rice plants (Karthik et al. 2014). For coping with drought stress, significant variation in metabolic status was observed among different species of sorghum. Samsorg 17, a highly drought-tolerant variety, accumulated sugars and sugar alcohols, whereas free amino acids accumulated in the Samsorg 40 variety having less tolerance towards drought (Ogbaga et al. 2016). Similarly under salinity stress, marked accumulation of amino acids as well as soluble sugars were found in the barley roots of tolerant variety (Shelden et al. 2016). Induction of amino acids and carbohydrates is also caused by chilling stress. A substantial change in metabolic profiles of rice varieties (Japonica vs. Indica) was observed under chilling stress. The Japonica variety tolerated the chilling stress by activating antioxidant pathways through modulating key metabolites such as adenine, glutamate, glycine, 5-oxoproline, γ-glutamyglutamine and adenine dinucleotide (Yi Zhang et al. 2016). Moreover this stress also causes the activation of the glycolytic pathway, though the recovery phase resumes its normal activity. Cold stress in wheat and barley expedites amino acid accumulation, and GABA-shunt genes are being induced for promoting the conversion of glutamate to GABA (Mazzucotelli et al. 2006). It is well known that flavones or flavones-glycosides accumulate in cereal grains, which provide protection against various stresses (Cassi-Lit et al. 2007). For instance, plenty of flavones-glycosides are being produced in rice for protection against herbivores and abiotic stress (Iwata et al. 2012). On the other hand, an enhancement of tryptophan, phospholipids azelaic acid and 1,3-bezoxazin-4-ones was observed in maize plants against herbivore tolerance (Marti et al. 2013). During plant pathogen interaction, metabolites related to resistance also accumulate. A large variety of metabolites, including coumaroylagmatine and coumaroyl putrescine, accumulate in the tolerant variety of wheat for conferring tolerance against fusarium head blight (Kage et al. 2017). Moreover, assessment of these compounds and their placement in the metabolic pathways helped in identifying an important gene, agmatine coumaroyl transferase (ACT) (Kumar et al. 2017).

13.3.1.2 Metabolomic Analysis of Gene-Edited Cereal Crops

The European Court of Justice (ECJ) in 2018 ruled that genome-edited crops should also be passed through the same stringent regulations as GM crops (Bobek 2018). However, according to the practitioners of the field, absence of foreign DNA in the crop should alleviate the need for lengthy regulation processes. Most countries that are interested in using this technology will develop and release a plethora of new alleles. Considering the limitations of assessment of gene editing through sequencing and potential similarity to existing alleles make it difficult to detect these changes. It is difficult to distinguish small deletions or single base changes from normal variation within a genotype. So there is a need that metabolomics to be used in a more prominent manner so that assessment of compositional changes can be possible. According to Fraser et al. (2020), metabolomics should be applied in gene-edited crops in three ways: (1) for the detection of gene-edited crops; (2) for the characterization of the unexpected changes that can affect safety, and (3) building on the track record for the regularity assessment of metabolically engineered crops. However, according to Fedorova and Herman (2020), (1) metabolomics cannot distinguish between mutations caused by genome editing or classical mutagenesis techniques; (2) application of metabolomics for searching possible compositional differences within crops using the least likely technique is not reasonable; and (3) similar to GM crops, repeating the onerous regulation approach for genome-edited crops is unlikely to have different results.

So far very few reports have been published regarding the metabolic analysis of genome-edited cereal crops. For the analysis of leaf senescence in rice plants, the LC-MS technique was used to analyze the metabolites generated in the CRISPR–Cas9-based knockout lines of the *OsR498G010613100* gene in the two cultivars of rice, and significant enrichment of genes and metabolites involved in the flavonoid pathway was found (Xue et al. 2021).

13.4 CONCLUDING REMARKS

Gene editing is an effective tool for crop improvement. CRISPR–Cas9 is a quite straightforward effective editing technique for multiple and precise modifications and is economical too. By using this technique, several traits such as higher yield, grain quality, nutritional benefits, disease resistance, and biotic and abiotic stress tolerance have been introduced in cereal crops. For the safety assessment of genetically edited crops, metabolomics is quite helpful, as it provides relevant information about the metabolites. Although metabolic profiling of plants has greatly accelerated in the last few years, chemical complexity of metabolites, as well as the dynamic range of concentrations are the major challenges faced by metabolomics. In this regard, advances in the analytical platform played a major role in addressing these challenges. As suggested by the foodomics strategy, it is necessary that the data generated by metabolomics be processed, integrated and interpreted along with the data of other high-throughput techniques such as transcriptomics and proteomics so that the biological significance and effect of detected changes can be easily understood.

REFERENCES

Aharoni, Asaph, and Gad Galili. 2011. "Metabolic engineering of the plant primary–secondary metabolism interface." *Current Opinion in Biotechnology* 22 (2): 239–244.

Alok, Anshu, Dulam Sandhya, Phanikanth Jogam, Vandasue Rodrigues, Kaushal K Bhati, Himanshu Sharma, and Jitendra Kumar. 2020. "The rise of the CRISPR/Cpf1 system for efficient genome editing in plants." *Frontiers in Plant Science* 11: 264.

Alseekh, Saleh, and Alisdair R Fernie. 2018. "Metabolomics 20 years on: what have we learned and what hurdles remain?" *The Plant Journal* 94 (6): 933–942.

Ansari, Waquar A, Sonali U Chandanshive, Vacha Bhatt, Altafhusain B Nadaf, Sanskriti Vats, Jawahar L Katara, Humira Sonah, and Rupesh Deshmukh. 2020. "Genome editing in cereals: approaches, applications and challenges." *International Journal of Molecular Sciences* 21 (11): 4040.

Beddington, John R, Mohammed Asaduzzaman, Fernandez A Bremauntz, Megan E Clark, Marion Guillou, Molly M Jahn, Lin Erda, Tekalign Mamo, Nguyen Van Bo, and Carlos A Nobre. 2011. "Achieving food security in the face of climate change: summary for policy makers from the Commission on Sustainable Agriculture and Climate Change."

Beleggia, Romina, Domenico Rau, Giovanni Laidò, Cristiano Platani, Franca Nigro, Mariagiovanna Fragasso, Pasquale De Vita, Federico Scossa, Alisdair R Fernie, and Zoran Nikoloski. 2016. "Evolutionary metabolomics reveals domestication-associated changes in tetraploid wheat kernels." *Molecular Biology and Evolution* 33 (7): 1740–1753.

Bobek. 2018. "Position paper on the opinion of advocate general Bobek delivered on 18th January 2018 in case C-528/16."

Cassi-Lit, Merdelyn T, Gregory J Tanner, Murali Nayudu, and Malcolm Whitecross. 2007. "Isovitexin-2'-O-β-[6-O-E-p-coumaroylglucopyranoside] from UV-B irradiated leaves of rice, *Oryza sativa* L. inhibits fertility of *Helicoverpa armigera*." *Photochemistry and photobiology, 83*(5), 1167–1173.

Chen, Wei, Yanqiang Gao, Weibo Xie, Liang Gong, Kai Lu, Wensheng Wang, Yang Li, Xianqing Liu, Hongyan Zhang, and Huaxia Dong. 2014. "Genome-wide association analyses provide genetic and biochemical insights into natural variation in rice metabolism." *Nature Genetics* 46 (7): 714–721.

Chen, Wei, Wensheng Wang, Meng Peng, Liang Gong, Yanqiang Gao, Jian Wan, Shouchuang Wang, Lei Shi, Bin Zhou, and Zongmei Li. 2016. "Comparative and parallel genome-wide association studies for metabolic and agronomic traits in cereals." *Nature Communications* 7 (1): 1–10.

Christian, Michelle, Tomas Cermak, Erin L Doyle, Clarice Schmidt, Feng Zhang, Aaron Hummel, Adam J Bogdanove, and Daniel F Voytas. 2010. "Targeting DNA double-strand breaks with TAL effector nucleases." *Genetics* 186 (2): 757–761.

Connorton, James M, Eleanor R Jones, Ildefonso Rodríguez-Ramiro, Susan Fairweather-Tait, Cristobal Uauy, and Janneke Balk. 2017. "Wheat vacuolar iron transporter TaVIT2 transports Fe and Mn and is effective for biofortification." *Plant Physiology* 174 (4): 2434–2444.

Dixon, Richard A, David R Gang, Adrian J Charlton, Oliver Fiehn, Harry A Kuiper, Tracey L Reynolds, Ronald S Tjeerdema, Elizabeth H Jeffery, J Bruce German, and William P Ridley. 2006. "Applications of metabolomics in agriculture." *Journal of Agricultural and Food Chemistry* 54 (24): 8984–8994.

FAO. 2009. "Proceedings of the expert meeting on how to feed the world in 2050."

Fedorova, Maria, and Rod A Herman. 2020. "Obligatory metabolomic profiling of gene-edited crops is risk disproportionate." *The Plant Journal* 103 (6): 1985–1988.

Fernandez, Olivier, Maria Urrutia, Stéphane Bernillon, Catherine Giauffret, François Tardieu, Jacques Le Gouis, Nicolas Langlade, Alain Charcosset, Annick Moing, and Yves Gibon. 2016. "Fortune telling: metabolic markers of plant performance." *Metabolomics* 12 (10): 1–14.

Fernie, Alisdair R, and Nicolas Schauer. 2009. "Metabolomics-assisted breeding: a viable option for crop improvement?" *Trends in Genetics* 25 (1): 39–48.

Fiaz, Sajid, Shakeel Ahmad, Mehmood Ali Noor, Xiukang Wang, Afifa Younas, Aamir Riaz, Adeel Riaz, and Fahad Ali. 2019. "Applications of the CRISPR/Cas9 system for rice grain quality improvement: perspectives and opportunities." *International Journal of Molecular Sciences* 20 (4): 888.

Foley, Jonathan A, Navin Ramankutty, Kate A Brauman, Emily S Cassidy, James S Gerber, Matt Johnston, Nathaniel D Mueller, Christine O'Connell, Deepak K Ray, and Paul C West. 2011. "Solutions for a cultivated planet." *Nature* 478 (7369): 337–342.

Fraser, Paul D, Asaph Aharoni, Robert D Hall, Sanwen Huang, James J Giovannoni, Uwe Sonnewald, and Alisdair R Fernie. 2020. "Metabolomics should be deployed in the identification and characterization of gene-edited crops." *The Plant Journal* 102 (5): 897–902.

Frederich, Michel, Bernard Pirotte, Marianne Fillet, and Pascal De Tullio. 2016. "Metabolomics as a challenging approach for medicinal chemistry and personalized medicine." *Journal of Medicinal Chemistry* 59 (19): 8649–8666.

Gao, Caixia. 2015. "Genome editing in crops: from bench to field." *National Science Review* 2 (1): 13–15.

———. 2018. "The future of CRISPR technologies in agriculture." *Nature Reviews Molecular Cell Biology* 19 (5): 275–276.

Gao, L., Cox, D. B., Yan, W. X., Manteiga, J. C., Schneider, M. W., Yamano, T., ... & Zhang, F. (2017). Engineered Cpf1 variants with altered PAM specificities. *Nature biotechnology, 35*(8), 789–792.

Gaudelli, Nicole M, Alexis C Komor, Holly A Rees, Michael S Packer, Ahmed H Badran, David I Bryson, and David R Liu. 2017. "Programmable base editing of A• T to G• C in genomic DNA without DNA cleavage." *Nature* 551 (7681): 464–471.

Godfray, H Charles J, John R Beddington, Ian R Crute, Lawrence Haddad, David Lawrence, James F Muir, Jules Pretty, Sherman Robinson, Sandy M Thomas, and Camilla Toulmin. 2010. "Food security: the challenge of feeding 9 billion people." *Science* 327 (5967): 812–818.

Gong, Liang, Wei Chen, Yanqiang Gao, Xianqing Liu, Hongyan Zhang, Caiguo Xu, Sibin Yu, Qifa Zhang, and Jie Luo. 2013. "Genetic analysis of the metabolome exemplified using a rice population." *Proceedings of the National Academy of Sciences* 110 (50): 20320–20325.

Guijas, Carlos, J Rafael Montenegro-Burke, Benedikt Warth, Mary E Spilker, and Gary Siuzdak. 2018. "Metabolomics activity screening for identifying metabolites that modulate phenotype." *Nature Biotechnology* 36 (4): 316–320.

Hess, Gaelen T, Josh Tycko, David Yao, and Michael C Bassik. 2017. "Methods and applications of CRISPR-mediated base editing in eukaryotic genomes." *Molecular Cell* 68 (1): 26–43.

Hu, Chaoyang, Takayuki Tohge, Shen-An Chan, Yue Song, Jun Rao, Bo Cui, Hong Lin, Lei Wang, Alisdair R Fernie, and Dabing Zhang. 2016. "Identification of conserved and diverse metabolic shifts during rice grain development." *Scientific Reports* 6 (1): 1–12.

Iwata, Hikaru, Amèlia Gaston, Arnaud Remay, Tatiana Thouroude, Julien Jeauffre, Koji Kawamura, Laurence Hibrand-Saint Oyant, Takashi Araki, Béatrice Denoyes, and Fabrice Foucher. 2012. "The TFL1 homologue KSN is a regulator of continuous flowering in rose and strawberry." *The Plant Journal* 69 (1): 116–125.

Jinek, Martin, Krzysztof Chylinski, Ines Fonfara, Michael Hauer, Jennifer A Doudna, and Emmanuelle Charpentier. 2012. "A programmable dual-RNA–guided DNA endonuclease in adaptive bacterial immunity." *Science* 337 (6096): 816–821.

Johnson, Caroline H, Julijana Ivanisevic, and Gary Siuzdak. 2016. "Metabolomics: beyond biomarkers and towards mechanisms." *Nature Reviews Molecular Cell Biology* 17 (7): 451–459.

Kage, Udaykumar, Shailesh Karre, Ajjamada C Kushalappa, and Curt McCartney. 2017. "Identification and characterization of a fusarium head blight resistance gene Ta ACT in wheat QTL-2 DL." *Plant Biotechnology Journal* 15 (4): 447–457.

Karthik, L, Gaurav Kumar, Tarun Keswani, Arindam Bhattacharyya, S Sarath Chandar, and KV Bhaskara Rao. 2014. "Protease inhibitors from marine actinobacteria as a potential source for antimalarial compound." *PloS One* 9 (3): e90972.

Kim, Hye Kyong, Young Hae Choi, and Robert Verpoorte. 2010. "NMR-based metabolomic analysis of plants." *Nature Protocols* 5 (3): 536–549.

Kim, Hyeran, Sang-Tae Kim, Jahee Ryu, Beum-Chang Kang, Jin-Soo Kim, and Sang-Gyu Kim. 2017. "CRISPR/Cpf1-mediated DNA-free plant genome editing." *Nature Communications* 8 (1): 1–7.

Kim, Yang-Gyun, Jooyeun Cha, and Srinivasan Chandrasegaran. 1996. "Hybrid restriction enzymes: zinc finger fusions to Fok I cleavage domain." *Proceedings of the National Academy of Sciences* 93 (3): 1156–1160.

Kumar, Rakesh, Abhishek Bohra, Arun K Pandey, Manish K Pandey, and Anirudh Kumar. 2017. "Metabolomics for plant improvement: status and prospects." *Frontiers in Plant Science* 8: 1302.

Labadie, Marc, Guillaume Vallin, Aurélie Petit, Ludwig Ring, Thomas Hoffmann, Amèlia Gaston, Aline Potier, Wilfried Schwab, Christophe Rothan, and Béatrice Denoyes. 2020. "Metabolite quantitative trait loci for flavonoids provide new insights into the genetic architecture of strawberry (Fragaria × ananassa) fruit quality." *Journal of Agricultural and Food Chemistry* 68 (25): 6927–6939.

Lawrenson, Tom, Oluwaseyi Shorinola, Nicola Stacey, Chengdao Li, Lars Østergaard, Nicola Patron, Cristobal Uauy, and Wendy Harwood. 2015. "Induction of targeted, heritable mutations in barley and Brassica oleracea using RNA-guided Cas9 nuclease." *Genome Biology* 16 (1): 1–13.

Li, Jian-Feng, Julie E Norville, John Aach, Matthew McCormack, Dandan Zhang, Jenifer Bush, George M Church, and Jen Sheen. 2013. "Multiplex and homologous recombination–mediated genome editing in Arabidopsis and *Nicotiana benthamiana* using guide RNA and Cas9." *Nature Biotechnology* 31(8): 688–691.

Li, Jun, Xiangbing Meng, Yuan Zong, Kunling Chen, Huawei Zhang, Jinxing Liu, Jiayang Li, and Caixia Gao. 2016. "Gene replacements and insertions in rice by intron targeting using CRISPR–Cas9." *Nature Plants* 2 (10): 1–6.

Li, Kun, Weiwei Wen, Saleh Alseekh, Xiaohong Yang, Huan Guo, Wenqiang Li, Luxi Wang, Qingchun Pan, Wei Zhan, and Jie Liu. "Large-scale metabolite QTL analysis provides new insights for high-quality maize improvement." *The Plant Journal*, 99(2), 216–230.

Liu, Fang, Pandi Wang, Nini Tian, Xiaojuan Xiong, Ping Fu, and Gang Wu. 2018. "LC-MS based secondary metabolic profile of BraLTP2 overexpressing Brassica napus." *Oil Crop Science* 3 (3): 141.

Macovei, Anca, Neah R Sevilla, Christian Cantos, Gilda B Jonson, Inez Slamet-Loedin, Tomáš Čermák, Daniel F Voytas, Il-Ryong Choi, and Prabhjit Chadha-Mohanty. 2018. "Novel alleles of rice eIF4G generated by CRISPR/Cas9-targeted mutagenesis confer resistance to rice tungro spherical virus." *Plant Biotechnology Journal* 16 (11): 1918–1927.

Marti, Guillaume, Matthias Erb, Julien Boccard, Gaetan Glauser, Gwladys R Doyen, Neil Villard, Christelle AM Robert, Ted CJ Turlings, Serge Rudaz, and Jean-Luc Wolfender. 2013. "Metabolomics reveals herbivore-induced metabolites of resistance and susceptibility in maize leaves and roots." *Plant, Cell & Environment* 36 (3): 621–639.

Matres, Jerlie Mhay, Julia Hilscher, Akash Datta, Victoria Armario-Nájera, Can Baysal, Wenshu He, Xin Huang, Changfu Zhu, Rana Valizadeh-Kamran, and Kurniawan R Trijatmiko. 2021. "Genome editing in cereal crops: an overview." *Transgenic Research* 30 (4): 461–498.

Mazzucotelli, Elisabetta, Alfredo Tartari, Luigi Cattivelli, and Giuseppe Forlani. 2006. "Metabolism of γ-aminobutyric acid during cold acclimation and freezing and its relationship to frost tolerance in barley and wheat." *Journal of Experimental Botany* 57 (14): 3755–3766.

Najera, Victoria Armario, Richard M Twyman, Paul Christou, and Changfu Zhu. 2019. "Applications of multiplex genome editing in higher plants." *Current Opinion in Biotechnology* 59: 93–102.

Obata, Toshihiro, and Alisdair R Fernie. 2012. "The use of metabolomics to dissect plant responses to abiotic stresses." *Cellular and Molecular Life Sciences* 69 (19): 3225–3243.

Obata, Toshihiro, Sandra Witt, Jan Lisec, Natalia Palacios-Rojas, Igor Florez-Sarasa, Salima Yousfi, Jose Luis Araus, Jill E Cairns, and Alisdair R Fernie. 2015. "Metabolite profiles of maize leaves in drought, heat, and combined stress field trials reveal the relationship between metabolism and grain yield." *Plant Physiology* 169 (4): 2665–2683.

Ogbaga, Chukwuma C, Piotr Stepien, Beth C Dyson, Nicholas JW Rattray, David I Ellis, Royston Goodacre, and Giles N Johnson. 2016. "Biochemical analyses of sorghum varieties reveal differential responses to drought." *PloS One* 11 (5): e0154423.

Oikawa, Akira, Fumio Matsuda, Miyako Kusano, Yozo Okazaki, and Kazuki Saito. 2008. "Rice metabolomics." *Rice* 1 (1): 63–71.

Okazaki, Yozo, and Kazuki Saito. 2016. "Integrated metabolomics and phytochemical genomics approaches for studies on rice." *GigaScience* 5 (1): s13742-016-0116-7.

Oliver, Stephen G, Michael K Winson, Douglas B Kell, and Frank Baganz. 1998. "Systematic functional analysis of the yeast genome." *Trends in Biotechnology* 16 (9): 373–378.

Pott, Delphine M, Sonia Osorio, and José G Vallarino. 2019. "From central to specialized metabolism: an overview of some secondary compounds derived from the primary metabolism for their role in conferring nutritional and organoleptic characteristics to fruit." *Frontiers in Plant Science* 10: 835.

Prado, Jose Rafael, Gerrit Segers, Toni Voelker, Dave Carson, Raymond Dobert, Jonathan Phillips, Kevin Cook, Camilo Cornejo, Josh Monken, and Laura Grapes. 2014. "Genetically engineered crops: from idea to product." *Annual Review of Plant Biology* 65: 769–790.

Puchta, Holger. 2005. "The repair of double-strand breaks in plants: mechanisms and consequences for genome evolution." *Journal of Experimental Botany* 56 (409): 1–14.

Ricroch, Agnès, Pauline Clairand, and Wendy Harwood. 2017. "Use of CRISPR systems in plant genome editing: toward new opportunities in agriculture." *Emerging Topics in Life Sciences* 1 (2): 169–182.

Ricepedia. 2020. "The global staple." *Ricepedia.* http://ricepedia.org/rice-as-food/the-global-staple-rice-consumers#.

Saito, Kazuki, and Fumio Matsuda. 2010. "Metabolomics for functional genomics, systems biology, and biotechnology." *Annual Review of Plant Biology* 61: 463–489.

Sarwar, Muhammad Haroon, Muhammad Farhan Sarwar, Muhammad Sarwar, Niaz Ahmad Qadri, and Safia Moghal. 2013. "The importance of cereals (Poaceae: Gramineae) nutrition in human health: a review." *Journal of Cereals and Oilseeds* 4 (3): 32–35.

Scossa, Federico, Yariv Brotman, Francisco de Abreu e Lima, Lothar Willmitzer, Zoran Nikoloski, Takayuki Tohge, and Alisdair R Fernie. 2016. "Genomics-based strategies for the use of natural variation in the improvement of crop metabolism." *Plant Science* 242: 47–64.

Shelden, Megan C, Daniel A Dias, Nirupama S Jayasinghe, Antony Bacic, and Ute Roessner. 2016. "Root spatial metabolite profiling of two genotypes of barley (*Hordeum vulgare* L.) reveals differences in response to short-term salt stress." *Journal of Experimental Botany* 67 (12): 3731–3745.

Shi, Jinrui, Huirong Gao, Hongyu Wang, H Renee Lafitte, Rayeann L Archibald, Meizhu Yang, Salim M Hakimi, Hua Mo, and Jeffrey E Habben. 2017. "ARGOS 8 variants generated by CRISPR-Cas9 improve maize grain yield under field drought stress conditions." *Plant Biotechnology Journal* 15 (2): 207–216.

Sun, Yongwei, Xin Zhang, Chuanyin Wu, Yubing He, Youzhi Ma, Han Hou, Xiuping Guo, Wenming Du, Yunde Zhao, and Lanqin Xia. 2016. "Engineering herbicide-resistant rice plants through CRISPR/Cas9-mediated homologous recombination of acetolactate synthase." *Molecular Plant* 9 (4): 628–631.

Tilman, David, Christian Balzer, Jason Hill, and Belinda L Befort. 2011. "Global food demand and the sustainable intensification of agriculture." *Proceedings of the National Academy of Sciences* 108 (50): 20260–20264.

Tohge, Takayuki, Federico Scossa, and Alisdair R Fernie. 2015. "Integrative approaches to enhance understanding of plant metabolic pathway structure and regulation." *Plant Physiology* 169 (3): 1499–1511.

Valdés, Alberto, Carolina Simó, Clara Ibáñez, and Virginia García-Cañas. 2013. "Foodomics strategies for the analysis of transgenic foods." *TrAC Trends in Analytical Chemistry* 52: 2–15.

Venkatesh, Tyamagondlu V, Alexander W Chassy, Oliver Fiehn, Sherry Flint-Garcia, Qin Zeng, Kirsten Skogerson, and George G Harrigan. 2016. "Metabolomic assessment of key maize resources: GC-MS and NMR profiling of grain from B73 hybrids of the nested association mapping (NAM) founders and of geographically diverse landraces." *Journal of Agricultural and Food Chemistry* 64 (10): 2162–2172.

Vickers, Neil J. 2017. "Animal communication: when i'm calling you, will you answer too?" *Current Biology* 27 (14): R713–R715.

Wang, Meng, Shubin Wang, Zhen Liang, Weiming Shi, Caixia Gao, and Guangmin Xia. 2018. "From genetic stock to genome editing: gene exploitation in wheat." *Trends in Biotechnology* 36 (2): 160–172.

Wang, Wei, Qianli Pan, Fei He, Alina Akhunova, Shiaoman Chao, Harold Trick, and Eduard Akhunov. 2018. "Transgenerational CRISPR-Cas9 activity facilitates multiplex gene editing in allopolyploid wheat." *The CRISPR Journal* 1 (1): 65–74.

Wen, Weiwei, Min Jin, Kun Li, Haijun Liu, Yingjie Xiao, Mingchao Zhao, Saleh Alseekh, Wenqiang Li, Francisco de Abreu e Lima, and Yariv Brotman. 2018. "An integrated multi-layered analysis of the metabolic networks of different tissues uncovers key genetic components of primary metabolism in maize." *The Plant Journal* 93 (6): 1116–1128.

Wen, Weiwei, Dong Li, Xiang Li, Yanqiang Gao, Wenqiang Li, Huihui Li, Jie Liu, Haijun Liu, Wei Chen, and Jie Luo. 2014. "Metabolome-based genome-wide association study of maize kernel leads to novel biochemical insights." *Nature Communications* 5 (1): 1–10.

Wen, Weiwei, Haijun Liu, Yang Zhou, Min Jin, Ning Yang, Dong Li, Jie Luo, Yingjie Xiao, Qingchun Pan, and Takayuki Tohge. 2016. "Combining quantitative genetics approaches with regulatory network analysis to dissect the complex metabolism of the maize kernel." *Plant Physiology* 170 (1): 136–146.

Xie, Li-Juan, Qin-Fang Chen, Mo-Xian Chen, Lu-Jun Yu, Li Huang, Liang Chen, Feng-Zhu Wang, Fan-Nv Xia, Tian-Ren Zhu, and Jian-Xin Wu. 2015. "Unsaturation of very-long-chain ceramides protects plant from hypoxia-induced damages by modulating ethylene signaling in Arabidopsis." *PLoS Genetics* 11 (3): e1005143.

Xue, Jiao, Dongbai Lu, Shiguang Wang, Zhanhua Lu, Wei Liu, Xiaofei Wang, Zhiqiang Fang, and Xiuying He. 2021. "Integrated transcriptomic and metabolomic analysis provides insight into the regulation of leaf senescence in rice." *Scientific Reports* 11 (1): 1–12.

Yang, Bei, Xiaosa Li, Liqun Lei, and Jia Chen. 2017. "APOBEC: from mutator to editor." *Journal of Genetics and Genomics* 44 (9): 423–437.

Yang, Li, Kui-Shan Wen, Xiao Ruan, Ying-Xian Zhao, Feng Wei, and Qiang Wang. 2018. "Response of plant secondary metabolites to environmental factors." *Molecules* 23 (4): 762.

Zafar, Kashaf, Khalid EM Sedeek, Gundra Sivakrishna Rao, Muhammad Zuhaib Khan, Imran Amin, Radwa Kamel, Zahid Mukhtar, Mehak Zafar, Shahid Mansoor, and Magdy M Mahfouz. 2020. "Genome editing technologies for rice improvement: progress, prospects, and safety concerns." *Frontiers in Genome Editing* 2: 5.

Zaib, Sania, Imtiaz Ahmad, and Samina N Shakeel. 2020. "Modulation of barley (*Hordeum vulgare*) defense and hormonal pathways by *Pseudomonas* species accounted for salinity tolerance." *Pakistan Journal of Agricultural Sciences* 57 (6), 1469–1481.

Zetsche, Bernd, Jonathan S Gootenberg, Omar O Abudayyeh, Ian M Slaymaker, Kira S Makarova, Patrick Essletzbichler, Sara E Volz, Julia Joung, John Van Der Oost, and Aviv Regev. 2015. "Cpf1 is a single RNA-guided endonuclease of a class 2 CRISPR-Cas system." *Cell* 163 (3): 759–771.

Zhang, Dandan, Zhenxiang Li, and Jian-Feng Li. 2016. "Targeted gene manipulation in plants using the CRISPR/Cas technology." *Journal of Genetics and Genomics* 43 (5): 251–262.

Zhang, Yangfan, Mark P Polinski, Phillip R Morrison, Colin J Brauner, Anthony P Farrell, and Kyle A Garver. 2019. "High-load reovirus infections do not imply physiological impairment in salmon." *Frontiers in Physiology* 10: 114.

Zhang, Yi, Da Li, Dingbo Zhang, Xiaoge Zhao, Xuemin Cao, Lingli Dong, Jinxing Liu, Kunling Chen, Huawei Zhang, and Caixia Gao. 2018. "Analysis of the functions of Ta GW 2 homoeologs in wheat grain weight and protein content traits." *The Plant Journal* 94 (5): 857–866.

Zhang, Yi, Zhen Liang, Yuan Zong, Yanpeng Wang, Jinxing Liu, Kunling Chen, Jin-Long Qiu, and Caixia Gao. 2016. "Efficient and transgene-free genome editing in wheat through transient expression of CRISPR/Cas9 DNA or RNA." *Nature Communications* 7 (1): 1–8.

Zhang, Yi, Karen Massel, Ian D Godwin, and Caixia Gao. 2018. "Applications and potential of genome editing in crop improvement." *Genome Biology* 19 (1): 1–11.

Zhou, Junhui, Zhao Peng, Juying Long, Davide Sosso, Bo Liu, Joon-Seob Eom, Sheng Huang, Sanzhen Liu, Casiana Vera Cruz, and Wolf B Frommer. 2015. "Gene targeting by the TAL effector PthXo2 reveals cryptic resistance gene for bacterial blight of rice." *The Plant Journal* 82 (4): 632–643.

14 Multiplexed Genome Editing in Cereals
Trait Improvement Using CRISPR/Cas9

Fozia Saeed, Sherien Bukhat, Tariq Shah,
Sumaira Rasul, Habib-ur-Rehman Athar,
and Hamid Manzoor

CONTENTS

14.1 INTRODUCTION

Genetic diversity is critical for the development of new plant varieties. Gene diversification for improved genetic architecture of crops has been performed for years through traditional plant breeding techniques. Traditional plant breeding started with the time-consuming selection of naturally occurring variants, but the recent developments in genome editing techniques have superseded the impediments of traditional breeding approaches and commenced a new era of plant genetic improvement. The sophistication of genome editing has significantly increased during the last few years due to the emergence of engineered site-specific nucleases (SSNs), which allow the introduction of transgenes at predetermined locations in the genome. These SSNs like zinc-finger nucleases (ZFNs), transcriptional activator-like effector nucleases (TALENS), and the clustered regularly interspaced short palindromic repeats (CRISPR)-associated endonuclease Cas9 (CRISPR/Cas9) produce a double-strand break (DSB) within the target DNA, which is usually corrected by the cell's own repair mechanism of non-homologous end joining (NHEJ) or homologous recombination (HR) (Miglani 2017). NHEJ is an error-prone mechanism and leaves short insertion and deletions that result in the efficient knockout of targeted genes. Whereas, the HR pathway results in more precise gene insertion or gene replacement events (Zhu et al. 2017). Gene editing techniques that can manipulate multiple targets are extremely useful. The great advantage of multiplex genome editing over traditional laborious and complex breeding programs is that it can target various sites simultaneously, resulting in novel lines with multiple mutations in one generation. Several different methods have been developed to manipulate the multiple genes in plants and the emergence of these SSNs has enabled the same technique to be used in crop improvement. For this purpose, the CRISPR/Cas9 system is a game-changing tool for crop genome editing to produce new safer genotypes. Multiplexed CRISPR technologies, which allow for the simultaneous expression of multiple gRNAs or Cas enzymes, each of which targets a different site, facilitates the powerful biological engineering applications (Cong et al. 2013a).

Cereals are a staple food in our daily intake supplying over 20% of all calorie consumption by global population. Therefore, cereals are most important to global food security. Maintaining a constant supply of cereals while enhancing their nutritional quality and combating climate change is difficult and demands the use of a variety of innovative agriculture breeding programs. Multiplex CRISPR-mediated genome editing is a new innovation with wide-ranging applications, acquiring new opportunities to advance novel varieties with improved yield and superiority. This chapter discusses the fundamentals of CRISPR/Cas9 tool and its implementation in multiplex genome editing. This chapter also presents the recent progress in the CRISPR/Cas9-based multiplex genome editing technique to develop novel traits for cereal crops with enhanced yield, stress tolerance, and better nutritional quality. Details on the possible challenges and limitations in implementing this system are also discussed.

14.2   MULTIPLEX GENOME EDITING: AN OVERVIEW

Multiplex genome editing (MGE) refers to the process using nucleases to precisely and simultaneously introduce two or more DSBs at target sites within the same genome in a single-round of mutagenesis (Cong et al. 2013b). Multiplex genome editing involves rapid desired change in distinct genes or quantitative trait loci (QTLs), leading to the formation of novel plant genetic variants with multiple mutations. The ability to make such genomic variations will permit scientists to better understand and investigate relationships between members of a gene family with duplicative functions, as well as examine epistatic interactions between strongly linked genetic pathways (Xing et al. 2014). Moreover, these multiple genetic edits or modifications may provide a practical framework for metabolic pathway engineering in crops. Numerous favorable crop developments can be facilitated by multiplex genome editing, such as the development of high-yielding agricultural crops and improved resistance to diseases (Gao et al. 2020; Mubarik et al. 2021), as well as the ability to avoid extensive separation in vegetatively propagating crops like sugarcane (Kannan et al. 2018). Unlike conventional random mutation, multiplexed genome-engineering technologies are adapted to precisely target particular genetic loci and generate the desired variations or mutations (Belhaj et al. 2015). Previously, it took years and many cycles of crossing and screening to accomplish specific mutated genes in an elite germplasm. Furthermore, conventional breeding has often become more complex due to linkage pull that hinders the transfer of genes. For example, creating high-yielding genetic makeups with superior quality traits have been regarded as a challenging task for plant breeders who use traditional plant breeding methods. Multiplexed genome editing methods have become more user friendly, especially since the emergence and adaptation of the CRISPR/Cas system (Cong et al. 2013a). Now, researchers can construct patterned genetic circuits that govern cellular functions or regulate metabolic processes by simultaneously editing, activating, and downregulating multitarget genes by producing different gRNAs and Cas protein in situ (Nielsen and Voigt 2014; Jakočiūnas et al. 2015; Gander et al. 2017).

14.3 DEVELOPMENT OF VECTORS FOR MULTIPLEXING

Numerous implementations, including genome editing to rewire metabolic processes and promoter engineering to introduce quantitative traits, necessitate simultaneous editing of multiple locations in a single targeted site or multiple targeted sites in a genome. It may be easy to design a single gRNA that targets multiple sites when various units of a gene family are targeted. Genome engineering, on the other hand, frequently involves many different genes from different families with no highly conserved (Lowder et al. 2015) regions in common. This necessarily requires a different approach known as multiplex CRISPR-based genome engineering, which involves simultaneously editing multitarget sites with multiple gRNAs. Individual gRNA transcription units powered either by U3 or U6 promoters can be stacked together in one construct to accomplish multiplex genome editing (Ma et al. 2015). However, using a U3 or U6 promoter in the same construct multiple times can cause gRNA expression levels to fluctuate and transgene suppression in plants (Ma et al. 2015). Furthermore, having a lot of gRNA transcription units may make cloning more challenging to the use of the same promoters and terminators over and over again. Moreover, if virus-based vectors are being used to deliver constructs, packaging all of the components becomes difficult due to the vector's cargo limit (Ali et al. 2015; Cody et al. 2017). As a result, more compact multiplex approaches are frequently chosen.

There are two types of compact multiplex CRISPR systems currently available: (1) single transcriptional unit (STU) multiplex systems (Figure 14.1) and (2) two transcriptional unit (TTU) multiplex systems (Figure 14.2). The TTU multiplex structure is further classified into two categories: (a) combined dual-promoter system, in which the Cas genes are transcribed with an RNA Pol II promoter as well as the gRNAs are transcribed with an RNA Pol III promoter (Figure 14.2); (b) dual promoter systems, in which both the gRNAs and Cas genes are expressed from two separate RNA Pol II promoters. One promoter and one terminator control both the gRNAs and Cas gene in the STU multiplex CRISPR/C as system. The STU system is further categorized into two parts: (a) intron-based STU (iSTU) and (b) RNA-processing enzyme-based STU (rpeSTU). STU systems are advantageous for applications requiring inducible or tissue-specific expression, and also CRISPR-mediated gene transcriptional regulation in crops (Lowder et al. 2015; Zhong et al. 2020).

Single transcriptional unit (STU) multiplex system

(A) RNA enzyme processing dependent STU system

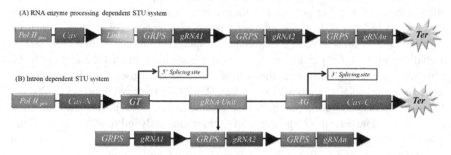

(B) Intron dependent STU system

FIGURE 14.1 Construct design of a single transcriptional unit (STU) multiplex system. Pol II refers to the RNA polymerase II promoter, while Pol III refers to the RNA polymerase III promoter. Linkers are nonfunctional DNA sequences that join two nearby functional DNA components. Abbreviations: Cas, CRISPR-related proteins; Cas-C, Cas protein's C-terminal region; Cas-N, Cas protein's N-terminal region; GRPS, guide RNA processing system, gRNA, single guide RNA; ter, terminator; pro, promoter.

Two transcriptional unit (TTU) multiplex system

(A) Single promoter

(B) Dual promoter

(C) Mixed dual promoter

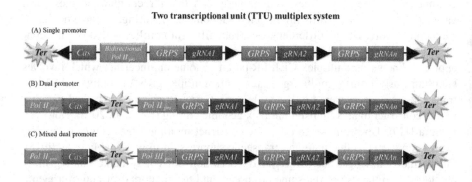

FIGURE 14.2 Construct design of a two transcriptional unit (TTU) multiplex system. Pol II refers to the RNA polymerase II promoter, while Pol III refers to the RNA polymerase III promoter; linkers are nonfunctional DNA sequences that join two nearby functional DNA components. Abbreviations: Cas, CRISPR-related proteins; Cas-C, Cas protein's C-terminal region; Cas-N, Cas protein's N-terminal region; GRPS, guide RNA processing system; gRNA, single guide RNA; ter, terminator; pro, promoter.

A multiplex CRISPR system's success is dependent on the precise handling and release of single gRNAs from a polycistronic transcript. Post-transcriptionally, polycistronic mRNA consisting of multiple gRNAs could be processed into the single gRNAs through RNA-cleaving enzymes. Several RNA-cleaving enzymes, including tRNA processing enzymes, CRISPR-associated RNA endoribonuclease (Csy4) from *Pseudomonas aeruginosa*, and HH or HDV dual ribozyme (HH-HDV) were shown to work for multiplexed genome engineering. Plants have been able to express up to 24 gRNAs from a single construct, though at a lower efficiency (Stuttmann et al. 2021). Because an incorrect concentration of every gRNA–Cas complex can result in decreased editing events (Stuttmann et al. 2021), it seems that composition of the gRNA–Cas complex within cell is critical for effective genetic manipulation by using multiplex CRISPR/Cas systems. Although using the strong ubiquitous promoters makes it feasible to express gRNAs at high rates, cohabitation of various gRNAs in a cell dilutes the saturation of every gRNA–Cas complex (Wang et al. 2015; McCarty et al. 2020). As a result, while the overall concentration of the gRNA–Cas complex may stay constant, the workable potency of every gRNA–Cas complex variant might decrease. To improve the effectiveness of multiplex editing, strong ubiquitous expression of the Cas gene is needed to raise the workable potency of every gRNA–Cas complex in plant cells (Wang et al. 2015; Hajiahmadi et al. 2019). Choosing the right Cas protein, such as SpCas9 for multiplex editing (Figure 14.1); RNA polymerase II-based promoter to express multiplex gRNAs; and gRNA processing system that is based on either HH-HDV, Csy4, or tRNA enzyme is a general guideline for the design of multiplex CRISPR constructs.

14.4 CRISPR/CAS9-MEDIATED MULTIPLEX GENOME EDITING IN CEREALS

CRISPR/Cas9 is simpler and more efficient than most other genome engineering tools, because of the design's simplicity, high editing fidelity, and multiplexing capability (Zhu et al. 2020). Using several sgRNAs and only a single Cas protein facilitates and improves the editing of various genes. The simplicity of plasmid development in CRISPR/Cas9 trials allows for the investigation of the role of several related genes. Efforts have been made to strengthen the CRISPR/Cas9 process in order to simplify and improve the effectiveness of engineering multiple genes. The primary goal has been to create a simple and quick toolbox for inserting several sgRNA-expression cassettes into a single binary vector. To insert a differential expression cassette containing sgRNAs for multiple targets in one construct, several rounds of a routine cloning procedure can be used, but this is time-consuming. Golden Gate cloning is a more sophisticated cloning process that relies on using type-IIS restriction endonuclease in DNA assembly. Non-palindromic sticky ends in plasmid DNA carrying sgRNAs have been generated using Golden Gate cloning. The Multiplex CRISPR/Cas9 system has been frequently used to produce knockout mutants, gene replacement, gene corrections, prime editing, and base editing in different cereal crops (Ma et al. 2015; Sun et al. 2016; Huang et al. 2018; Li et al. 2018; Wang, Simmonds et al. 2018). In one early example, two gRNAs were used for simultaneous modification of two targets in the acetolactate synthase gene (*ALS1*) in rice, resulting in the effective amino acid substitutions *S627* and *W548L* (Sun et al. 2016). The rice genes *OsGSTU*, *OsMRP15*, and *OsAnP*, which are involved in anthocyanin accumulation and transport, were all mutated at the same time in a purple-leafed rice line to produce genetic variants having green leaves (Ma et al. 2015). Qi et al. (2016) adapted and introduced maize glycine-tRNA to design multiple tRNA-gRNA units under the regulation of the maize U6 promoter for the manufacturing of various gRNAs. This study demonstrates specific gene targeting by two separate gRNAs to use a tRNA base approach that significantly improves mutagenesis efficiency in corn. In maize, Feng et al. (2016) used a CRISPR/Cas9 construct to simultaneously mutate the three rice genes: young seedling albino (*YSA*), stromal processing peptidase (*SPP*), and rice outermost cell-specific gene 5 (*ROC5*). Experiment results have revealed that various applications and processes are gradually appearing. In this research field, enormous technological improvements have recently occurred, by using more diverse genetic, epigenetic, and transcriptional manipulations for future research.

14.5   GENE EDITING AND IMPROVEMENT OF CEREALS: CRISPR/CAS9 PERSPECTIVE

The agriculture sector is facing numerous problems as a result of different abiotic and biotic factors, in addition to the goal of enhancing yield and quality. Multiplexed genome editing with CRISPR/Cas9 gives scientists the ability to decode difficult biological issues by targeting multiple genes at the same time. The multiplexed CRISPR/Cas9 system has been widely used in various crop species. This includes rice, wheat, maize, soybean, barley, sorghum, potato, tomato, rapeseed, tobacco, cotton, Arabidopsis, banana, and apple. The most common application of this multiplexed genome editing tool is to create gene knockouts alleles through insertions or deletions (indels) that result in frameshift mutations and the emergence of stop codon. Among the many crops edited by the multiplex CRISPR/Cas9 system, this chapter focuses mainly on cereal crops.

14.6   IMPROVEMENT OF YIELD AND RELATED TRAITS

Cereals are the most important food crop worldwide. High-yielding ability is a most essential trait for cereal breeding programs and crop quality traits drive farmers' adoption of new genotypes. High-yielding cultivars with better quality are now more appealing to consumers, which increases the marketing value for the important cereal crop. Therefore, multiple QTLs must be introduced into an elite genotype to generate a novel genetic variant in terms of plant quality, yield, and nutritive values with the aim of securing the food supply. Numerous multiplex genome editing studies have been successfully conducted to change the yield-attributing characteristics and nutritional properties in commercially important cereal crops. An excellent example of multiplexed genome editing influencing cereal development targeted 11 members of a rice FT-like gene family of a floral regulator, with one construct constituting eight gRNAs but the other comprising three. Mutations have been found in ten genes, with the most simultaneous alterations in one line being seven, giving rise to an early maturing genetic makeup (Ma et al. 2015). Xu et al. (2016) used CRISPR/Cas9-based multiplex editing of grain width 2 (*GW2*, *GW5*, *GW6*, and *GS3*) negative regulators, which encode grain mass, to significantly improve grain size and weight (Xu et al. 2016). Increases of 8.47% and 11.69% were seen in the *gw5tgw6* double mutant in grain length and grain width, respectively, and total grain width all increased by 12.68%. Also, four *gw2gw5tgw6* triples mutants demonstrated an increase of 20% to 30% in the very same parameters, suggesting that genetic variations in such QTLs resulted in the expected multiplicative impacts (Xu et al. 2016). In an attempt to generate early maturing cultivars, three floral-repressor genes— *Hd2*, *Hd4*, and *Hd5*—have been mutated in rice plants (Li et al. 2017). The authors utilized five gRNAs for successful editing of single sites in *Hd5* and multiple sites in

Hd4 and *Hd2* in this experiment. Fourteen of the 18 recovered lines had mutations in all three genes, the majority of which were single nucleotide indels. According to the authors, all the triple mutants flowered up to 40 days prior to the wild-type rice plants. Lacchini et al. (2020) used multiplexed genome editing of the *OsGS3*, *OsGW2*, and *OsGn1a* genes to improve the shape of rice grains. Another recent multiplex genome engineering experiment on Nippon bare rice resulted in a combined effect of abiotic stress tolerance and yield improvement. *OsMYB30* (a cold-tolerance gene), *GS3* (a grain-size gene), and *OsPIN5b* (a panicle-length gene) were all targeted, and thus the mutants showed increased cold tolerance, grain size, and panicle length (Zeng et al. 2019). In wheat, multiple members (100 members) of α-gliadin gene family were mutated with two different consensus gRNAs, yielding a variety of multiplex mutants, including a line with simultaneous genetic variations in 35 specific genes (Sánchez-León et al. 2018). The resulting transgene-free, low-gluten wheat lines are used as a starting point for introducing this attribute to other elite wheat genotypes. Knockout mutants of wheat *TaGW2*, *TaMLO*, and *TaLpx-1* genes were created via multiplexed genome editing and exhibited improved seed size as well as thousand kernel weight (Wang, Pan et al. 2018). Multiple members of a *k1C* gene family, which encodes the main seed storage protein a-kafirin, were targeted with only one gRNA in sorghum, tending to result in 72 plants by 26 events of mutations, including the one with concurrent genetic variations in 11 distinct *k1C* genes and the other in the *k1C*3–9 group. The resulting (T1 as well as T2) seeds had lower levels of α-kafirin, and the preferred T2 events had substantially higher grain lysine content and protein digestibility (Huang et al. 2018). All these reports illustrating the use of CRISPR/Cas9 to obtain ideotypes through targeting multiple genes simultaneously in cereal crops is expected to grow rapidly in the future. Table 14.1 lists the targeted genes and traits in various cereal crops that have been modified using CRISPR/Cas9-based multiplexed genome editing techniques to construct new varieties with higher yield and nutrient content.

14.7 IMPROVEMENT OF CEREAL DEFENSE RESPONSES

Various biotic and abiotic factors directly threaten food and nutrition security by reducing cereal crop production. A large number of multiplexed genome engineering experiments in cereal crops have been reported, with the goal of targeting multiple environmental stresses to improve defense response (Table 14.2). The primary goal of cereal crop genetic improvement is to increase food productivity and nutritional security for the world's population boom. Among the traits targeted by the multiplexed CRISPR/Cas9 system in cereal crops is improved resistance to diseases caused by bacterial, viral, and fungal pathogens. Moreover, concerns about genome editing aimed at improving herbicide tolerance were also top of the list. Recently, Oliva et al. (2019) used a CRISPR/Cas9-based multiplex strategy to make a mutated Kitaake line of rice, and mutated lines of mega varieties IR-64 and Ciherang-Sub1 with various combinations of genetic variations in the effector-binding elements (EBEs) of *OsSWEET14*, *OsSWEET13*, and *OsSWEET11* promoters imparted resistance toward a great amount of *Xoo*-tested strains. *Xoo* strains bearing various TALE-infecting rice lines with mutations in the associated EBEs of *OsSWEET* promoter resulted in a

TABLE 14.1

Summary of the Efforts Made with Multiplex Genome-Editing to Revolutionize Crop Improvement Activities in Cereals

Plant species	Targeted genes	Gene function	Reference
Rice	FT family genes	Early maturation	(Ma et al. 2015)
	OsGW2, *OsGW5*, and *OsTGW6*	Increased grain size	(Xu et al. 2016)
	GS3, *Gn1a*, *DEP1*, and *IPA1*	Improved grain length, yield with dwarf stature	(Li et al. 2016)
	OsGS3, *OsGW2*, and *OsGn1a*	Increasing gain size	(Zhou et al. 2019)
	GS3 and *Gn1a*	Increased gain yield	(Shen et al. 2018)
	Gn1a and *OsDEP1*	Improved yield	(Huang et al. 2018)
	Group I (*PYL1–PYL6*, *PYL12*) and Group II (*PYL7–PYL11*, *PYL13*)	Improved growth and productivity	(Miao et al. 2018)
	OsPIN5b, *GS3*, and *OsMYB30*	Increased grain size, panicle length, and cold tolerance	(Zeng et al. 2019)
	GS3, *GW2*, and *GN1A*	Increased gain size	(Lacchini et al. 2020)
	GS3, *GW2*, and *Gn1a*	Increased grain size, number, width, and weight	(Zhou et al. 2019)
	FWL (FW 2.2-like) and *OsFWL4*	Increased tiller number and yield	(Gao et al. 2020)
	Multiple sites	Domestication of wild relative (*O. alta*)	(Yu et al. 2021)
Wheat	Two sites at *α-gliadin* gene	Wheat with low-gluten content	(Sánchez-León et al. 2018)
	TaGW2, *TaLpx-1*, and *TaMLO*	Increased grain weight and seed size	(Wang et al. 2018)
Durum wheat	*α-Amylase trypsin* inhibitor genes	Decreased allergen proteins	(Camerlengo et al. 2020)
Maize	*ZmIPK*, *ZmIPK1A*, and *ZmMRP4*	Synthesis of phytic acid	(Liang et al. 2014)
Barley	Two *HvPM19* paralogs	Hormone signaling	(Lawrenson et al. 2015)
Sorghum	Twenty *K1C* paralogs	Improved digestibility and protein quality	(Li et al. 2018)

suitable disease interaction in several cases. Recently, a blight resistance diagnostic kit with double- or triple-knockout mutated lines was offered to assess that which *SWEET* gene seems to be the target of a specific *Xoo* strain (Eom et al. 2019). Such rice lines with several interruptions in various S-genes caused by genome engineering strategies are possible genetic assets for fighting bacterial leaf blight diseases. Similarly, Xu et al. (2019) created mutations in two EBEs of the S genes (*OsSWEET11* and *OsSWEET14*) in rice cv. Kitaake. The mutants that resulted were stable and constantly showed broad-spectrum resistance for many bacterial leaf blight strains

TABLE 14.2
List of Notable Multitarget Genome Editing Studies for Improving Cereal Defense Responses

Plant species	Targeted genes	Gene function	Reference
Rice	*OsSWEET11* and *OsSWEET14*	Rice with broad-spectrum disease resistance	(Xu et al. 2019)
	TMS5, *Pi21*, and *Xa13*	Yield enhancement and disease resistance	(Draz et al. 2019)
	OsACC-T1, *OsCDC48-T3*, *OsALS-T1*, *OsDEP1-T2*, *OsNRT1.1B-T1*, and *OsDEP1-T1*	Resistance to herbicides	(Li et al. 2018)
	Three sgRNAs that target the *FTIP1e* and *ALS* genes	Herbicide resistance	(Shimatani et al. 2017)
	SWEET11, *SWEET13*, and *SWEET14*	Developing broad-spectrum disease resistance	(Oliva et al. 2019)
	Various loci of *OsACC* gene	Herbicide resistance	(Liu et al. 2020)
	OsPIN5b, *GS3*, and *OsMYB30*	Improved yield and cold stress tolerance	(Zeng et al. 2019)
	Various loci at *OsALS1*	Increased herbicide resistance	(Kuang et al. 2020)
Wheat	Three homologs of *EDR1*	Resistance to powdery mildew	(Zhang et al. 2017)
	TaERF3 and *TaDREB2*	Drought tolerance	(Kim et al. 2018)
Cotton	*AV2/AV1*, *AC2/AC3*, and *AC1/AC4*	Broad-spectrum resistance to cotton leaf curl virus	(Mubarik et al. 2021)
Maize	*ZmALS1* and *ZmALS2*	Herbicide tolerance	(Jiang et al. 2020)

(Oliva et al. 2019). In rice, Draz et al. (2019) created genetic variations in *Pi21*, *TMS5*, and *Xa13* genes. Triple T1 mutants exhibited thermosensitive genic male sterility (TGMS), increased blast, and bacterial blight resistance. Multiplex genome editing will open the way for long-term disease resistance development by approaching further S genes in various plant species. Recently, CRISPR/Cas9-mediated multiplexed genome editing was successfully applied to enhance resistance against cotton leaf curl viruses (CLCuVs) by simultaneously targeting the six genes of CLCuV in cotton plants (Mubarik et al. 2021). Powdery mildew, a biotrophic fungal pathogen, is amongst the most common diseases severely restricting wheat production (Draz et al. 2019). To solve the problem, three homoeologs of *TaEDR1* were simultaneously targeted by Zhang et al. (2017) to enhance the resistance against powdery mildew in Bread wheat. Herbicide tolerance is a widely used trait of cereals to achieve a long-term weed resistance. In this regard, herbicide tolerant maize variety was developed by Jiang et al. (2020) through prime-editing performed to modulate acetolactate synthase (*ZmALS1* and *ZmALS2*) genes and produce S621I/W542L double mutation lines that confer tolerance against multiple herbicides. The CRISPR/Cas9 system is being constantly improved in order to maximize the efficiency and specificity of multiplexed gene targeting. To overcome the issue of resistance to environmental stresses and fungal and bacterial disease breakdown over time, the aforementioned strategies could be used to develop tolerance by targeting multiple genes. Table 14.1 lists the targeted genes and traits in various cereal crops that have been modified using CRISPR/Cas9-mediated multiplexed genome editing technology to improve defense response and increase resistance against biotic and abiotic factors and disease resistance caused by microbial pathogens.

14.8 CHALLENGES AND OUTLOOK OF MULTIPLEXING

Multiplex CRISPR-mediated genetic manipulation is a gene regulatory tool that enables users to simultaneously activate or suppress the expression of several genes. Multiplex CRISPR/Cas9 allows scientists to reformat and reconfigure chromosomal segments according to their needs. CRISPR/Cas9-mediated multiplexed gnome editing has been used to edit a variety of traits in cereal crops, ranging from agronomically important traits to enhanced cereal quality and health and nutritional benefits for consumers. The multiplexed CRISPR/Cas9 system has also eliminated difficulties associated with the simultaneous creation of mutations into different genes and has provided immediate access toward the complex genome sequences of polyploid cereal crops like hexaploid wheat. All of these advancements made it much easier to use multiplex genome editing technologies to improve plant biology understanding by disrupting gene function and deciphering gene functions in metabolic processes.

Despite the versatility of multiplexed CRISPR/Cas9 techniques, there are significant biological/technical challenges ahead. The current challenges in creating long arrays of gRNAs and predicting how these gRNAs will act in living cells are now the most crucial aspects from an engineering perspective. One major issue in using CRISPR multiplexing to cut several genetic loci at the same time seems to be the introduction of unwanted chromosomal rearrangement. Another major issue is that as the total count of gRNAs in a cell grows, they must become competitive for a shrinking pool of endonucleases. This competition alters the efficiency of each gRNA, a phenomenon known as retroactivity. Moreover, Cas9 leads to off-target bindings as the number of targets grows (Abdelrahman et al. 2021). These obstacles can possibly be alleviated by reducing the repetitive sequences in sgRNA, using a strong promoter to drive the Cas9 genes and using a conditional gRNA construct. Furthermore, advances in basic biology and genomics, chemistry, and computation will help to alleviate the bottlenecks in CRISPR/Cas9-based multiplexing. By overcoming these limitations of CRISPR/Cas9-based multiplexing for the advancement of agriculturally important crops, the production of new and healthier food can be increased. Additionally, implementation of this technique is essential for understanding the dynamics of various metabolic pathways by exploring the enzyme functionalities, interactions between different pathways, gene regulation, and signal transduction in response to different environmental stresses. Such advanced knowledge with new modeling techniques and a transferable editing system will accelerate the development of crop varieties that resist routine transformation and help in discoveries/development of new elite crop germplasm.

REFERENCES

Abdelrahman, M., Z. Wei, J. S. Rohila and K. Zhao (2021). "Multiplex genome-editing technologies for revolutionizing plant biology and crop improvement." *Front Plant Sci* 12, 721203

Ali, Z., A. Abul-faraj, L. Li, N. Ghosh, M. Piatek, A. Mahjoub, M. Aouida, A. Piatek, N. J. Baltes, D. F. Voytas, S. Dinesh-Kumar and M. M. Mahfouz (2015). "Efficient virus-mediated genome editing in plants using the CRISPR/Cas9 system." *Mol Plant* 8(8): 1288–1291.

Belhaj, K., A. Chaparro-Garcia, S. Kamoun, N. J. Patron and V. Nekrasov (2015). "Editing plant genomes with CRISPR/Cas9." *Curr Opin Biotechnol* 32: 76–84.

Camerlengo, F., A. Frittelli, C. Sparks, A. Doherty, D. Martignago, C. Larré, R. Lupi, F. Sestili and S. Masci (2020). "CRISPR-Cas9 multiplex editing of the α-amylase/trypsin inhibitor genes to reduce allergen proteins in durum wheat." *Front Sustain Food Syst* 4: 104.

Cody, W. B., H. B. Scholthof and T. E. Mirkov (2017). "Multiplexed gene editing and protein overexpression using a tobacco mosaic virus viral vector." *Plant Physiol* 175(1): 23–35.

Cong, L., F. Ran, D. Cox, S. Lin, R. Barretto and N. Habib (2013a). "Ingeniería de genoma multiplex utilizando sistemas CRISPR/Cas." *Ciencias* 339: 819–823.

Cong, L., F. A. Ran, D. Cox, S. Lin, R. Barretto, N. Habib, P. D. Hsu, X. Wu, W. Jiang, L. A. Marraffini and F. Zhang (2013b). "Multiplex genome engineering using CRISPR/Cas systems." *Science* 339(6121): 819–823.

Draz, I. S., S. M. Esmail, M. A. E.-H. Abou-Zeid and T. A. E.-M. Essa (2019). "Powdery mildew susceptibility of spring wheat cultivars as a major constraint on grain yield." *Ann Agric Sci* 64(1): 39–45.

Eom, J.-S., D. Luo, G. Atienza-Grande, J. Yang, C. Ji, J. C. Huguet-Tapia, S. N. Char, B. Liu, H. Nguyen and S. M. Schmidt (2019). "Diagnostic kit for rice blight resistance." *Nat Biotechnol* 37(11): 1372–1379.

Feng, C., J. Yuan, R. Wang, Y. Liu, J. A. Birchler and F. Han (2016). "Efficient targeted genome modification in maize using CRISPR/Cas9 system." *J Genet Genom* 43(1): 37–43.

Gander, M. W., J. D. Vrana, W. E. Voje, J. M. Carothers and E. Klavins (2017). "Digital logic circuits in yeast with CRISPR-dCas9 NOR gates." *Nat Commun* 8(1): 1–11.

Gao, Q., G. Li, H. Sun, M. Xu, H. Wang, J. Ji, D. Wang, C. Yuan and X. Zhao (2020). "Targeted mutagenesis of the rice FW 2.2-like gene family using the CRISPR/Cas9 system reveals OsFWL4 as a regulator of tiller number and plant yield in rice." *Int J Mol Sci* 21(3), 809

Hajiahmadi, Z., A. Movahedi, H. Wei, D. Li, Y. Orooji, H. Ruan and Q. Zhuge (2019). "Strategies to increase on-target and reduce off-target effects of the CRISPR/Cas9 system in plants." *Int J Mol Sci* 20(15), 3719

Huang, L., R. Zhang, G. Huang, Y. Li, G. Melaku, S. Zhang, H. Chen, Y. Zhao, J. Zhang and Y. Zhang (2018). "Developing superior alleles of yield genes in rice by artificial mutagenesis using the CRISPR/Cas9 system." *The Crop Journal* 6(5): 475–481.

Jakočiūnas, T., I. Bonde, M. Herrgård, S. J. Harrison, M. Kristensen, L. E. Pedersen, M. K. Jensen and J. D. Keasling (2015). "Multiplex metabolic pathway engineering using CRISPR/Cas9 in Saccharomyces cerevisiae." *Metab Eng* 28: 213–222.

Jiang, Y.-Y., Y.-P. Chai, M.-H. Lu, X.-L. Han, Q. Lin, Y. Zhang, Q. Zhang, Y. Zhou, X.-C. Wang, C. Gao and Q.-J. Chen (2020). "Prime editing efficiently generates W542L and S621I double mutations in two ALS genes in maize." *Genome Biol* 21(1): 257.

Kannan, B., J. H. Jung, G. W. Moxley, S. M. Lee and F. J. P. b. j. Altpeter (2018). "TALEN-mediated targeted mutagenesis of more than 100 COMT copies/alleles in highly polyploid sugarcane improves saccharification efficiency without compromising biomass yield." *Plant Biotechnol J* 16(4): 856–866.

Kim, D., B. Alptekin and H. Budak (2018). "CRISPR/Cas9 genome editing in wheat." *Funct Integr Genomics* 18(1): 31–41.

Kuang, Y., S. Li, B. Ren, F. Yan, C. Spetz, X. Li, X. Zhou and H. Zhou (2020). "Base-editing-mediated artificial evolution of OsALS1 in planta to develop novel herbicide-tolerant rice germplasms." *Mol Plant* 13(4): 565–572.

Lacchini, E., E. Kiegle, M. Castellani, H. Adam, S. Jouannic, V. Gregis and M. M. Kater (2020). "CRISPR-mediated accelerated domestication of African rice landraces." *PLoS One* 15(3): e0229782.

Lawrenson, T., O. Shorinola, N. Stacey, C. Li, L. Østergaard, N. Patron, C. Uauy and W. Harwood (2015). "Induction of targeted, heritable mutations in barley and Brassica oleracea using RNA-guided Cas9 nuclease." *Genome Biol* 16: 258.

Li, A., S. Jia, A. Yobi, Z. Ge, S. J. Sato, C. Zhang, R. Angelovici, T. E. Clemente and D. R. Holding (2018). "Editing of an alpha-kafirin gene family increases, digestibility and protein quality in sorghum." *Plant Physiol* 177(4): 1425–1438.

Li, C., Y. Zong, Y. Wang, S. Jin, D. Zhang, Q. Song, R. Zhang and C. Gao (2018). "Expanded base editing in rice and wheat using a Cas9-adenosine deaminase fusion." *Genome Biol* 19(1): 59.

Li, M., X. Li, Z. Zhou, P. Wu, M. Fang, X. Pan, Q. Lin, W. Luo, G. Wu and H. Li (2016). "Reassessment of the four yield-related genes Gn1a, DEP1, GS3, and IPA1 in rice using a CRISPR/Cas9 system." *Front Plant Sci* 7: 377.

Li, X., W. Zhou, Y. Ren, X. Tian, T. Lv, Z. Wang, J. Fang, C. Chu, J. Yang and Q. Bu (2017). "High-efficiency breeding of early-maturing rice cultivars via CRISPR/Cas9-mediated genome editing." *J Genet Genomics* 44(3): 175–178.

Liang, Z., K. Zhang, K. Chen and C. Gao (2014). "Targeted mutagenesis in *Zea mays* using TALENs and the CRISPR/Cas system." *J Genet Genomics* 41(2): 63–68.

Liu, X., R. Qin, J. Li, S. Liao, T. Shan, R. Xu, D. Wu and P. Wei (2020). "A CRISPR-Cas9-mediated domain-specific base-editing screen enables functional assessment of ACCase variants in rice." *Plant Biotechnol J* 18(9): 1845–1847.

Lowder, L. G., D. Zhang, N. J. Baltes, J. W. Paul 3rd, X. Tang, X. Zheng, D. F. Voytas, T. F. Hsieh, Y. Zhang and Y. Qi (2015). "A CRISPR/Cas9 toolbox for multiplexed plant genome editing and transcriptional regulation." *Plant Physiol* 169(2): 971–985.

Ma, X., Q. Zhang, Q. Zhu, W. Liu, Y. Chen, R. Qiu, B. Wang, Z. Yang, H. Li, Y. Lin, Y. Xie, R. Shen, S. Chen, Z. Wang, Y. Chen, J. Guo, L. Chen, X. Zhao, Z. Dong and Y. G. Liu (2015). "A robust CRISPR/Cas9 system for convenient, high-efficiency multiplex genome editing in monocot and dicot plants." *Mol Plant* 8(8): 1274–1284.

McCarty, N. S., A. E. Graham, L. Studená and R. Ledesma-Amaro (2020). "Multiplexed CRISPR technologies for gene editing and transcriptional regulation." *Nat Commun* 11(1): 1281.

Miao, C., L. Xiao, K. Hua, C. Zou, Y. Zhao, R. A. Bressan and J. K. Zhu (2018). "Mutations in a subfamily of abscisic acid receptor genes promote rice growth and productivity." *Proc Natl Acad Sci U S A* 115(23): 6058–6063.

Miglani, G. S. (2017). "Genome editing in crop improvement: Present scenario and future prospects." *J Crop Improv* 31(4): 453–559.

Mubarik, M. S., X. Wang, S. H. Khan, A. Ahmad, Z. Khan, M. W. Amjid, M. K. Razzaq, Z. Ali and M. T. Azhar (2021). "Engineering broad-spectrum resistance to cotton leaf curl disease by CRISPR-Cas9 based multiplex editing in plants." *GM Crops Food*, 12(2), 1–12.

Nielsen, A. A., & Voigt, C. A. (2014). Multi-input CRISPR/C as genetic circuits that interface host regulatory networks. *Molecular systems biology*, 10(11), 763.

Oliva, R., C. Ji, G. Atienza-Grande, J. C. Huguet-Tapia, A. Perez-Quintero, T. Li, J. S. Eom, C. Li, H. Nguyen, B. Liu, F. Auguy, C. Sciallano, V. T. Luu, G. S. Dossa, S. Cunnac, S. M. Schmidt, I. H. Slamet-Loedin, C. Vera Cruz, B. Szurek, W. B. Frommer, F. F. White and B. Yang (2019). "Broad-spectrum resistance to bacterial blight in rice using genome editing." *Nat Biotechnol* 37(11): 1344–1350.

Qi, W., T. Zhu, Z. Tian, C. Li, W. Zhang and R. Song (2016). "High-efficiency CRISPR/Cas9 multiplex gene editing using the glycine tRNA-processing system-based strategy in maize." *BMC Biotechnol* 16(1): 1–8.

Sánchez-León, S., J. Gil-Humanes, C. V. Ozuna, M. J. Giménez, C. Sousa, D. F. Voytas and F. Barro (2018). "Low-gluten, nontransgenic wheat engineered with CRISPR/Cas9." *Plant Biotechnol J* 16(4): 902–910.

Shen, L., C. Wang, Y. Fu, J. Wang, Q. Liu, X. Zhang, C. Yan, Q. Qian and K. Wang (2018). "QTL editing confers opposing yield performance in different rice varieties." *J Integr Plant Biol* 60(2): 89–93.

Shimatani, Z., S. Kashojiya, M. Takayama, R. Terada, T. Arazoe, H. Ishii, H. Teramura, T. Yamamoto, H. Komatsu, K. Miura, H. Ezura, K. Nishida, T. Ariizumi and A. Kondo (2017). "Targeted base editing in rice and tomato using a CRISPR-Cas9 cytidine deaminase fusion." *Nat Biotechnol* 35(5): 441–443.

Stuttmann, J., K. Barthel, P. Martin, J. Ordon, J. L. Erickson, R. Herr, F. Ferik, C. Kretschmer, T. Berner, J. Keilwagen, S. Marillonnet and U. Bonas (2021). "Highly efficient multiplex editing: One-shot generation of 8× *Nicotiana benthamiana* and 12× Arabidopsis mutants." *Plant J* 106(1): 8–22.

Sun, Y., X. Zhang, C. Wu, Y. He, Y. Ma, H. Hou, X. Guo, W. Du, Y. Zhao and L. Xia (2016). "Engineering herbicide-resistant rice plants through CRISPR/Cas9-mediated homologous recombination of acetolactate synthase." *Mol Plant* 9(4): 628–631.

Wang, W., Q. Pan, F. He, A. Akhunova, S. Chao, H. Trick and E. Akhunov (2018). "Transgenerational CRISPR-Cas9 activity facilitates multiplex gene editing in allopolyploid wheat." *CRISPR J* 1(1): 65–74.

Wang, W., J. Simmonds, Q. Pan, D. Davidson, F. He, A. Battal, A. Akhunova, H. N. Trick, C. Uauy and E. Akhunov (2018). "Gene editing and mutagenesis reveal inter-cultivar differences and additivity in the contribution of TaGW2 homoeologues to grain size and weight in wheat." *Theor Appl Genet* 131(11): 2463–2475.

Wang, Z. P., H. L. Xing, L. Dong, H. Y. Zhang, C. Y. Han, X. C. Wang and Q. J. Chen (2015). "Egg cell-specific promoter-controlled CRISPR/Cas9 efficiently generates homozygous mutants for multiple target genes in Arabidopsis in a single generation." *Genome Biol* 16(1): 144.

Xing, H.-L., L. Dong, Z.-P. Wang, H.-Y. Zhang, C.-Y. Han, B. Liu, X.-C. Wang and Q.-J. Chen (2014). "A CRISPR/Cas9 toolkit for multiplex genome editing in plants." *BMC Plant Biol* 14(1): 1–12.

Xu, R., Y. Yang, R. Qin, H. Li, C. Qiu, L. Li, P. Wei and J. Yang (2016). "Rapid improvement of grain weight via highly efficient CRISPR/Cas9-mediated multiplex genome editing in rice." *J Genet Genomics* 43(8): 529–532.

Xu, W., W. Song, Y. Yang, Y. Wu, X. Lv, S. Yuan, Y. Liu and J. Yang (2019). "Multiplex nucleotide editing by high-fidelity Cas9 variants with improved efficiency in rice." *BMC Plant Biol* 19(1): 511.

Yu, H., T. Lin, X. Meng, H. Du, J. Zhang, G. Liu, M. Chen, Y. Jing, L. Kou, X. Li, Q. Gao, Y. Liang, X. Liu, Z. Fan, Y. Liang, Z. Cheng, M. Chen, Z. Tian, Y. Wang, C. Chu, J. Zuo, J. Wan, Q. Qian, B. Han, A. Zuccolo, R. A. Wing, C. Gao, C. Liang and J. Li (2021). "A route to de novo domestication of wild allotetraploid rice." *Cell* 184(5): 1156-1170.e1114.

Zeng, Y., J. Wen, W. Zhao, Q. Wang and W. Huang (2019). "Rational improvement of rice yield and cold tolerance by editing the three genes OsPIN5b, GS3, and OsMYB30 With the CRISPR-Cas9 system." *Front Plant Sci* 10: 1663.

Zhang, Y., Y. Bai, G. Wu, S. Zou, Y. Chen, C. Gao and D. Tang (2017). "Simultaneous modification of three homoeologs of TaEDR1 by genome editing enhances powdery mildew resistance in wheat." *Plant J* 91(4): 714–724.

Zhong, Z., S. Liu, X. Liu, B. Liu, X. Tang, Q. Ren, J. Zhou, X. Zheng, Y. Qi and Y. Zhang (2020). "Intron-based single transcript unit CRISPR systems for plant genome editing." *Rice (N Y)* 13(1): 8.

Zhou, J., X. Xin, Y. He, H. Chen, Q. Li, X. Tang, Z. Zhong, K. Deng, X. Zheng, S. A. Akher, G. Cai, Y. Qi and Y. Zhang (2019). "Multiplex QTL editing of grain-related genes improves yield in elite rice varieties." *Plant Cell Rep* 38(4): 475–485.

Zhu, C., L. Bortesi, C. Baysal, R. M. Twyman, R. Fischer, T. Capell, S. Schillberg and P. Christou (2017). "Characteristics of genome editing mutations in cereal crops." *Trends Plant Sci* 22(1): 38–52.

Zhu, H., Li, C., & Gao, C. (2020). Applications of CRISPR–Cas in agriculture and plant biotechnology. *Nature Reviews Molecular Cell Biology, 21*(11), 661–677.

Index

A

ABA, *see* Abscisic acid
Abiotic stress, 51
Abiotic stress and plant metabolism, 142–143
    drought, 143
    flood, 144–145
    heavy metals, 145
    salinity, 143–144
    temperature, 144
    UV light, 144
Abiotic stress-tolerant plants, 145
    conventional techniques, 146, 147
    CRISPR/Cas 9, 148–150
    transgenic approaches, 146, 148
Abscisic acid (ABA), 41
ACC, *see* 1-Aminocyclopropane-1-carboxylic
    acid
AFLP, *see* Amplified fragment length
    polymorphisms
Aggressivity (A), 15
Aging, 69
    mechanism, 73
    phenotype, 72–73
Agrobacterium, 251, 252, 255
    gene transformation, 281–282, 283
Agrobacterium tumefaciens, 281, 282
    transformation
        with ballistic transfection, 284–285
        with pollination, 284
        using silicon carbide fibers, 285
Agronomic strategies, 86
AGROS8 gene, 168
Allium fistulosum, 183
Alternative splicing (AS), 50
Amino acid biosynthesis genes, 98–99
Amino acids, 71
1-Aminocyclopropane-1-carboxylic (ACC) acid,
    41, 222
Ammonia volatilization, 90
Amplified fragment length polymorphisms
    (AFLP), 164
Antioxidant system, 40–41
APCI, *see* Atmospheric pressure chemical
    ionization
API, *see* Atmospheric pressure ionization
Arabidopsis thaliana, 176, 177, 213, 215
Area time equivalent ratio (ATER), 14–15
ARGOS8 gene, 260, 289
Arthrobacter globiformis, 255
AS, *see* Alternative splicing
Ascorbate, 224, 225

B

Barley
    botany, 114
    landraces, 53
    origin, domestication, distribution, and
        spread, 115
    taxonomy, 112
Barley *vs.* wheat, yield in dry environments
    adaptation, 51
        apical development and tiller appearance,
            53–54
        morphological characters, 51–53
        plant ideotype and grain yield, 54–55
    breeding, 41–43
        genetic variation, growth stages, 43–45
        genotyping and phenotyping, 45–46
        nanotechnology, 46
    genetics, 46–47
        basis, 47
        functional validation, 47–48
        genetic engineering, 50–51
        genomic analyses, 48–50
    morphological and physiological
        characteristics, 56, 57
    physiological and biochemical reactions,
        34–36
        oxidative status, 40–41
        photosynthesis, 35, 37–39
        water and nutrient relations, 39–40
Biofortification, 86
Biolistic transformation, 276
Biological nitrogen fixation (BNF), 6, 7
Biotic constraints, 126
Biotic stress, 183, 249
BNF, *see* Biological nitrogen fixation
Boron, 87, 88
Brown planthopper (BPH) infection, 186

C

Calcium ammonium nitrate (CAN), 94
Carotenoids, 225
Caryopsis, 37
Cassava (*Manihot esculenta*), 23–25

Asian Development Bank studies, 83
ATER, *see* Area time equivalent ratio
Atmospheric pressure chemical ionization
    (APCI), 181
Atmospheric pressure ionization (API), 180, 181
Autumn season sugarcane, 21

Printed in the United States
by Baker & Taylor Publisher Services